Martian Impact Craters

Our exploration of Mars has revealed a world as fascinating as Earth, with a changing climate, giant volcanoes, former oceans, polar ice caps, and numerous impact craters. This book provides a comprehensive summary of the morphology and distribution of meteorite craters on Mars, and the wealth of information these can provide on the crustal structure, surface geology, climate, and evolution of the planet. The chapters present highly illustrated case studies of landforms associated with impact craters to highlight their morphological diversity, using high-resolution images and topographic data to compare these features with those on other bodies in the Solar System. Including research questions to inspire future work, this book will be valuable for researchers and graduate students interested in impact craters (both terrestrial and extra-terrestrial) and Mars geology, as well as planetary geologists, planetary climatologists, and astrobiologists.

PETER J. MOUGINIS-MARK is a planetary geomorphologist and is currently an Emeritus Researcher at the Hawaii Institute of Geophysics and Planetology, University of Hawaii at Manoa. He specializes in impact craters and volcanoes and has published over 140 peer-reviewed research papers on these topics.

JOSEPH M. BOYCE is an affiliate faculty member at the University of Hawaii at Manoa and was previously a NASA Program Scientist on fifteen space missions from 1979 to 2002, many of which were to Mars, including the first Mars landers (Viking 1 and 2) and the first Mars rover (Pathfinder). He received NASA's Exceptional Service Medal in 2002. Asteroid 1978 VO 5 BOYCE is named in honor of his scientific work on impact craters.

Cambridge Planetary Science

Series Editors:

Jim Bell, Leigh Fletcher, David Jewitt, Ralph Lorenz, Francis Nimmo, David Vokrouhlický

Books in the Series:

1. Jupiter: The Planet, Satellites and Magnetosphere†
   Edited by Bagenal, Dowling, and McKinnon
   978-0-521-03545-3
2. Meteorites: A Petrologic, Chemical and Isotopic Synthesis†
   Hutchison
   978-0-521-03539-2
3. The Origin of Chondrules and Chondrites†
   Sears
   978-1-107-40285-0
4. Planetary Rings†
   Esposito
   978-1-107-40247-8
5. The Geology of Mars: Evidence from Earth-Based Analogs†
   Edited by Chapman
   978-0-521-20659-4
6. The Surface of Mars†
   Carr
   978-0-521-87201-0
7. Volcanism on Io: A Comparison with Earth†
   Davies
   978-0-521-85003-2
8. Mars: An Introduction to its Interior, Surface and Atmosphere†
   Barlow
   978-0-521-85226-5
9. The Martian Surface: Composition, Mineralogy and Physical Properties
   Edited by Bell
   978-0-521-86698-9
10. Planetary Crusts: Their Composition, Origin and Evolution†
    Taylor and McLennan
    978-0-521-14201-4
11. Planetary Tectonics†
    Edited by Watters and Schultz
    978-0-521-74992-3
12. Protoplanetary Dust: Astrophysical and Cosmochemical Perspectives†
    Edited by Apai and Lauretta
    978-0-521-51772-0
13. Planetary Surface Processes
    Melosh
    978-0-521-51418-7
14. Titan: Interior, Surface, Atmosphere and Space Environment
    Edited by Müller-Wodarg, Griffith, Lellouch, and Cravens
    978-0-521-19992-6

15. Planetary Rings: A Post-Equinox View (Second edition)
Esposito
978-1-107-02882-1

16. Planetesimals: Early Differentiation and Consequences for Planets
Edited by Elkins-Tanton and Weiss 978-1-107-11848-5

17. Asteroids: Astronomical and Geological Bodies
Burbine
978-1-107-09684-4

18. The Atmosphere and Climate of Mars
Edited by Haberle, Clancy, Forget, Smith, and Zurek
978-1-107-01618-7

19. Planetary Ring Systems
Edited by Tiscareno and Murray
978-1-107-11382-4

20. Saturn in the 21st Century
Edited by Baines, Flasar, Krupp, and Stallard
978-1-107-10677-2

21. Mercury: The View after MESSENGER
Edited by Solomon, Nittler, and Anderson
978-1-107-15445-2

22. Chondrules: Records of Protoplanetary Disk Processes
Edited by Russell, Connolly Jr., and Krot
978-1-108-41801-0

23. Spectroscopy and Photochemistry of Planetary Atmospheres and Ionospheres
Krasnopolsky
978-1-107-14526-9

24. Remote Compositional Analysis: Techniques for Understanding Spectroscopy, Mineralogy, and Geochemistry of Planetary Surfaces
Edited by Bishop, Bell III, and Moersch
978-1-107-18620-0

25. Meteoroids: Sources of Meteors on Earth and Beyond
Edited by Ryabova, Asher, and Campbell-Brown
978-1-108-42671-8

26. Meteorite Mineralogy
Rubin and Ma
978-1-108-48452-7

27. Vesta and Ceres: Insights into the Dawn of our Solar System
Edited by Marchi, Raymond, and Russell
978-1-108-47973-8

28. Ganymede
Edited by Volwerk, McGrath, Jia and Spohn
978-1-108-83295-3

29. Martian Impact Craters: A Morphological Perspective
Mouginis-Mark and Boyce
978-1-107-15491-9

† Reissued as a paperback

MARTIAN IMPACT CRATERS

A Morphological Perspective

PETER J. MOUGINIS-MARK
University of Hawaii at Manoa

JOSEPH M. BOYCE
University of Hawaii at Manoa

CAMBRIDGE
UNIVERSITY PRESS

CAMBRIDGE
UNIVERSITY PRESS

Shaftesbury Road, Cambridge CB2 8EA, United Kingdom

One Liberty Plaza, 20th Floor, New York, NY 10006, USA

477 Williamstown Road, Port Melbourne, VIC 3207, Australia

314–321, 3rd Floor, Plot 3, Splendor Forum, Jasola District Centre, New Delhi – 110025, India

103 Penang Road, #05–06/07, Visioncrest Commercial, Singapore 238467

Cambridge University Press is part of Cambridge University Press & Assessment,
a department of the University of Cambridge.

We share the University's mission to contribute to society through the pursuit of
education, learning and research at the highest international levels of excellence.

www.cambridge.org
Information on this title: www.cambridge.org/9781107154919

DOI: 10.1017/9781316651162

When citing this work, please include a reference to the DOI 10.1017/9781316651162

First published 2025

Printed in the United Kingdom by CPI Group Ltd, Croydon CR0 4YY

Cover image: The relatively fresh crater Bacolor is ~20 km in diameter and is a classic
Type-1 double layered ejecta crater, located at 33.01oN 118.58oE. The Thermal Emission
Imaging System (THEMIS) on the Mars Odyssey spacecraft imaged Bacolor crater at both
visible and infrared wavelengths, which were used here to produce this color image.
NASA/JPL-Caltech/Arizona State University.

A catalogue record for this publication is available from the British Library

*A Cataloging-in-Publication data record for this book is available from the
Library of Congress*

ISBN 978-1-107-15491-9 Hardback

CONTENTS

	Preface	*page* xi
	List of Abbreviations	xiii
1	Impact Crater Formation	1
	1.1 Introduction	1
	1.2 Contact and Compression	2
	1.3 Excavation	2
	1.4 Modification	3
	1.5 Ejecta Emplacement	4
	1.6 Crater Geometry	6
	1.7 Summary	10
2	Morphology of Craters throughout the Solar System	12
	2.1 Introduction	12
	2.2 Earth	12
	2.3 Craters on Volatile-Poor Bodies – The Moon and Mercury	15
	2.3.1 Interior Features	15
	2.3.2 Exterior Features	21
	2.4 Craters on Volatile-Rich Bodies – Ganymede, 1 Ceres, and Charon	23
	2.4.1 Interior Features	23
	2.4.2 Exterior Features	25
	2.5 Craters on Bodies with an Atmosphere – Venus and Titan	27
	2.5.1 Interior Features	27
	2.5.2 Exterior Features	27
	2.6 Summary	29
3	Knowing Mars	30
	3.1 Introduction	30
	3.2 Data Types	30
	3.3 Ages of Martian Rocks	34
	3.4 Physiography of Mars	35
	3.5 Major Variations with Time	39
	3.5.1 Potential Northern Ocean	39
	3.5.2 Obliquity Changes	39
	3.5.3 Distribution of Near-Surface Ice	40
	3.6 Human-Scale Morphology of Mars	41
	3.7 Identifying the Youngest Craters on Mars	43
	3.7.1 Superposed Craters	43
	3.7.2 Pitted Floor Material	43
	3.7.3 Depth/Diameter Ratio	43
	3.7.4 Impact Melt	44
	3.7.5 Ejecta Rays	44
	3.7.6 Thermal Inertia	44
	3.8 Summary	45

4 Fresh Crater Interiors 46
4.1 Introduction 46
4.2 Floor Features 46
4.2.1 Bowl-Shaped Craters 46
4.2.2 Flat-Floored Craters 46
4.2.3 Central Peaks 47
4.2.4 Peak Rings 49
4.2.5 Central Pits 49
4.2.6 Concentric Craters 53
4.3 Other Interior Features 53
4.3.1 Pitted Terrain 53
4.3.2 "Swirl" Texture 54
4.4 Wall and Rim Features 56
4.4.1 Terraces 56
4.4.2 Rim Uplift 60
4.4.3 Impact Melt 60
4.5 Summary 62

5 Modified Crater Interiors 63
5.1 Introduction 63
5.2 Floor Features 63
5.2.1 Dunes 63
5.2.2 Flat Floors 63
5.2.3 Chaotic/Fractured Texture 64
5.2.4 Concentric Crater Fill 64
5.2.5 Exposures of Megabreccia 65
5.2.6 Massive Sedimentary Deposits 65
5.2.7 Deltaic Fans 69
5.2.8 Hydrothermal Deposits 73
5.3 Wall Features 73
5.3.1 Gullies 73
5.3.2 Debris Flows 74
5.4 Polar Craters 75
5.5 Degradation of Large Craters and Basins 75
5.5.1 Schiaparelli and Huygens Basins 75
5.5.2 Argyre Basin 77
5.5.3 Isidis Basin 77
5.5.4 Hellas Basin 77
5.5.5 Utopia Basin 78
5.6 Summary 79

6 Mars Crater Exterior Features – Major Types of Ejecta Blanket 80
6.1 Introduction 80
6.2 Single, Double, and Multi-layered Ejecta 80
6.3 Radial Extent of Ejecta 89
6.4 Influence of Pre-existing Topography on Ejecta Emplacement 89
6.5 Ejecta Thickness 91
6.6 Distal Ramparts 93
6.7 Secondary Craters 95
6.8 Summary 99

7 Mars Crater Exterior Features – Other Ejecta Attributes 100
7.1 Introduction 100
7.2 Small, Fresh Craters 100

7.3 Radial Ejecta Craters 101
7.4 LARLE Craters 102
7.5 Pedestal Craters 103
7.6 Self-Secondary Craters 104
7.7 Lobate Flows 107
7.8 Oblique Impacts 107
7.9 Crater Rays 109
7.10 Highly Fluid Ejecta 109
7.11 Pits on the Ejecta Layers 109
7.12 "Dewatering" of Ejecta Blankets 111
7.13 Ejecta from Large Craters 111
7.14 Ejecta Partially Buried by the Vastitas Borealis Formation 116
7.15 Summary 116

8 Ejecta Flow Features 117
8.1 Introduction 117
8.2 Thermal Properties of Ejecta? 117
8.3 Characteristics of Distal Ejecta 117
8.4 Clues to Ejecta Layer Emplacement 119
8.5 Early Phases of Modification of DLE Ejecta 126
8.6 Summary 127

9 Remaining Puzzles Concerning Martian Craters 128
9.1 Introduction 128
9.2 Are Similar-Sized Craters Created Equal? 128
9.3 Is There an Elevation Effect for Ejecta Fluidization? 129
9.4 Correlation between Interior and Exterior Crater Morphologies 130
9.5 Could Fluidized Ejecta Craters Indicate Time-Varying Amounts of Volatiles? 130
9.6 What Happens to Ejecta After Emplacement? 134
9.7 Zunil Crater: Two Ejecta Types at One Crater 136
9.8 How Might Some of the Questions about Fluidized Ejecta Be Answered? 142
9.9 Concluding Thoughts 146

Appendix 148
References 150
Index 160

Color plates are to be found between pages 82 and 83

PREFACE

Impact craters are the only planetary landform for which the final shape can be predicted from basic physics, so that the diversity of morphologies and geometries can provide information about the target and the post-impact evolution of the crater. The physical processes by which an impact crater formed on a planetary surface can be investigated with laboratory experiments, field observations of high-energy (i.e., nuclear) tests, and numerical simulations, but the real test comes in trying to understand landforms on a planet. Thus, Mars forms an interesting set of challenges as the planet displays a great diversity in rock types, elevation, and the effects of climatic changes over geologic time. One particularly intriguing feature of Mars craters is the presence of lobate ejecta that appears to have been emplaced as a fluidized flow during crater formation. No similar ejecta pattern is observed either on the Moon or Mercury, although similar deposits have been recognized on the icy moons Ganymede and Charon. Teasing out the inter-relationships between these diverse attributes of Martian craters that range in diameter from a few hundred meters to a few hundred kilometers presents a fascinating challenge to the planetary geologist.

The authors have collectively spent over 100 years studying the geomorphology of Mars craters, dating back to the earliest days of the *Viking Orbiter* missions in 1976 and the selection of the *Viking Lander* landing sites. Despite having published over a hundred papers and abstracts on the subject, they, and their colleagues, still have not solved many of the questions that were first raised during the *Viking Orbiter* missions back in the 1970s! This book sets out some of the challenges in interpreting these craters and is focused on the morphology of the relatively fresh craters on Mars as they provide the best clues to the cratering process and cover many of the discoveries since the *Viking Orbiter* missions. In writing this book, there has inevitably been a long gestation period, starting in 2015 and ending in the summer of 2024; a strong effort has been made to keep track of new ideas that have been proposed during this time, but inevitably some of the most recent research may not be given as much attention here as it will ultimately warrant. First and foremost, our intent has been to provide copious illustrations of various attributes of Martian craters so that the book serves as a teaser for future planetary geomorphologists to know both what is known and what is unknown about Mars craters. We hope that graduate students and post-doctoral researchers looking for specific topics on Mars craters, or researchers keen to gain a better understanding of the literature, will find this to be a valuable book. We wish them success in finding the answers that have been so elusive to us.

We want to thank the many people who made the book possible through the acquisition of such fine data sets. The efforts are particularly noteworthy of Michael Carr as the Viking Imaging System Team Leader, Philip Christensen as the Principal Investigator (PI) of the THEMIS instrument, Alfred McEwen as the PI of the HiRISE camera, and Michael Malin as the PI of the CTX camera. Without these images, we would still be looking through the haze of Martian dust seen at low resolution. Similarly, David Smith as the PI of the MOLA team provided valuable insights into the topography of Mars. The JMARS Team at Arizona State University is thanked for their production of a web interface that made cross-checking the names and locations of all the landforms discussed in this book so much easier. The work of Harold Garbeil was much appreciated for the generation of the digital elevation models used throughout the book to produce the oblique computer-generated images.

Finally, this book is dedicated to our good friend, Nadine Barlow (below photo), who sadly passed away in 2020 during the early formulation of this book. Nadine was a leader in the collection of morphometric data of impact craters throughout the Solar System, with a particular interest in Mars. Many of her ground-breaking investigations are cited here, particularly in Chapters 4 and 6 where we discuss the global distribution of features associated with craters. We are delighted that there is now a 87 km diameter crater located in the northern portion of Sinai Planum on Mars (11.65°S, 270.09°E) named in her honor. At the time of her passing, Nadine was a professor in the Department of Physics and Astronomy at Northern Arizona University, but her career spanned several positions including research conducted at the NASA Johnson Space Center, the Lunar and Planetary Institute in Houston, the University of Central Florida, and the U.S. Geological Survey in Flagstaff, Arizona. At all these venues, she specialized in creating catalogs of planetary craters that subsequently helped guide many other researchers of planetary craters. Her insights, good humor, and professional approach to all the topics that she tackled will be greatly missed.

Figure P.1 Nadine Barlow in 2013. Reproduced with permission of Northern Arizona University.

ABBREVIATIONS

CaSSIS	Colour and Stereo Surface Imaging System
CCF	concentric crater fill
CODEX	Chemistry, Organics, and Dating Experiment
CRISM	Compact Reconnaissance Imaging Spectrometer
CTX	Context Camera
d/D	depth/diameter ratio
DLE	double-layered ejecta
DEM	digital elevation model
EM	ejecta mobility
GRS	Gamma Ray Spectrometer
GSFC	Goddard Space Flight Center
HiRISE	High Resolution Imaging Science Experiment
HRSC	High Resolution Stereo Camera
JMARS	Java Mission-planning and Analysis of Remote Sensing
JPL	Jet Propulsion Laboratory
LARLE	low-aspect-ratio layered ejecta
L-Band	24-cm wavelength radar
LRO	Lunar Reconnaissance Orbiter
LROC	Lunar Reconnaissance Orbiter Camera
MARSIS	Mars Advanced Radar for Subsurface and Ionospheric Sounding
MCC	Mars Color Camera
MFF	Medusae Fossae Formation
MLE	multi-layered ejecta
MOC	Mars Orbiter Camera
MOLA	Mars Orbiter Laser Altimeter
MRO	Mars Reconnaissance Orbiter
NASA	National Aeronautical and Space Administration
OMEGA	Observatoire pour la Mineralogie, l'Eau, les Glaces et l'Activite
SAR	synthetic aperture radar
SHARAD	Shallow Radar
SIR-A	Shuttle Imaging Radar experiment-A
SLA	single-layered ejecta
SNC	Shergottite, Nakhlite, and Chassignite meteorites
THEMIS	Thermal Emission Imaging System
TIR	thermal infrared
VBF	Vastitas Borealis Formation
VIS	visible wavelengths
WAC	Wide-Angle Camera

1

Impact Crater Formation

1.1 INTRODUCTION

Craters were first discovered on the Moon by Galileo in 1610, but he only described these features as "circular spots." In 1791, the astronomer J. H. Schröter first used the term "crater" to describe lunar craters. Because little was known about the physics of impact crater formation, at that time most astronomers believed these features to be extinct volcanoes. Even at the dawn of the Space Age, some scientists believed that lunar craters were volcanic in origin (cf., Ronca, 1966; Wells and Fielder, 1967; Green, 1971) because almost all the morphological features in and around the craters appeared to agree with a volcanic origin. Initial clues favoring an impact origin had been obtained a few years earlier with the identification of shock metamorphic features in rocks found at Meteor crater in Arizona, USA (Shoemaker, 1963). Acceptance of the idea that craters resulted from the collision of large pieces of debris from space with a planetary surface was reached only after high-resolution images and *in situ* investigation by the Apollo astronauts in the late 1960s and early 1970s. Hyper-velocity impact experiments (Gault et al., 1968), the analysis of nuclear and chemical explosions (Roddy, 1977), and field studies of impact craters (Shoemaker, 1963) provided valuable additional insights into the physics of impact events.

In this chapter, we introduce the mode of formation of craters on planetary surfaces to set the stage for comparisons of crater morphology throughout the Solar System (Chapter 2) and on Mars specifically (Chapters 4 to 8). We will see that there are some common features of craters across the Solar System and some significant differences in their morphology that are influenced by the material within which the crater formed as well as the presence/absence of an atmosphere and the gravity of the body. Mars also has some unusual properties that are climate-related, with the distribution of ice in the near-surface layers varying across the planet as a function of time and location, and these properties are introduced in Chapter 3. The book concludes in Chapter 9 with a discussion of some of the issues pertaining to Martian crater morphology that remain unresolved, as well as some suggestions for how further advances in our knowledge might be achieved.

Impact craters are the only planetary landform for which the final shape can be predicted from basic physics, and so the diversity of morphologies and geometries provide information about the target and the post-impact evolution of the crater. A few craters, such as Meteor crater in Arizona, have been extensively studied via drilling (Roddy et al., 1975) or ground-penetrating radar (Pilon et al., 1991) to better understand the three-dimensional structure of craters. Some excellent texts have documented the different stages of formation of hypervelocity impact craters, to which the reader is referred for a more detailed explanation. For example, Melosh (1982, 1989, 2011) provided a comprehensive review of the physics of impact and explosion cratering. More recently, Osinski and Pierazzo (2013) included in their book some comparisons of lunar and Martian craters to terrestrial examples such as Lonar crater in India and Meteor crater, as well as considered the geological effects associated with impact cratering on Earth. The 24 km diameter Nordlinger Ries crater in Germany has also been investigated as a terrestrial analog to Martian craters, aided by the relatively young age (~14.8 Ma) of the crater that allows parts of the ejecta blanket to be investigated in the field (Hörz and Banholzer, 1980; Sturm et al., 2013, 2015).

The physical processes by which an impact crater formed on a planetary surface can be investigated with laboratory experiments, field observations of high-energy (i.e., nuclear) tests, and from numerical simulations. In this chapter, we are planetary agnostics; the basic process of a projectile hitting a planetary surface, the excavation and relaxation of the crater cavity, and the emplacement of the ejecta are similar whether the target is the Moon, Venus, Mercury, Mars, or one of the moons of the outer Solar System planets. True, the details of the amount of energy partitioned into the excavation of the crater and the resultant ejecta will be different depending on such attributes as the velocity at which the projectile hits (which in part is determined by the gravity of the impacted body), the composition of the projectile (for example, an iron or a carbonaceous meteorite), and even the angle at which the projectile hits the surface, but the basic physics as described here will be the same.

When a projectile strikes the surface of a planet or moon at high velocity (typically many kilometers per second), an orderly sequence of events is initiated that produces a crater. Although this is a continuous process, to aid in the interpretation of impact craters presented in subsequent chapters, it helps to describe the overall process by focusing on distinct stages, each dominated by different physical processes (Gault et al., 1968; Melosh, 1989): contact/compression, excavation, and modification. The excavation of the crater cavity produces a copious volume of material that is then thrown onto the surrounding terrain or, particularly on bodies with a low gravity, thrown off the body to escape into space. The processes associated with ejecta emplacement can be predicted from numerical simulations (e.g., Oberbeck and Morrison, 1973; Oberbeck,

1975; Oberbeck et al., 1975), and these can serve as a basis for the analysis of ejecta blankets on airless worlds such as the Moon. In turn, as subsequent chapters focus on Martian ejecta deposits, the differences between ejecta on Mars and the Moon (or Mercury) can be identified.

1.2 CONTACT AND COMPRESSION

At the beginning of the process of forming an impact crater, when the projectile first touches the target, the contact and compression stage typically lasts only briefly; a few times longer than the time required for the impacting projectile to traverse its own diameter into the surface of the target. The incoming projectile can encounter the surface at a speed as high as 70 km/sec, but commonly the velocity is dependent on the gravity of the planet, which causes the projectile to strike Mars with a "modal impact velocity" of ~12 km/sec. For the Moon, the modal impact velocity is typically ~20 km/s due to proximity to Earth. For Mercury, modal impact velocity is ~42.5 km/s because the planet is close to the Sun.

At first contact, a system of shock waves forms that provides the mechanism for transferring the kinetic energy of the impacting projectile into the target. During this stage, the projectile plunges into the target generating a strong shock wave as material in both the projectile and target are compressed. This slows the projectile and compresses it, as well as compressing and accelerating the target material downward. The shock wave in the object eventually reaches the projectile's trailing edge releasing the pressure as the surface of the compressed projectile expands upward. A wave of pressure relief, a "rarefaction wave," propagates back toward the interface between the object and the target. During decompression of the projectile, heat deposited during the irreversible compression process may cause the projectile to change phases to either liquid or vapor. The projectile generally penetrates the target less than its diameter below the original surface before it is destroyed as the decompression wave rips it apart. The projectile typically retains about half of the total initial energy if the compressibility and density of the target and projectile are similar. The balance of energy moves into the target where it eventually is expended in opening the crater and heating the target.

The contact and compression stage is generally accompanied by high-velocity jets of highly shocked material that form where strongly compressed material is close to the free surface. These jets may exceed the impact velocity by a factor greater than five. In vertical impacts, these jets account for less than 10% of the projectile's mass, but jetting can be much more important in oblique impacts.

1.3 EXCAVATION

Following the contact and compression stage, the excavation stage begins with expansion of the shock wave. The basic shock wave geometry approximates a hemispherical shell of compressed material, expanding radially and distributing the kinetic energy over a steadily increasing mass of the target material. As a result, the shock pressure and energy drop as the shock wave spreads outward through the target. This shock wave eventually weakens into an elastic wave, while the crater itself is opened by the much slower excavation flow.

Target material engulfed by the shock wave is quickly released as the shock passes but still has a residual velocity of ~20% of the particle velocity in the shock wave. The residual velocity is caused by thermodynamic irreversibility in the shock compression of the target and eventually excavates the crater. This velocity field has a characteristic downward-outward-then-upward pattern that excavates and ejects target material from the crater cavity. Lines of flow ("streamlines") of target material coming out of the growing crater cut across the contours of maximum shock pressure. As a result, material ejected at any one time may contain material with a wide range of shock levels, although early in ejection process the fast ejecta contains a higher proportion of highly shocked material than the later slower ejecta. In addition, the mixture of vaporized projectile and target may expand rapidly out of the growing crater, forming a vapor plume. If massive enough, this plume may blow aside the surrounding atmosphere during ejecta emplacement or even play a role in impact erosion of the entire atmosphere (Walker, 1986; Kegerreis et al., 2020).

Near the free surface, the expanding shock wave encounters a special condition that causes the pressure at the surface to be always zero. This is because stress cannot be transferred across a free surface and hence drops to zero at that surface (e.g., faults cannot extend from the ground into the air). In impact events, this results in a thin layer of rock at the surface comprising only about 1% to 3% of the total ejected mass being ejected at very high velocity, sometimes nearly as high as the impact velocity. Because the surface rocks are not compressed to high pressure, small quantities of unshocked ejected rock may exceed the escape velocity of the planet. In the case of impact events on Mars, this material can have particular importance because some ejecta can find their way to Earth to be collected as meteorites. The Shergottite, Nakhlite, and Chassignite (SNC) meteorites (Treiman et al., 2000; McSween, 2002), of which at least 262 have been identified from at least 11 ejection events (Udry et al., 2020), are examples of this type of material.

The shape of the growing crater is at first hemispherical, with the depth and diameter growing at nearly the same rate. But soon the cavity stops increasing in depth, while the diameter continues to increase to form a shallow bowl-shaped depression with a diameter about three to four times its depth. This is called the "transient" crater. Material deeper than about one third of the transient crater depth is pushed downward into the target. Deformation of the surrounding rock accommodates the volume of this material. Material lying above this depth in the transient crater is ejected ballistically from beyond the transient cavity rim to form the continuous ejecta blanket (Oberbeck, 1975), thereby forming one of the most studied aspects of impact craters on Mars (see Chapters 6, 7, and 8). Ejecta blankets can be divided into continuous deposits close to the crater rim and discontinuous deposits at greater radial range. Secondary craters, formed by individual blocks of ejecta re-impacting the target to produce new small craters, populate most of the discontinuous ejecta blanket.

1.4 MODIFICATION

The modification stage of the cratering process starts shortly after excavation has opened the transient crater and the ejecta have been launched into ballistic trajectories (Melosh and Ivanov, 1999). The shapes of craters are modified by different processes. Some involve the direct effects of impact cratering at the end of the transient crater formation, while others such as gradation, infilling by lava or sediment, and isostatic adjustment processes operating over a longer timescale to modify the shapes of the craters (Craddock et al., 2018). Crater cavity modification lasts for about the same amount of time as the excavation phase. Major changes in motion occur both beneath and within the crater, as debris begins to move downward and back toward the center of the crater. This collapse is driven by gravity, although elastic rebound of the compressed, underlying rock may also play a role. The effects of this collapse are dependent on the size of the crater (Figure 1.1). In small craters ("simple craters"), this generally entails the collapse of the steep rim slopes inside the crater onto its floor and uplifted strata within the crater rim. This results in a shallow bowl-shaped depression, a "simple craters," whose diameter is about five times its depth. In larger craters ("complex craters"), collapse causes the floors to rise and the formation of central peaks or peak rings, central pits, and stepped terraces. Collectively, these processes result in the formation of shallow, broad, relatively low-relief depressions (Figure 1.1). Melt-rich crater-fill material may occur at various places within the crater, as ponds of material collect in local topographic depressions that are controlled by terrace blocks forming along listric faults (Figure 1.2). For transient crater depths of 1–10 km (i.e., final crater diameter in the range 10–100 km), the gravity collapse time is 10–30 seconds (Melosh and Ivanov, 1999).

Simple craters on Mars (Figure 1.3) are generally less than about 8 km diameter but can be larger on the Moon due to the lower gravity (Pike, 1980a; Boyce and Garbeil, 2007; Robbins and Hynek, 2012). When fresh, these craters possess an uplifted rim and are filled with an allochthonous breccia lens that comprises largely unshocked target materials, possibly mixed with impact melt-bearing lithologies (Shoemaker, 1963). The overall low shock level of material in the breccia lens suggests that it is formed due to slumping of the transient cavity walls and is not "fallback" material (Grieve and Cintala, 1981). Volume conservation suggests that the transient crater diameter of the crater is increased by ~15% from this collapse.

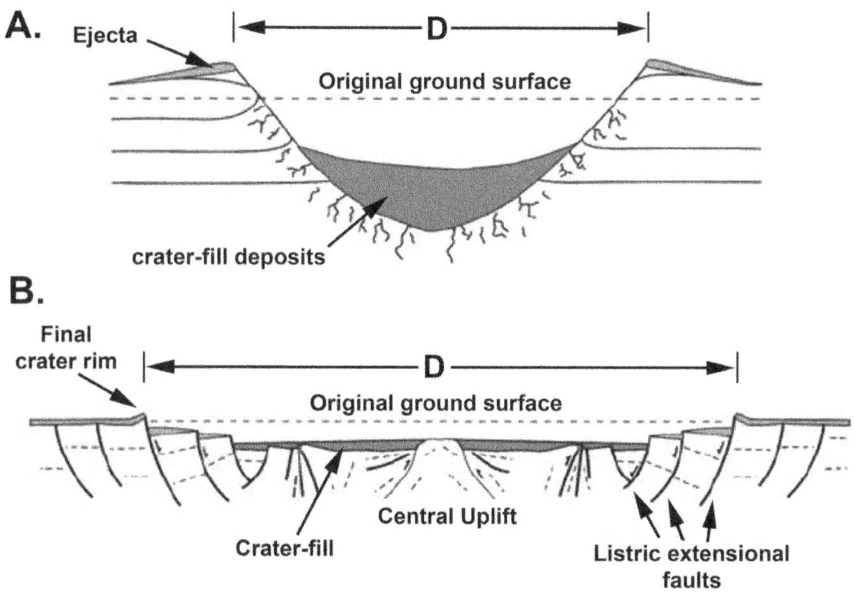

Figure 1.1 Typical cross-sections of simple (A) and complex (B) craters. "D" represents the final crater diameter which is defined as the diameter of the topographic rim that rises above the surface for simple craters or above the outermost slump block not covered by ejecta for a complex crater. Note that the crater-fill within the complex crater occurs not only around the central uplift but also on top of some of the terrace blocks that formed along the listric extensional faults. Reprinted from Osinski et al. (2011). "Impact ejecta emplacement on terrestrial planets," *Earth Planetary Science Letters*, vol. 310, pp. 167–181, Figure 1. With permission from Elsevier.

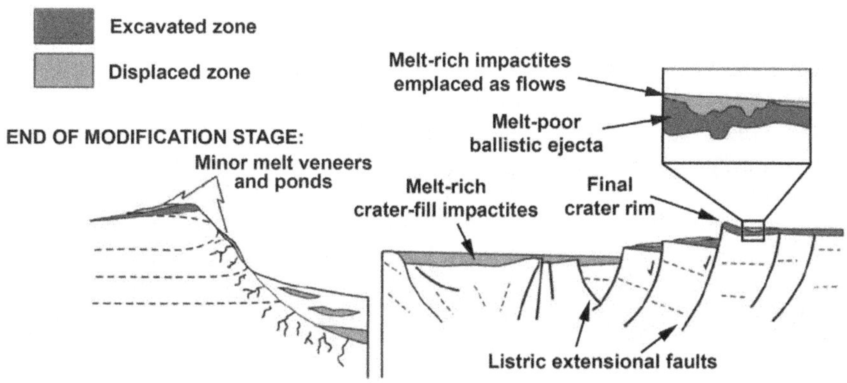

Figure 1.2 Representations of the final cross-section of simple craters (left) and complex craters (right), illustrating the relative positions of displaced and excavated materials. Reprinted from Osinski et al. (2011). "Impact ejecta emplacement on terrestrial planets," *Earth Planetary Science Letters*, vol. 310, pp. 167–181, Figure 7. With permission from Elsevier.
Note: A black and white version of this figure will appear in some formats. For the color version, please refer to the plate section.

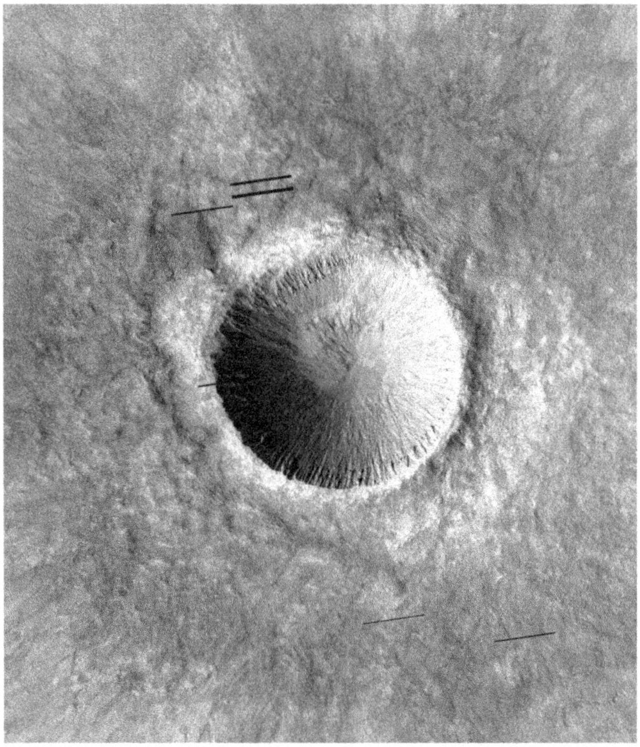

Figure 1.3 A fine example of a fresh simple crater on Mars located at 51.60°N, 333.02°E. This crater is ~1.2 km in diameter. The thin black lines on the ejecta blanket are data dropouts. HiRISE image ESP_027866_2320.

Figure 1.4 The lunar crater Moretus (70.63°S, 354.05°E) displays a fine example of a central peak. The crater is ~114 km in diameter. Illumination is from the left. LROC WAC mosaic. NASA/GSFC/ Arizona State University.

Complex craters display a much wider variety of interior features. These tend to transition to different types as a function of crater size and target material. Central peaks (Figure 1.4) result from the uplift of the floor after the passage of the shock wave, with the uplifted material "freezing" in place to form a central peak. A result of complex crater formation is that material from depth is brought to the surface. Thus, the materials identified within the central peak of a crater may well be different from the composition of the surrounding terrain or the material exposed on the crater rim. Central peaks are, therefore, probes into the subsurface geology of a planet (Pieters, 1982; Michalski and Niles, 2010; Caudill et al., 2012; Pan et al., 2015). For craters the size of Manicouagan (100 km) on Earth or Copernicus (92 km) on the Moon, the distribution and expected cooling rate of impact melt on the crater floor have allowed the time for central peak formation to be estimated to be ~100 seconds (Melosh and Ivanov, 1999).

The walls of the complex craters experience greater collapse than that seen in simple craters. These walls typically exceed the angle of repose in the transient cavity, so that large-scale slumping may occur immediately after the excavation stage to produce terrace blocks (Figure 1.5). Slumping occurs in complex craters along discrete faults, forming terraces whose widths are controlled by strength of the rock and the stresses as the floor rises (Melosh, 2011). Structural uplift of the rim crest is also a permanent feature of the cratering process. For Martian craters, Sturm et al. (2016) documented this phenomenon for 10 complex craters in the diameter range 8.2–53.0 km and found that the mean structural uplift at the final crater rim comprises ~80% of the total rim elevation, implying that the thickness of the ejecta on the rim is ~20% of the total (Figure 1.6). A similar amount of rim uplift was also reported by Sharpton (2014) for lunar craters.

The largest "craters" in the Solar System are known as multi-ringed basins, of which Orientale Basin (Figure 1.7) is one of the best-defined basins on the Moon (Head, 1974). Baker et al. (2011) gave a fine summary of the progression of morphologies with increasing crater diameter (Figure 1.8).

1.5 EJECTA EMPLACEMENT

The ejecta blankets of Martian craters are some of the most diverse and, potentially, most informative about the characteristics of the target material (Head and Roth, 1976; Carr et al., 1977; and many others). This is particularly true when impacts into targets containing volatiles are concerned (Gault and Greeley, 1978). The major types of ejecta blanket on Mars will be discussed in detail in Chapter 6. Absent of atmospheric effects, ejecta from any impact crater are thrown from the crater cavity in ballistic trajectories (Oberbeck, 1975; Oberbeck et al., 1975). Ejecta from close to the point of impact are ejected at the highest velocities and highest angles relative to the impacted surface (Figure 1.9). Ejected material will first hit the surrounding surface near the crater rim and then at progressively greater ranges. Importantly for the analysis of the layered ejecta surrounding craters on Mars, the re-impacting ejecta land at progressively higher velocities with increasing range due to their point of origin within the crater cavity. Some of the characteristic flow features associated with fluidized ejecta, such as the distal ramparts (see Section 6.6), may be a consequence of this velocity field (Boyce and Mouginis-Mark, 2025).

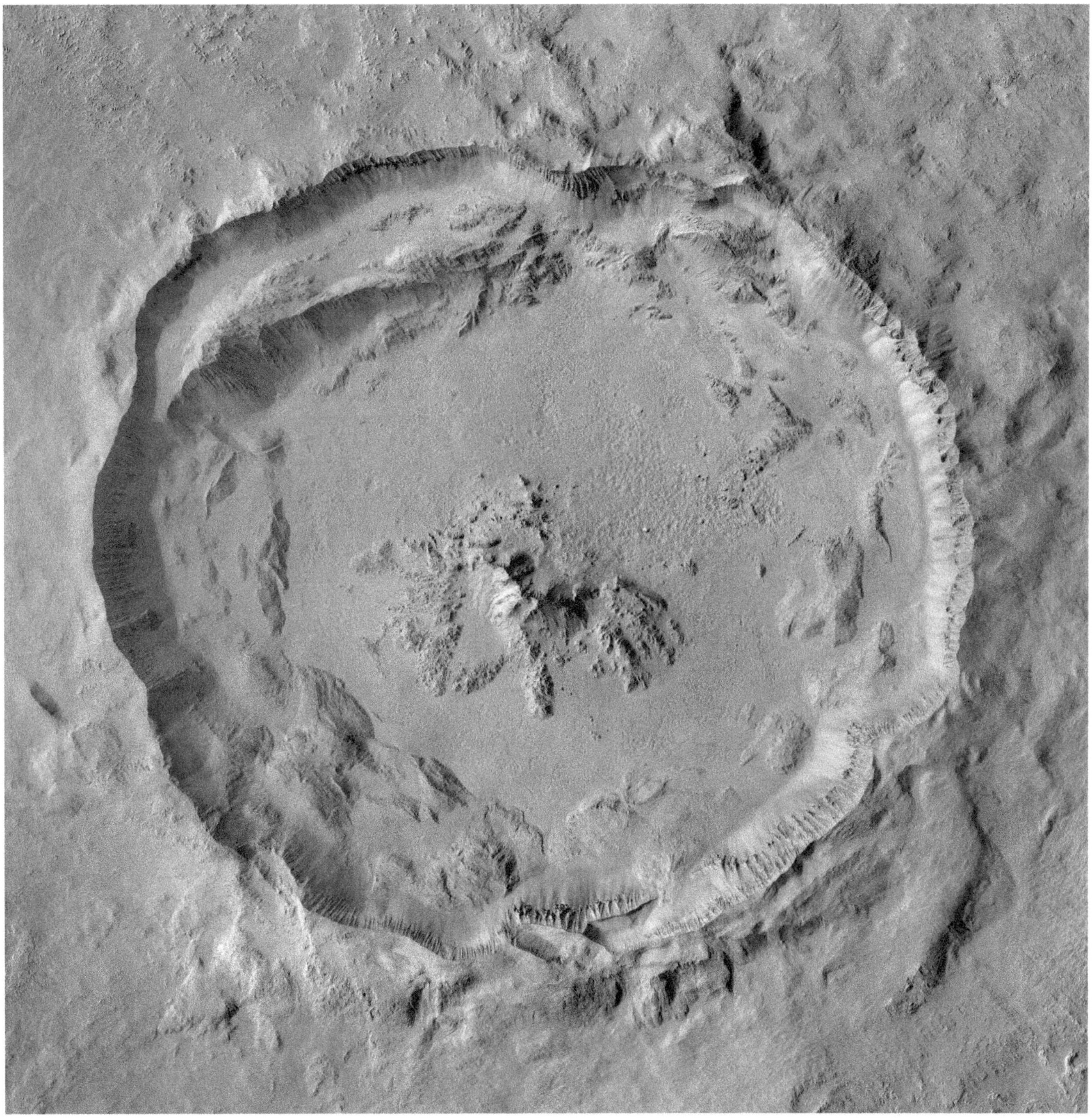

Figure 1.5 Tooting crater on Mars (23.21°N, 207.76°E) displays multiple terrace blocks on the inner western walls of the crater, as well as a very large central peak. Tooting crater is 28.0 km in diameter. CTX image P01_001538_2035.

As the crater size increases, the velocity at which ejecta re-impact the surface will be sufficiently high that the surrounding surface is excavated by secondary cratering. Material from these "secondary craters" will mix with the primary ejecta. Where individual ejecta blocks land sufficiently close to each other, they form a "herringbone" or dune-shaped pattern. The ejecta from the lunar crater Copernicus (Figure 1.10) provide fine examples of such secondary craters (Guest and Murray, 1971). Copernicus crater is estimated to be ~800 Ma old (Eberhardt et al., 1973), so that the slow rate of erosion on the Moon allows the analysis of these fine-scale features (Pieters et al., 1985).

During the ballistic emplacement of the ejecta, a considerable amount of the original surface is excavated (Oberbeck, 1975). Large fragments of the ejecta in low-angle-low-velocity trajectories are the first to impact the surface close to the crater rim. With increasing velocity, particles from closer to the point of impact attain a greater range from the still-forming primary crater and hit the surface at higher angles. Collectively, the ejecta sweeps up additional material from the pre-existing terrain such that the ejecta blanket contains more locally derived material with a greater radial distance from the impact site. The physical properties of the radial flow of the mixed primary and

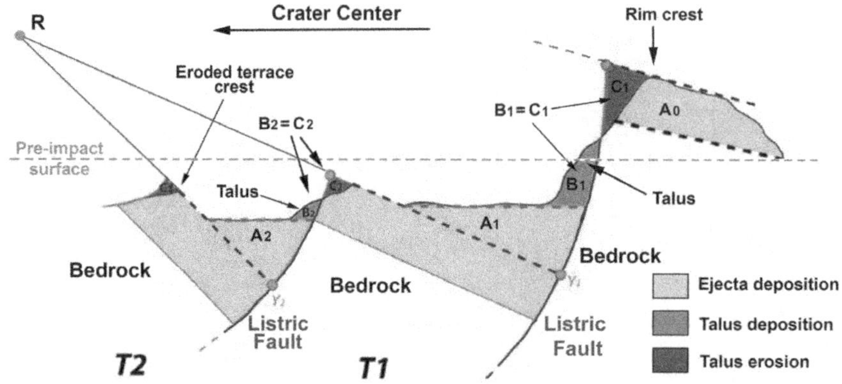

Figure 1.6 Illustration of how terrace blocks (T1 and T2) may form along listric faults and then may be eroded in a complex crater. Talus eroded from exposed blocks were rotated along these faults (C1 and C2), and then redeposited at the base of the blocks (B1 and B2). This diagram illustrates that one might find ejecta deposits (impact breccia and impact melt) at multiple places within the interiors of complex craters. Reprinted from Sturm et al. (2016) "Ejecta thickness and structural rim uplift measurements of Martian impact craters: Implications for the rim formation of complex impact craters," *Journal of Geophysical Research: Planets*, vol. 121, pp. 1026–1053, Figure 9. With permission of John Wiley and Sons, Inc. Note: A black and white version of this figure will appear in some formats. For the color version, please refer to the plate section.

Figure 1.7 Orientale Basin is the youngest of the large lunar basins (19.56°S, 266.25°E). The distinct outer ring is about 930 km in diameter from east to west. The black lines on the eastern and western rims are due to data dropouts. LROC WAC mosaic. NASA/GSFC/Arizona State University.

secondary material behind the ejecta curtain also change with radial distance from large relatively slow-moving particles close to the rim to finer particles moving more quickly.

In the case of the Moon, approximately half the volume of ejecta falls within one crater radius of the rim (McGetchin et al., 1973; Settle and Head, 1977). The ejecta blanket becomes thinner with radial distance and becomes increasingly discontinuous. Over 90% of the debris excavated from the crater cavity falls within ~4 radii of the crater rim. In the case of Mars, the ejecta may incorporate volatiles (water or ice) from multiple levels of the stratigraphic column of the target. These volatiles may provide a level of fluidization that allows the ejecta to flow across the surface as a late-stage part of the ejecta emplacement process. Thus, ejecta blankets may well be the "Rosetta Stone" of Martian geology when it comes to investigating target conditions at the time of impact. As we investigate ejecta blanket morphology in later chapters, it must be recognized that some of the material originated from within the crater cavity and that other material is reworked material incorporated from the surrounding terrain. The influence of the atmosphere should also be considered (Schultz and Gault, 1979), as aerodynamic drag may restrict the range of fine particles.

1.6 CRATER GEOMETRY

As described earlier, crater diameter is a critical parameter in determining what the interior features might be within a fresh impact crater (Turtle et al., 2005). The measurement of the diameter and other dimensions of craters (Figure 1.11) has become an important way in which the global population of craters can be investigated as well as quantify the degree of degradation of crater shape. As is the case on the Moon and Mercury, it is well known that Mars displays a spectrum of morphologies that increase in complexity as diameter increases (Hartmann, 1972; Cintala et al., 1976). Simple bowl-shaped craters are associated with the smallest craters. As diameter increases, the walls are sufficiently unstable so that they start to collapse, forming "terraces" on the walls that partially infill the crater floor, making the crater shallower. The simple-to-complex transition for Mars craters occurs at diameters between ~8 and ~11 km (Garvin and Frawley 1998; Robbins and Hynek, 2012). Central peaks and central pits both occur at larger crater diameters beyond this transition.

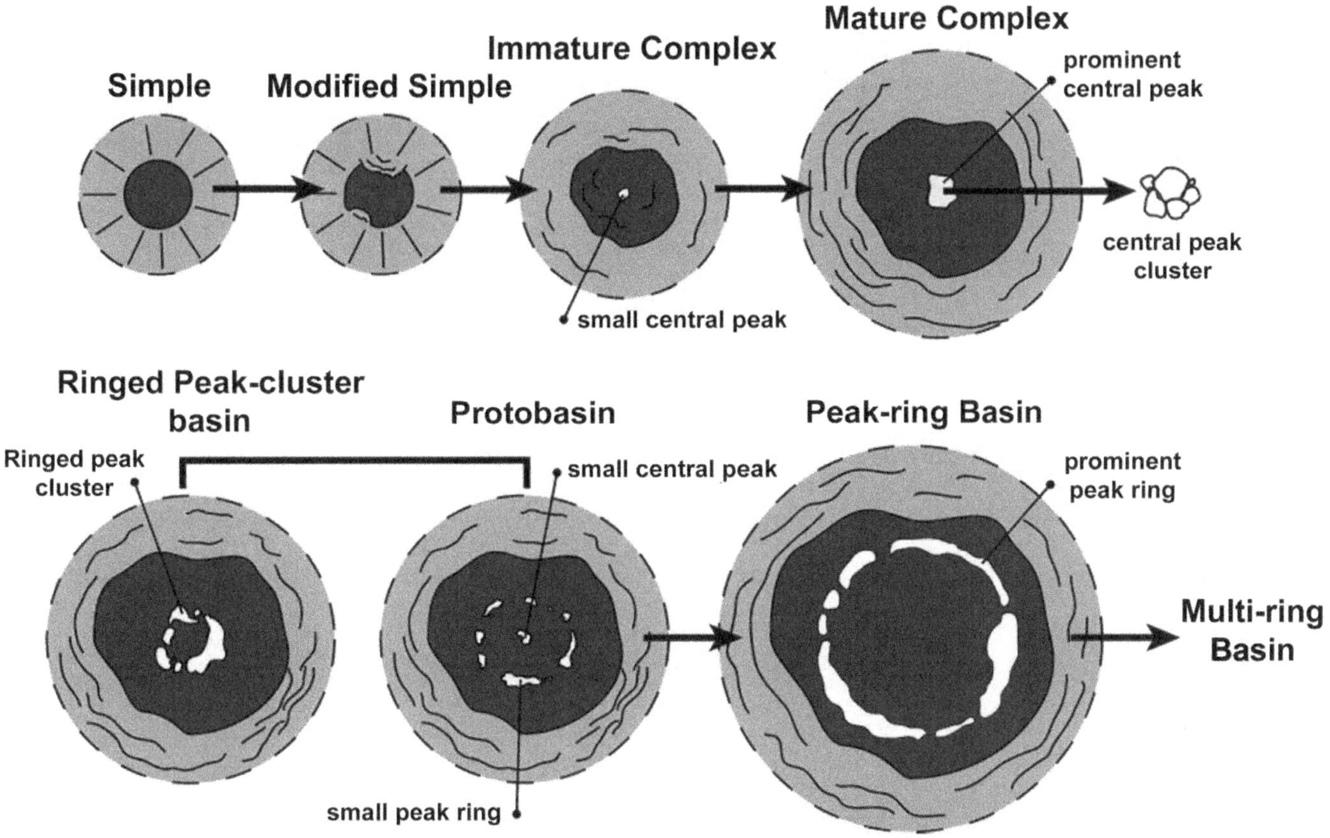

Figure 1.8 The complexity of fresh craters as a function of diameter. Simple craters exhibit smooth, featureless interiors (simple) to minor wall slumping (modified simple). Complex craters exhibit large slump deposits and rudimentary terracing with small central peaks (immature complex) to strong terracing and single central peaks to clusters of central peak elements (mature complex). The onset of basin morphologies occurs with a ringed arrangement of peak elements (ringed peak cluster) or the presence of both a small central peak and peak ring (proto-basin). Peak-ring basins have large, prominent peak rings with no central peaks. Multi-ring basins (not shown here) exhibit three or more rings and commonly an inner depression. Reprinted from Baker et al. (2011) "The transition from complex crater to peak-ring basin on Mercury: New observations from MESSENGER flyby data and constraints on basin formation models," *Planetary Space Science*, vol. 59, pp. 1932–1948, Figure 1. With permission from Elsevier.

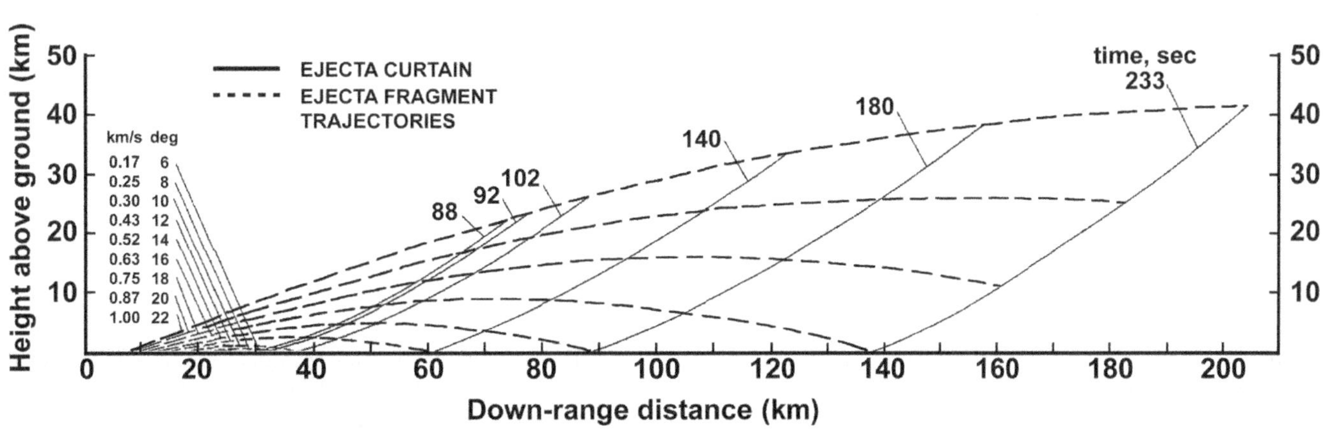

Figure 1.9 Calculated trajectories (dashed lines) and positions of ejecta at a given constant time (solid lines) after the impact of the body that produced the 92 km diameter Copernicus crater on the Moon. Note that the ejection angles vary from 22° close to the point of impact to 6° further from the impact. Over this range of angles, ejecta leave the cavity at speeds of 1.0 to 0.17 km/sec and will re-impact the surface at these speeds. Due to the higher surface gravity, flight times and down-range distances will be smaller on Mars for a comparable energy impact. Reprinted from Oberbeck (1975). "The role of ballistic erosion and sedimentation in lunar stratigraphy," *Reviews of Geophysics*, vol. 13, pp. 337–362, Figure 12. With permission from John Wiley and Sons, Inc.

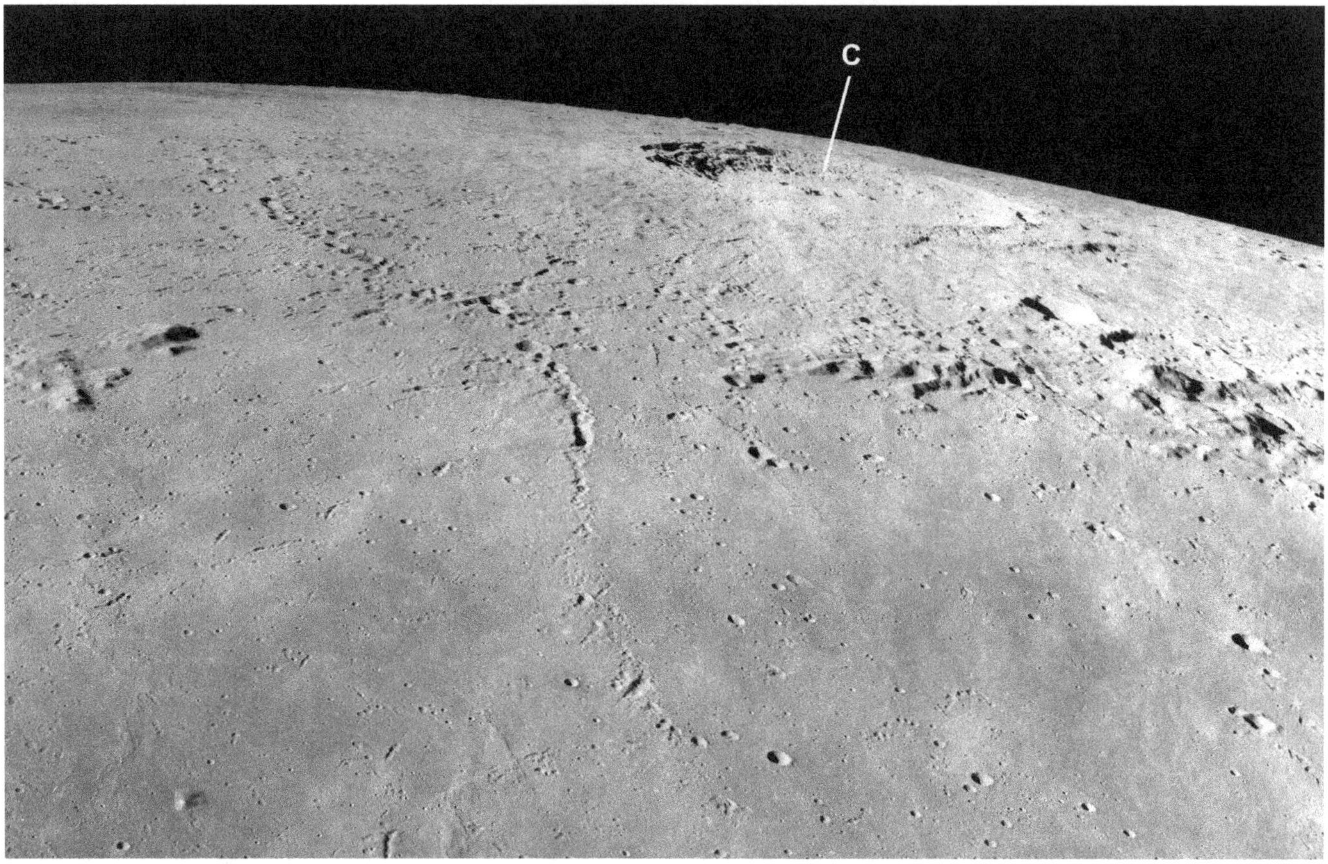

Figure 1.10 Ejecta northeast of Copernicus crater on the Moon (9.61°N, 339.92°E) clearly illustrate the role of secondary impact cratering and the formation of crater chains with the associated "V-shaped" deposits between individual craters. Copernicus crater (92 km in diameter) is near the horizon just right of the center in this image ("C"). In this view, the closest pit chain to the crater is ~170 km from the rim of Copernicus, and the secondary craters at the bottom of the image are ~243 km from the rim. The mounds at right and far left are topographic features that predate Copernicus crater. Apollo image AS17-M-2439.

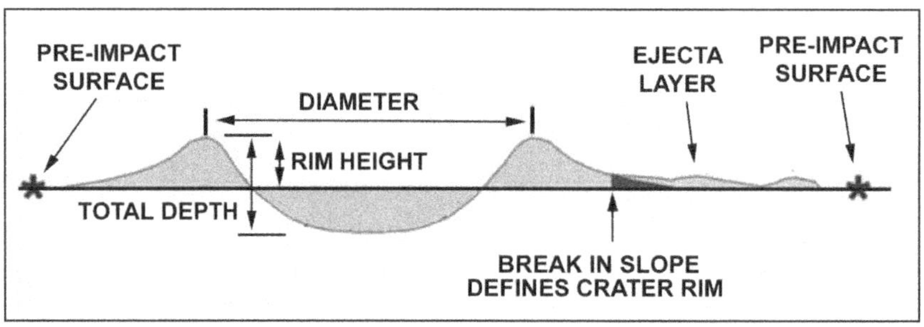

Figure 1.11 Idealized cross-section for an impact crater on Mars, showing the attributes of the cavity that are most often used in the geometric analysis of the crater. Reprinted from Mouginis-Mark et al. (2018). "Determination of Mars crater geometric data: Insights from high-resolution digital elevation models," *Meteoritics and Planetary Science*, vol. 53, pp. 726–740, Figure 1. Reprinted with permission from John Wiley and Sons.

The geometry of these attributes (Figure 1.11) has been investigated for Mars for more than 40 years (e.g., Cintala and Mouginis-Mark, 1980; Pike, 1980b; Garvin et al., 2000; Chappelow and Sharpton, 2002; Boyce et al., 2005; Tornabene et al., 2018a; Robbins et al., 2018a). The goal has often been focused on the search for the effects of subsurface volatiles (Mouginis-Mark and Hayashi, 1993; Barlow and Perez, 2003) as well as target strength (Boyce et al., 2006; Stewart and Valiant, 2006), or the geometric comparison of Martian craters with those on the Moon and Mercury (Malin and Dzurisin, 1977) to identify the effects of gravity or the atmosphere of Mars. Crater depth/diameter ratio (Figure 1.12) has been used most often (Daubar et al., 2014; Tornabene et al., 2018a), but other geometric attributes, such as rim height versus diameter (Craddock et al., 1997; Watters et al., 2015), central peak diameter versus crater diameter (Barlow et al., 2017), rim height versus depth (Figure 1.13; Watters et al., 2015), and central peak height versus rim height (Bennett and Bell, 2016), have all been investigated. Other relationships have addressed different attributes of the ejecta blanket, such as the "ejecta mobility," which is the maximum range of the ejecta from the crater rim (Mouginis-Mark, 1978; Barlow et al., 2014), the ejecta run-out distance versus the crater rim height (Mouginis-Mark and Baloga, 2006), and the "lobateness" of the ejecta layers (Barlow, 1994; Baratoux et al., 2002).

Measuring crater diameter is relatively straightforward, although some changes as the crater is eroded have been noted (Chappelow and Sharpton, 2002). But critical to the correct interpretation of any of these geometric relationships is the accurate measurement of height, be it the height of the rim crest above the crater floor, the height of the rim above the surrounding terrain, or the height of the distal rampart on the ejecta blanket. Over the decades, there have been many methods employed to measure heights with increasing levels of accuracy. For Mars, these have included Earth-based radar altimetry (Roth et al., 1989), shadow length measurements using known lighting geometry (Cintala and Mouginis-Mark, 1980), photometry of the surface from single spacecraft images (Davis and Soderblom, 1984), laser altimetry (Garvin and Frawley, 1998; Garvin et al., 2000; Robbins and Hynek, 2013), and measurement from digital elevation models derived from stereo images (Mouginis-Mark et al., 2018). Robbins et al. (2018a) provide a comprehensive summary of these different techniques.

Most often used today because of the global nature of the data set is the altimetry derived from the Mars Orbiter Laser Altimeter (MOLA) (Smith et al., 2001). However, even this data set has limited spatial resolution at the kilometer scale (Figure 1.14). Particularly near the Equator, the cross-track (roughly east-west) spacing of the laser profiles may be a few kilometers, while along-track (approximately north-south) the 300 m-spacing of the individual laser shots means that the highest points on the rim, or the lowest points on the floor, may be missed. In a comparison between MOLA and High-Resolution Stereo Camera (HRSC) digital elevation models, Robbins and Hynek (2013) found that crater diameters determined from MOLA data are reliable for diameters >10 km. For craters smaller than 1 km, the shadow measurement technique of Chappelow and Sharpton (2002) is preferred for the determination of crater depths compared either to altimetric or stereo measurements, both of which have lower spatial resolution (Herrick, 2013).

Complicating the determination of geometric ratios that involve crater depth and rim height are actual topographic variations of the rim crest, which will lead to varying estimations of the crater depth. Uneven crater floors can also

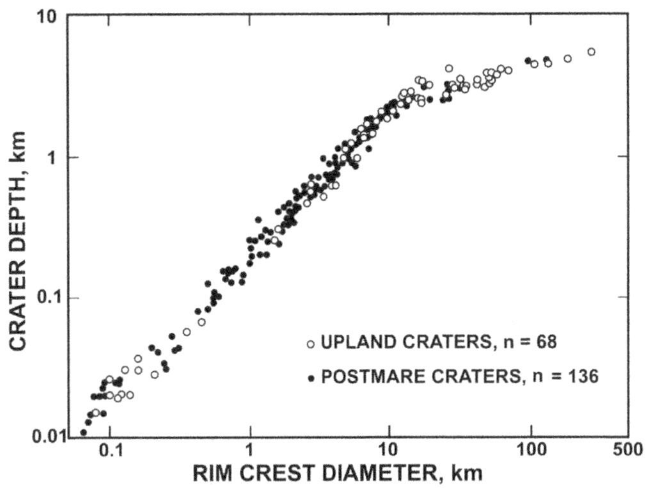

Figure 1.12 An example of a depth/diameter plot for fresh craters. This example is for 204 fresh lunar craters, compiled photogrammetrically from Apollo 15, 16, and 17 images. Note the transition from simple to complex craters at ~11 km in diameter. Reprinted from Pike (1974). "Depth/diameter relations of fresh lunar craters: Revisions from spacecraft data," *Geophysical Research Letters*, vol. 1, pp. 726–740, Figure 1, with permission from John Wiley and Sons.

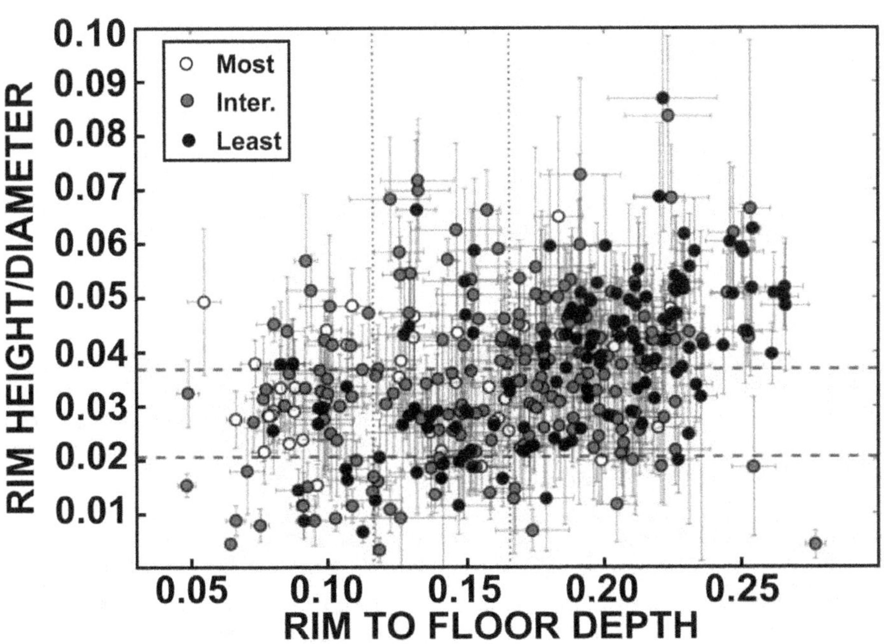

Figure 1.13 Diameter-normalized rim height (h/D) versus rim-to-floor depth (d/D) for a global distribution of Mars craters 25 m to 5 km diameter. White points represent the most modified craters, black points the least modified, and gray points the intermediate values. The purpose of this diagram is to investigate how the rims of craters degrade over time compared to the widening of the crater as erosion proceeds. Reprinted from Watters et al. (2015). "Morphometry of small recent impact craters on Mars: Size and terrain dependence, short-term modification," *Journal of Geophysical Research: Planets*, vol. 120, pp. 226–254, Figure 4, with permission from John Wiley and Sons.

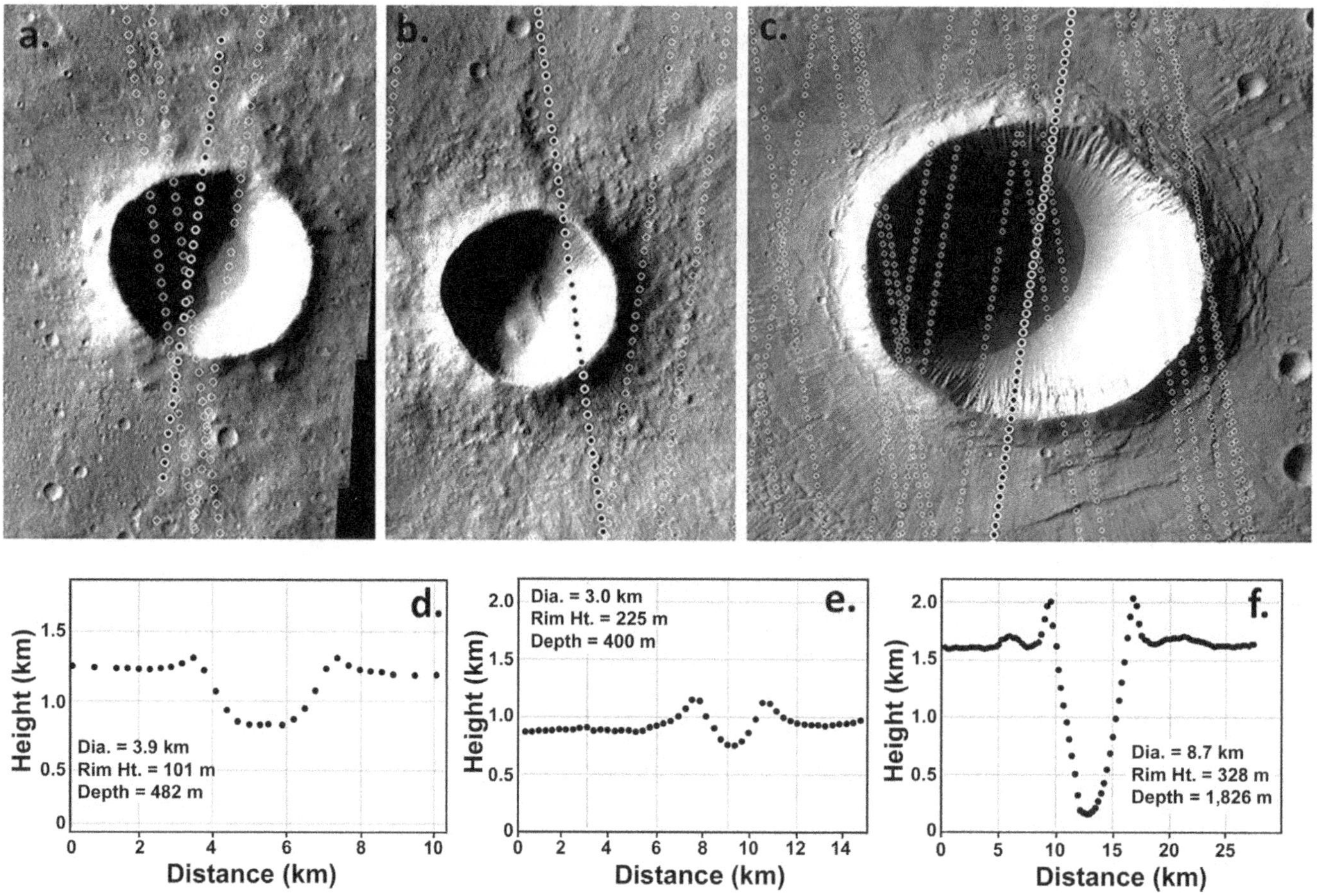

Figure 1.14 Examples of individual MOLA tracks across Mars craters of different diameters. Faint lines of dots indicate the locations of all the MOLA laser shots in the area, and black dots indicate where the data have been used for each profile. "Distance" is measured along the profile from the top of each image. (a) A 3.9 km diameter crater (15.21°N, 98.30°E) where four profiles cross the central part of the crater floor. THEMIS image V28627019. (b) A 4.1 km diameter crater (8.77°N, 181.34°E) where only one MOLA profile crosses the crater with the measurement falling on the inner wall rather than the crater floor. THEMIS image V53698016. (c) Multiple MOLA profiles cross this 9.3 km diameter crater (29.65°N, 116.55°E). THEMIS image V28876009. (d) to (f) show the corresponding topographic profiles across each crater. Reprinted from Mouginis-Mark et al. (2018). "Determination of Mars crater geometric data: Insights from high-resolution digital elevation models," *Meteoritics and Planetary Science*, vol. 53, pp. 726–740, Figure 2, with permission from John Wiley and Sons.

make these measurements less meaningful, because craters can often have tens to hundreds of meters of height variation across the floor due to slumping of the inner walls or to nonuniform uplift of the central peak. Originally recognized at lunar craters by Sharpton (2014), Mouginis-Mark et al. (2018) demonstrated that the elevation of crater rims on Mars can vary with azimuth by several hundred meters (Figure 1.15). A detailed discussion of this rim height variation is provided in Section 9.7. Thus, the simple relationship between crater depth and diameter may not be as simple as originally believed. Particularly given uneven crater floor topography, the actual rim height and the measured floor depth may vary significantly (perhaps ±10%). At Tooting crater, for example, Mouginis-Mark and Garbeil (2007) noted that the rim crest has a minimum height of 386 m above the surrounding terrain, a maximum height of 893 m, and an average height of 628 m (i.e., a variation of ~40% from the average). Taking 28.0 km as the diameter of the crater, the average rim height value gives a rim height/crater diameter ratio of 0.022. Making similar measurements for the depth of the floor below the rim crest gave a value of 1,690 m for the northern portion of the crater floor and 1,881 m for the southern part of the floor, giving an average of ~1,785 m and a range equivalent to ~11% of the measurements. Although not drastically different, the depth/diameter ratio may consequently vary from 0.058 to 0.065 depending upon where the measurements were made.

1.7 SUMMARY

In this chapter we have reviewed the stages of the impact cratering process including ejecta emplacement. Measurement of parameters such as the crater depth, diameter, and rim height may be used to compare craters from different locations on any planetary body. The basic classification of craters as either simple or complex, as well as the larger basin-sized craters, is a useful tool to investigate details of the morphology of specific craters. In Chapter 2 we take a

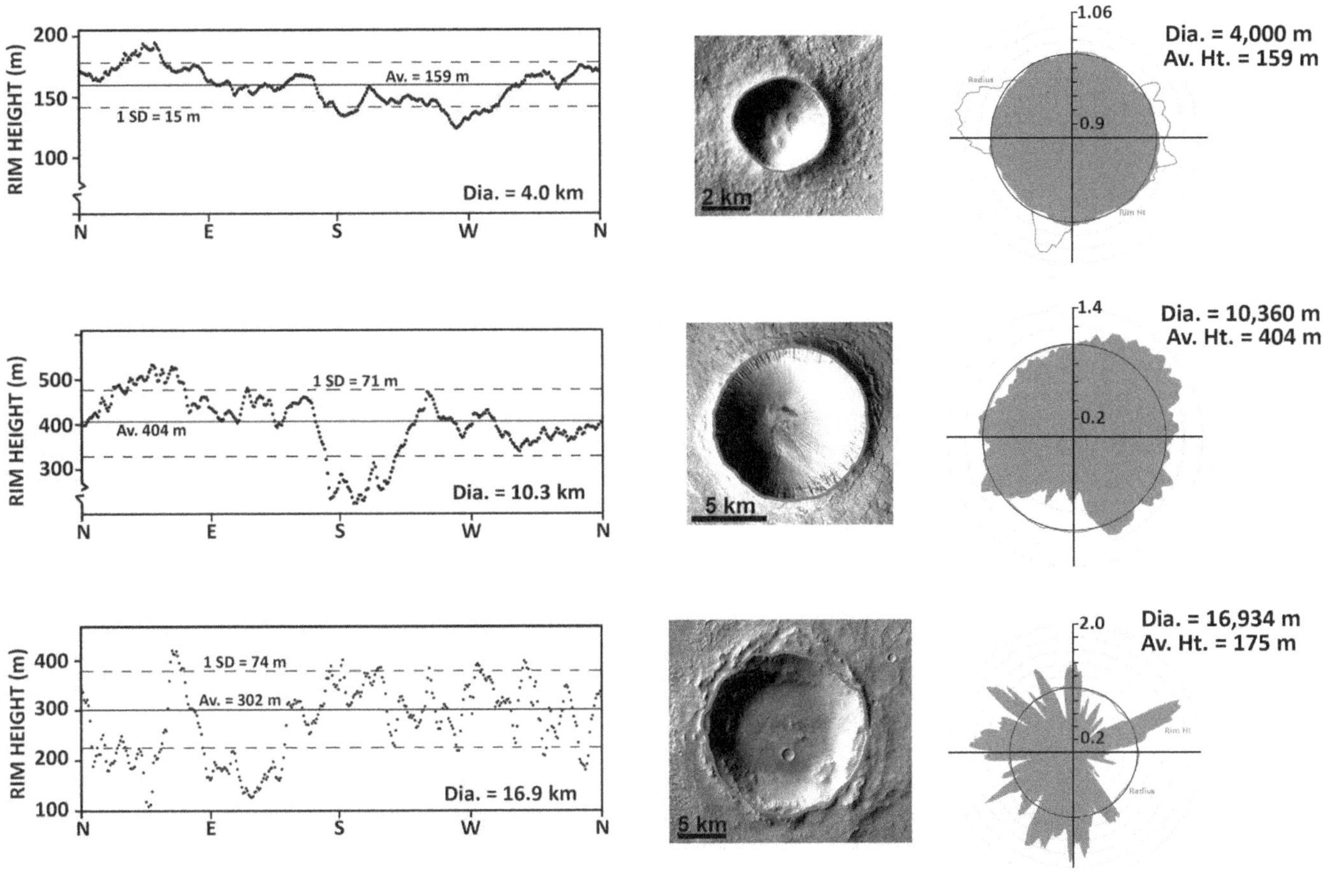

Figure 1.15 Variations in rim height for three Mars craters in the diameter range of 4.0–16.9 km. At left are traverses along the identifiable rim crater. North is at the far left and far right of each plot. Average (solid horizontal line) and one standard deviation values (dashed horizontal lines) for the 360 individual rim height measurements are also shown. Center images are CTX frames (Top: D22_035823_1870; middle: P15_007027_2099; bottom: D22_035823_1870) with the rim crest showing the trace of each measured rim crest. At right are azimuth plots showing the variation in crater rim height as a fraction of the average rim height for the three hundred and sixty 1° azimuth measurements. Notice that for the smallest crater (top), there is little azimuthal variation in rim height, but for the largest crater (bottom), there are large excursions from the average height.

tour through the Solar System to gain an overview of crater morphology, as seen on bodies with rocky or icy surfaces, as well as those with or without an atmosphere. This will form a reference point against which the morphology of impact craters on Mars can be intercompared. Not all places on Mars are the same, and so in Chapter 3 we will consider some of the spatial and temporal differences that may result in craters having very different morphologies even if they are the same size and age, as well as what might constitute a "fresh" or "pristine" crater on Mars.

2

Morphology of Craters Throughout the Solar System

2.1 INTRODUCTION

In Chapter 1 we reviewed the basic physics of the impact cratering process from the excavation of the crater cavity to the formation of an ejecta blanket. Various attributes of the crater geometry were also introduced by which craters can be quantitatively intercompared. This gives us a framework against which specific examples of craters on any planetary body can be investigated. While this book is focused on Mars, it is nevertheless valuable to first look at craters elsewhere in the Solar System where the environment is less complicated. Thus, in the subsequent chapters (Chapters 3 to 8), we will be able to recognize attributes of craters that are unique to Mars and, through their distribution across the planet, tell us more about the Red Planet.

This chapter reviews impact craters throughout the Solar System, looking first at craters formed on Earth, where we have the best field knowledge. We then investigate craters formed on airless rocky bodies (the Moon and Mercury), where the cratering process is not affected by atmospheric effects. We follow this with a glimpse of craters on volatile-rich bodies that also lack an atmosphere, specifically Ganymede, 1 Ceres, and Charon. Here the target material is most likely water ice. Finally, we examine craters formed on bodies with thick atmospheres (Venus and Titan) to see what landforms may have been formed by the interaction of the projectile and the ejecta with the atmosphere. In each of these environments (rocky airless, volatile-rich, and those with atmospheres), we first look at the interior of the crater and then move radially outward to the rim, the ejecta blanket, and finally the extended ejecta deposits beyond the continuous ejecta blanket.

Key landforms that are known to change between different planetary bodies are the interior features (e.g., central peaks, wall units, and the distribution of impact melt), as well as the ejecta deposits that lie beyond the rim crest. Different gravity fields also come into play, with a low gravity allowing ejecta to be thrown much further for a given energy than on a planet with Earth-like gravity. Crater shape is also affected by gravity, with the walls prone to collapse to form terraces at smaller crater diameters on planetary bodies with higher gravity. Interior walls also deform more easily on a body with a high gravity, so that the simple bowlshape of a small crater is more rapidly deformed by terrace formation on Venus than it would be on the Moon. A higher gravity will result in a smaller crater for the same impact energy since gravity will work against the opening of the crater during the excavation stage. Conversely, a higher gravity will also mean that a projectile of a given mass will hit the surface with greater kinetic energy than if the object were to hit a body with low mass, so that a larger crater may result.

2.2 EARTH

Terrestrial craters yield valuable information pertinent to the interpretation of craters formed on rocky planets. Osinski and Pierazzo (2013) have detailed the geology of several impact craters on Earth that provide insights into planetary impact craters. Meteor crater, Arizona (Figures 2.1 and 2.2), offers some of the best insights into impact craters on Earth (Roddy et al., 1975). This crater is ~1.2 km in diameter and ~175 m deep and is very young. Little erosion has been identified in the ejecta blanket of Meteor crater since its formation ~50,000 years ago (Nishiizumi et al., 1991; Grant and Schultz, 1993). Ground-penetrating radar (Pilon et al., 1991) reveals that the total uplift at the crater rim is ~55 m and the effects of this uplift can be detected to a radial distance of ~400 m from the rim. Kumar et al. (2010) developed a useful schematic cross-section for Meteor crater (Figure 2.3) that includes uplift of the rim material (the Kaibab Formation and the Coconino sandstone) that is relevant to the interpretation of rim-uplift on planetary craters.

Lonar crater, India, a ~1.9 km diameter crater formed within the Deccan traps, is a useful analog for small craters on the terrestrial planets and the Moon as it is one of only four known craters on Earth that formed in basaltic rock (Kumar, 2005; Maloof et al., 2010). The 14 km diameter Ries crater, Germany, exhibits fluidization and hydrothermal alteration of ejecta deposits (Newsom et al., 1986), as well as erosion of the basement rocks via surface flow of the ejecta (Hörz and Banholzer, 1980; Hörz et al., 1983). Manicouagan crater (~100 km in diameter) in Quebec, Canada (Figure 2.4), has been used for both structural comparisons with large lunar craters (Grieve and Head, 1983) as well as the analysis of the rate of cooling of impact melt generated during the impact event (Simonds et al., 1978). At a larger scale, the remnants of the ~200 km diameter Chicxulub crater in Mexico reveal details of the collapse of the transient crater cavity and the mode of formation of peak rings (Morgan et al., 2000; Gulick et al., 2013). The largest terrestrial crater is the 2.02 Ga Vredefort impact structure in South Africa (Henkel and Reimold, 1998). Considering that as much as ~6 km of vertical erosion has

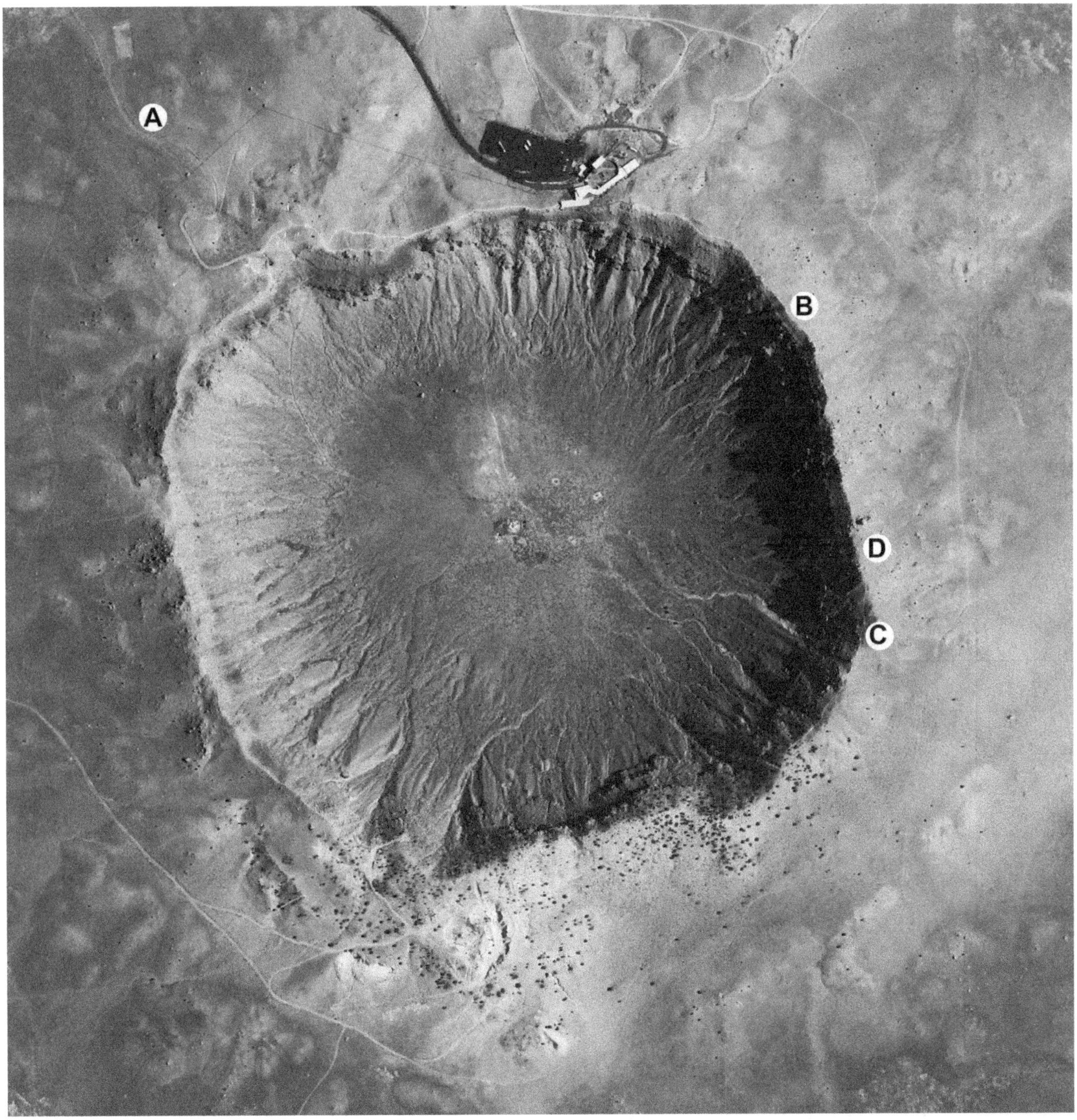

Figure 2.1 Meteor crater, Arizona (35.01°N, 111.00°W). The crater is 1.2 km in diameter. This air photograph serves as a great comparison to some of the images we will see of planetary craters, with the layered rim material, gullies on the inner walls, and some large ejecta blocks all visible. See Figure 2.2 for the ground photos identified at points A, B, C, and D. U.S. Geological Survey National Map Service image 2007158.

taken place here, scaling to the original final rim diameter provides an estimate of close to 300 km for the original final rim diameter (Therriault et al., 1997). Vredefort gives us some of the best indicators for the processes associated with basin formation on the other planets.

Due to erosion, it is difficult to assign the ~215 Ma Manicouagan crater to a particular morphological class of impact structure, but based on the available data, it is most likely that the pre-erosional form was that of a central peak crater or possibly a peak-ring basin. A reconstruction of the pre-erosion geology of Manicouagan has been used by Grieve and Head (1983) and revealed that, at the time of its formation, the Manicouagan crater may have been both dimensionally and possibly morphologically similar to the 92 km diameter lunar crater Copernicus.

Figure 2.2 Field photographs of Meteor crater, Arizona. See Figure 2.1 for image locations. Top left is a view of the crater's northern rim crest seen from the approach road. Top right shows the interior of crater. The crater is ~175 m deep. The bottom left image illustrates the inner southern wall. The bottom right illustrates one of the largest ejecta blocks on the SE rim. All photos by the authors. Note: A black and white version of this figure will appear in some formats. For the color version, please refer to the plate section.

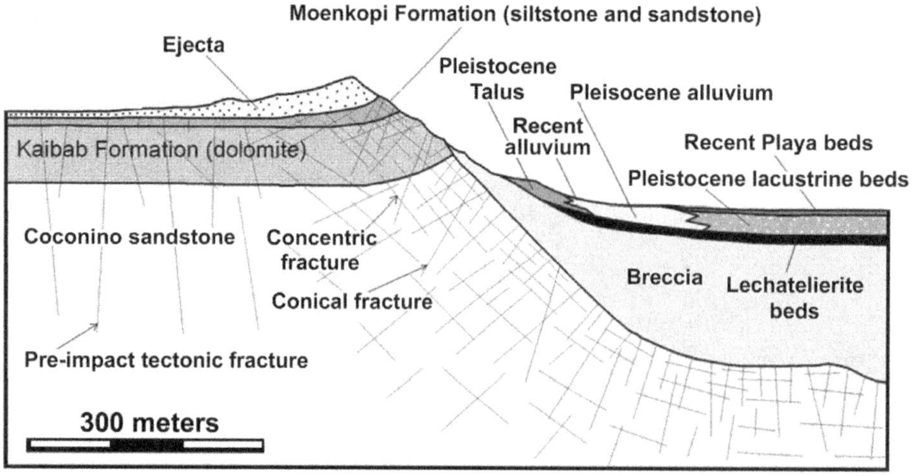

Figure 2.3 Schematic geologic cross-section of Meteor crater, showing various rock units, structures, and crater floor sediments. Reprinted from Kumar et al. (2010). "Erosional modification and gully formation at Meteor Crater, Arizona: Insights into crater degradation processes on Mars," *Icarus*, vol. 208, pp. 608–620, Figure 13, with permission from Elsevier.

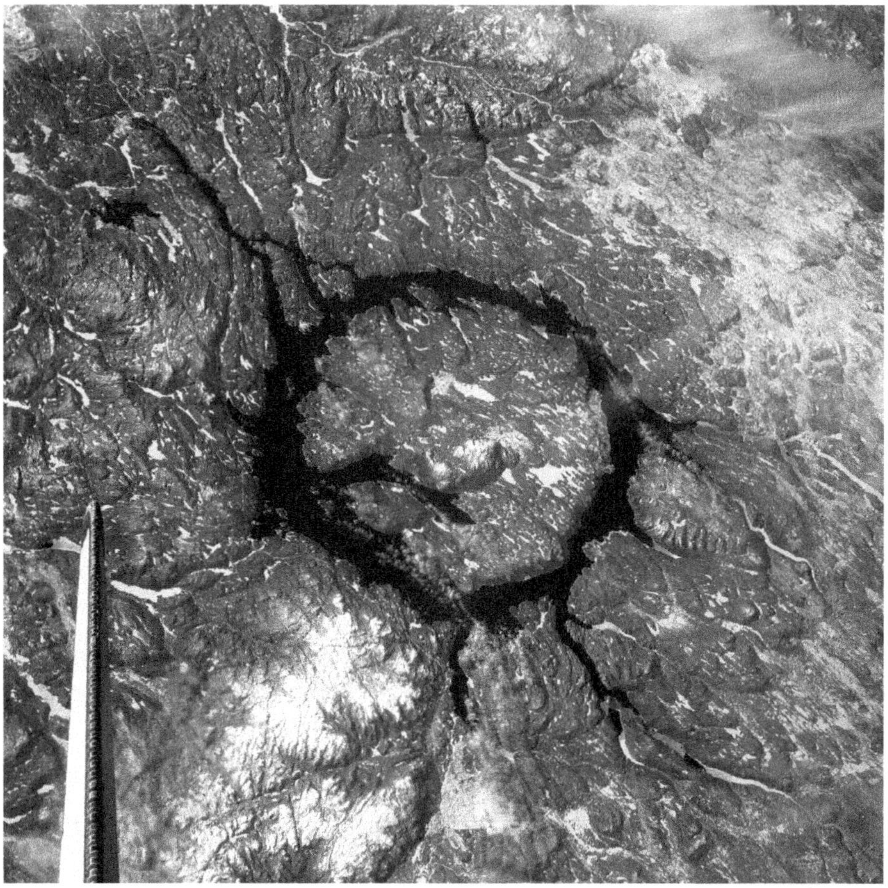

Figure 2.4 The eroded Manicouagan crater in Canada (51.10°N; 68.70°W) is ~215 million years old. The crater has a multi-ring structure ~100 km across and is filled by an annular lake ~70 km across. This lake surrounds an inner island plateau comprising the remnants of the central peak structure. Space Shuttle photo no. STS009-048-3139.

2.3 CRATERS ON VOLATILE-POOR BODIES – THE MOON AND MERCURY

The Moon and Mercury offer the possibility to see morphological changes in crater shape and features as a function of size unencumbered by the effects of an atmosphere or target volatiles. This comparison is particularly instructive as Mercury has a surface gravity approximately double that of the Moon. Mercury and Mars have comparable gravities. Thus, in this section, we review the characteristics of fresh craters on these airless rocky bodies to contrast them with craters formed in icy targets, or craters formed on planets or moons with an atmosphere. But watch out! Craters will appear to have different characteristics depending upon the lighting geometry of the images, as Figure 2.5 demonstrates. Fortunately, for the Moon in particular, numerous space missions have imaged lunar craters under a wide range of lighting and viewing geometries as well as different image spatial resolutions.

2.3.1 Interior Features

2.3.1.1 Simple Craters

Crater diameter is closely linked to the frequency of occurrence of interior features such as terraces on the inner walls, simple bowl-shaped or flat floors, and the possible occurrence of central peaks (Hartmann, 1972; Pike, 1977; Cintala et al., 1977). As we saw in Chapter 1, such structures are formed due to the collapse of the transient crater cavity under the gravity of the body. Smaller craters (Figure 2.6), typically only a few kilometers in diameter, possess fewer of these landforms, but their rate of occurrence increases with crater size. Xiao et al. (2014) investigated how gravity may influence the geometry of fresh complex craters during the impact excavation process on Mercury and the Moon. As a rule of thumb, as gravity increases, the walls of a crater become unstable at smaller crater diameters, so that a simple crater on the Moon may be larger than a complex crater of comparable diameter on Mercury.

2.3.1.2 Terraces

Terraces within the interior of complex craters appear to be the head scarps of landslides that carried large amounts of material from the unstable rim into the interior of the crater. Terraces may be up to a few kilometers wide, and the widths of terrace slumps of complex craters can be used to determine the effective strength of the target material during crater collapse (Leith and McKinnon, 1991). As a result of terrace formation, wall failure has significantly altered the geometry of large, fresh-appearing craters (Settle and Head, 1979). In general, terrace widths increase outward toward the rim for any given crater and in

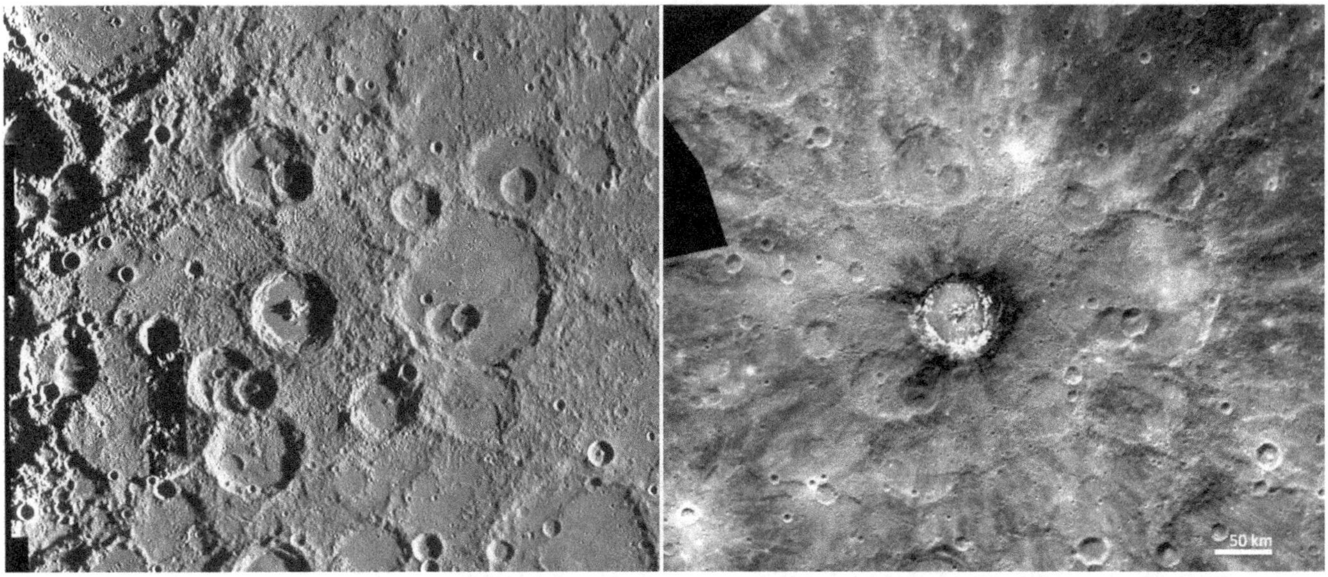

Figure 2.5 Planetary craters can be two-faced, at least when viewed under radically different lighting conditions! This side-by-side comparison of Basho crater (74.6 km diameter) on Mercury illustrates the importance of studying the same landform under different lighting conditions. At left we see a low-Sun image that highlights the subtle topography around the crater and suggests a very deep crater. While at right, the high-Sun view highlights albedo differences, with the bright crater rays and dark rim material (impact melt?) prominent. Image centered at 32.38°S, 189.55°E. NASA/JPL-Caltech Photojournal image PIA16343.

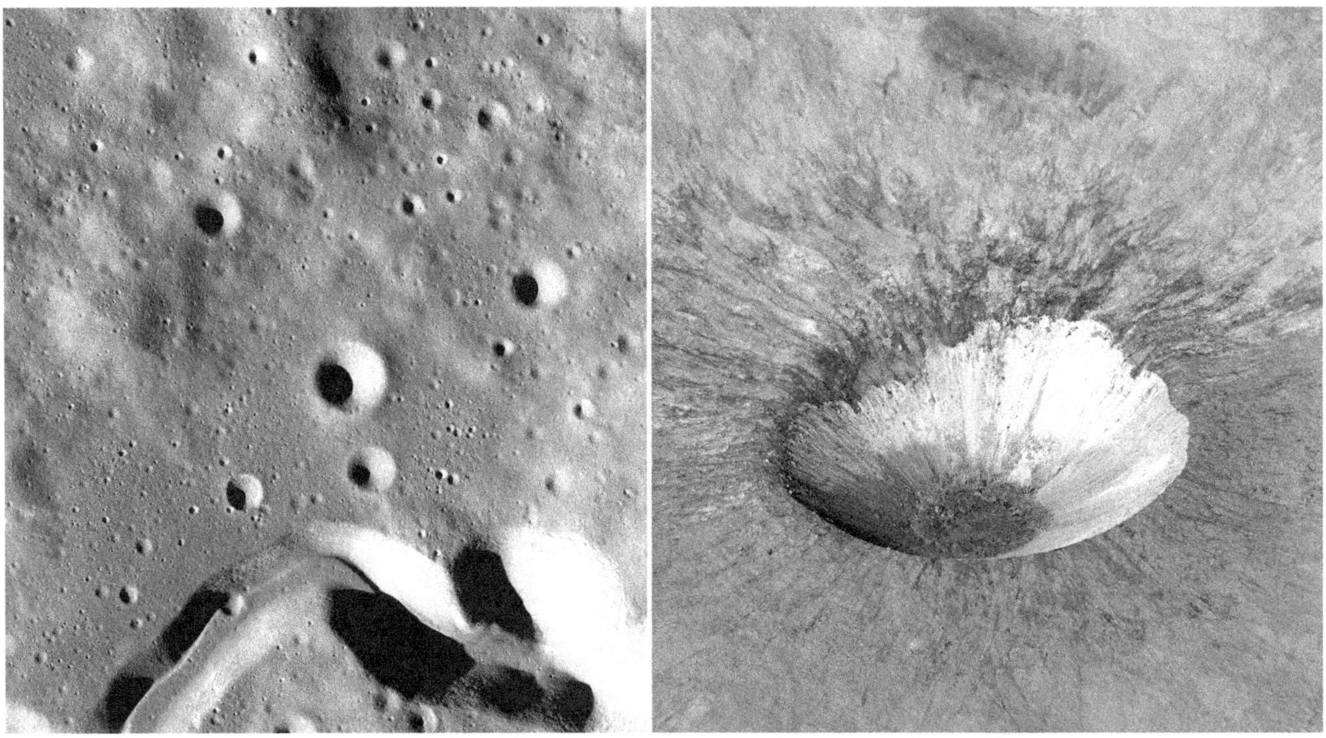

Figure 2.6 Left: Numerous simple craters are seen in this late afternoon (low Sun) image of the area just west of the crater Plato on the Moon (scene centered at 51.92°N, 346.97°E). The largest crater in this view is 2.0 km in diameter. The lighting direction is toward the right. LROC NAC PLATOEJCLOC_E514N3470, NASA/GSFC/Arizona State University. At right: An oblique view of Hell Q crater (33.00°S, 355.55°E), which is 3.7 km and one of the younger craters on the Moon. This simple bowl-shaped crater has been modified by material eroding off the crater walls to produce a small flat floor. The crater also displays near-pristine impact melt that lines the crater walls. Ejecta were thrown out to several crater radii, and dark impact-melt streamers that formed late in the impact process crossed over the earlier emplaced ejecta. LROC NAC image M1221369684LR, NASA/GSFC/Arizona State University.

general the outermost major terrace increases as a function of crater diameter. Several attributes of terraces provide clues to their mode of formation. In most cases, the toes of terraces fade smoothly into the solidified floor of the crater, without any sign of disruption by movement after the floor solidified. Terraces are often coherent blocks, showing no signs of internal deformation, suggesting that the landslides were large-scale rather than the incremental collapse of the inner crater wall. On Mercury (Figure 2.7) and the Moon (Figure 2.8), the terraces themselves are frequently veneered with impact melt that has ponded in closed depressions formed by the rotation of the head scarp and these ponds are relatively undeformed by subsequent terrace motion (Melosh, 1989, p. 142). Thus, terraces most likely formed as soon as the rim was formed during the crater excavation process and created local topographic obstacles to enable ponds of impact melt to form.

2.3.1.3 Central Structures (Central Peaks and Peak Rings)

Central peaks only form in craters larger than ~15–20 km in diameter on the Moon (Wood, 1973). The rock that forms the peak originates from the greatest depth of all the material found within the crater (Pieters, 1982; Song et al., 2013). This type of complex lunar crater morphology is most clearly seen at Tycho crater (85 km in diameter and 4.8 km deep) (Figures 2.9 and 2.10). It is estimated that Tycho crater is only ~109 ± 4 million years old from samples returned from the Apollo 17 landing site (Stöffler and Ryder, 2001) and so provides good insight into the original structure of a peak. Similar central peak structures to those seen on the Moon are also observed on Mercury, where a rocky target but higher gravity (by a factor of ~2x) compared to the Moon can make the peaks more complicated in outline (Figure 2.11). At larger crater diameters, the peak is replaced by a peak ring on both the Moon and Mercury (Figure 2.12).

2.3.1.4 Impact Melt

Impact melt is produced by the kinetic energy of the impact event heating the target rock (see section 1.4). Young lunar craters often have their inner walls draped with impact melt (Bray et al., 2010). Neish et al. (2014) discovered that 80% of complex craters and 60% of simple lunar craters

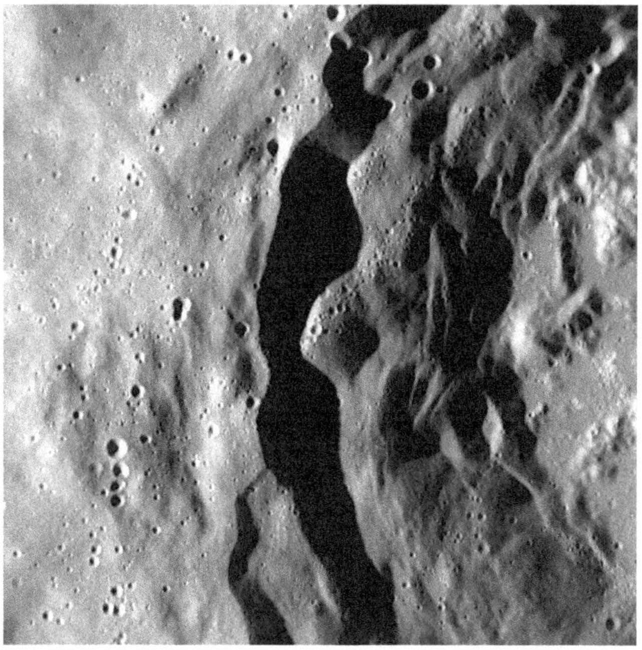

Figure 2.7 Inner wall of a 45 km diameter crater on Mercury (54.58°N, 314.23°E) showing well-preserved terraces and a hummocky floor. Contrast this morphology to that of the lunar crater Aristarchus (Figure 2.8). Illumination is from the left. Crater floor is at right. JPL Photojournal image PIA19195.

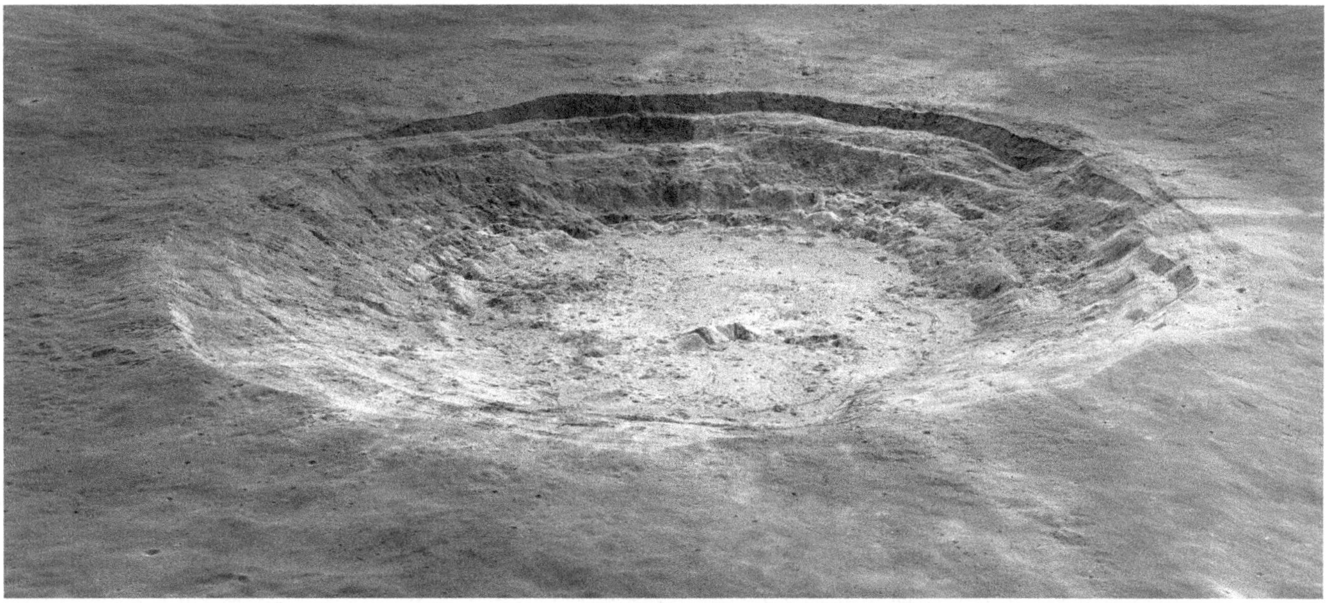

Figure 2.8 A spectacular view looking west across Aristarchus crater (40 km diameter) on the Moon (23.75°N, 312.52°E). In this image, the multiple terraces on the far wall, as well as the small central peak, are clearly visible. LROC image M1259297876LR, NASA/GSFC/Arizona State University.

Figure 2.9 LROC WAC mosaic of Tycho crater on the Moon, ~85 km in diameter, located at 43.28°S, 348.79°E. Even at this resolution, the flat, low-albedo ponds of impact melt are clearly visible. See also Figure 2.10 for more detailed views of this crater. NASA/GSFC/Arizona State University.

(Figure 2.13) have melt dispersed in directions that are coincident or nearly coincident with the lowest point of their rim, implying that pre-existing topography played a dominant role in melt emplacement beyond the crater rim crest. A clear view of the floor of a young impact crater can be gained at Jackson crater (Figure 2.14), where blocks of rock a few kilometers in width are draped by a smoother material that was once molten. Cooling cracks are also visible on the floor of Jackson crater. It has been proposed that few craters on Mars would display such preserved impact melt due to the presence of volatiles in the target at the time of crater formation (Kieffer and Simonds, 1980), although a few rare examples can be found, such as at the 28 km diameter Tooting crater (Mouginis-Mark and Boyce, 2012).

2.3.1.5 *Pits and Hollows*

The floors of numerous craters on Mercury display some features that are absent on the Moon; one example is the curious features called "pits" or "hollows" (Figure 2.15). These features are shallow, flat-floored irregular depressions notable for their relatively high reflectance and characteristic color. They are believed to have formed by degassing of volatiles through a combination of sublimation, outgassing, and/or pyroclastic

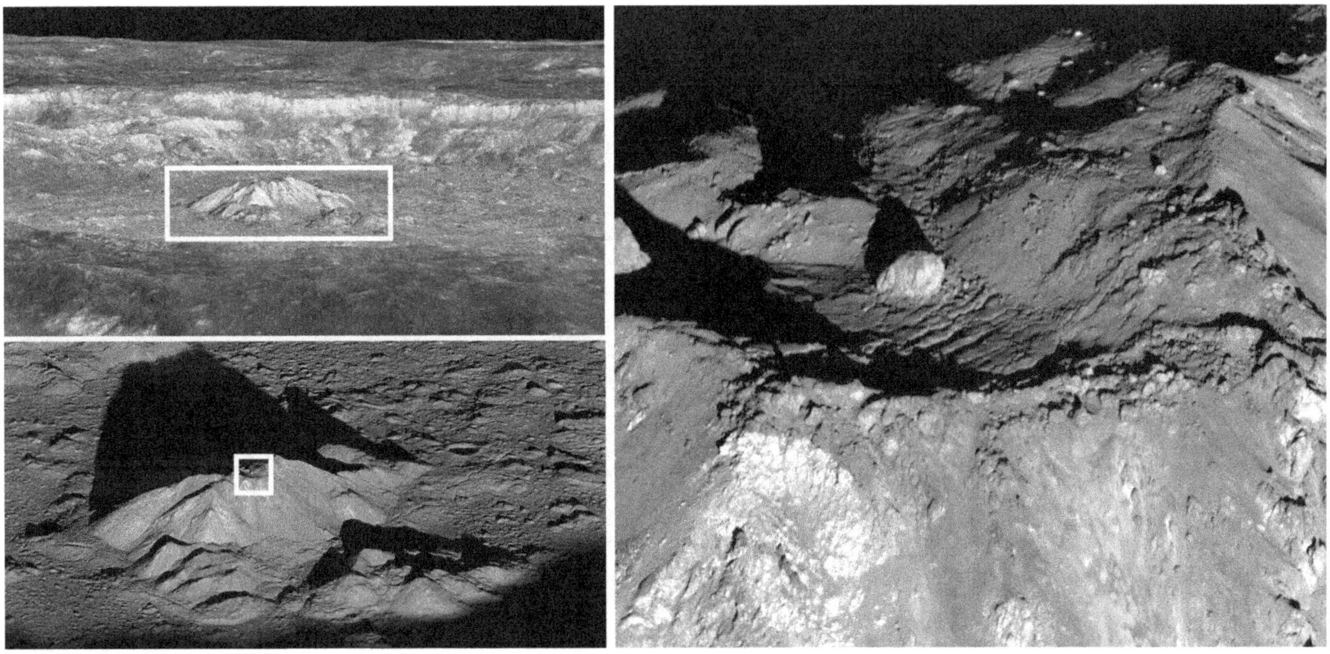

Figure 2.10 Top left: Oblique view looking across Tycho crater (~85 km diameter) on the Moon. From the crater floor to the crest of the west wall is >4,400 m. Box delineates the area shown at lower left. LROC frame M1167178525LR. Lower left: Closer oblique view of the central peak (taken under different lighting conditions). Insert marks the location of the scene at right. LROC image M162350671LR. Right: Oblique view of the summit area of Tycho crater's central peak, showing a boulder sitting close to the summit. This boulder is ~120 m in diameter and appears to be nestled in a chilled pond of impact melt that displays multiple cooling cracks. All images NASA/GSFC/Arizona State University.

Figure 2.11 Abedin crater (61.75°N, 349.34°E) on Mercury is ~116 km in diameter. This mosaic of oblique images highlights the impressive interior of the crater, which has a floor covered with impact melt. Cracks that formed as this impact melt cooled are visible to the left of the central peak. Somewhat intriguing is the shallow depression (possibly a volcanic vent) that lies within the central peak complex of the crater. JPL Photojournal image PIA19423.

Figure 2.12 At left, the peak-ring crater Michelangelo (230 km diameter) on Mercury (44.92°S, 250.22°E), imaged by MESSENGER, frame EN0241236952M. NASA/Johns Hopkins University Applied Physics Lab. At right, the Moon's 312 km diameter Schrodinger Basin (74.73°S, 132.92°E). The 2.5 km high peak ring is 150 km in diameter. NASA Scientific Visualization Studio image.

activity (Blewett et al., 2011, 2013). Hollows range from tens of meters to a few kilometers across and can occur not only on the crater floor but also on central peaks and can extend across the crater rim. Unlike the pits that are found within fresh Martian craters, which are interpreted to form during the cratering process (see Section 4.3.1), the Mercury hollows are

Figure 2.13 The low reflectance material above the step, bright, inner wall of this young unnamed lunar crater (80.03°N, 278.89°E) displays multiple horizontal cracks where the impact melt hardened and then slowly fell into the crater cavity at the bottom of the image. Illumination is from the bottom of this image, which is ~1.2 km in width. LROC NAC image M125130801R. NASA/GSFC/Arizona State University.

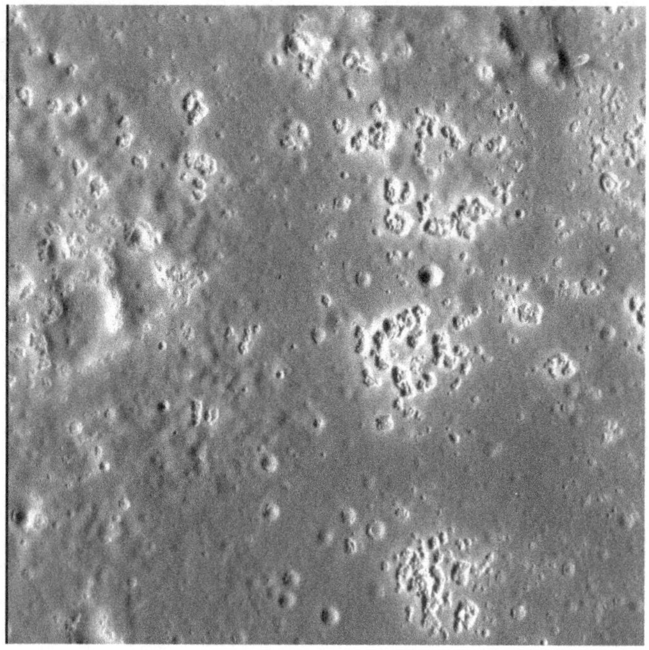

Figure 2.15 Parts of the floor of the 129 km diameter Zeami Basin (2.96°S, 212.59°E) on Mercury are peppered by hollows, which superficially resemble pitted materials seen within Mars craters. These examples are more irregular in shape and can be isolated depressions that do not possess rims, and so most likely have a different origin compared to the Martian examples. Image width is ~20 km. Illumination is from the left. JPL Photojournal image PIA19267.

Figure 2.14 Impact melt on the floor of Jackson crater (71.3 km diameter) on the lunar far side (22.05°N, 196.68°E). Illumination is from the right. Note the multiple fractures that occur on top of some of the mounds, which suggest that these mounds are covered by melt. Image width is ~6 km. LROC image M1139975629RE. NASA/GSFC/Arizona State University.

Figure 2.16 Low-Sun image of a fresh 550 m diameter crater in the lunar highlands. Image width is 950 m. Small craters reflect the varying geology found in target rocks that form the top few tens of meters. This crater (at 51.42°S, 68.73°E) is surrounded by a very large number of boulders that could indicate that the substrate was relatively strong and did not break apart during cavity formation. Illumination is from the left. JPL Photojournal image PIA14421.

believed to have formed by sublimation and erosion over an extended period of time (Blewett et al., 2011).

2.3.2 Exterior Features

2.3.2.1 Ejecta Blocks and Impact Melt

The formation of ejecta blankets was discussed in Chapter 1, where it was noted that the material from the cavity becomes mixed with the surrounding terrain as ejecta re-impact the surface. Thus, when we look at ejecta blankets in planetary images, we need to consider what material originated from within the crater cavity and how local material may have been incorporated in the final deposit. Ejecta blocks are common on the rims of fresh lunar craters (Figure 2.16). During crater formation, impact melt can also escape the crater cavity to be distributed within the ejecta blanket (Hawke and Head, 1977; Ashley et al., 2012). The volume of melt can be quite large, with the Al-Tusi pond (which is adjacent to and influenced by King crater) and the floor of King crater (~76 km in diameter) estimated to be ~72 km^3 (Ashley et al., 2012). For lunar craters smaller than 5 km in diameter, impact melt deposits occur at twice as many nonmare craters as at comparable mare craters (Stopar et al., 2014). Impact melt can be seen not only within the crater interior but also beyond the rim crest of small craters on Mercury (Figure 2.17).

Figure 2.17 Waters crater (15 km diameter) on Mercury (8.96°S, 254.55°E) displays an unusual dark flow feature on its southern rim, which is most likely impact melt. Superficially this might be confused with the fluidized ejecta seen around many fresh craters on Mars. The fine fingering of the margins of the flow aids in the correct interpretation. MESSENGER image 2641461. NASA/Johns Hopkins University Applied Physics Lab.

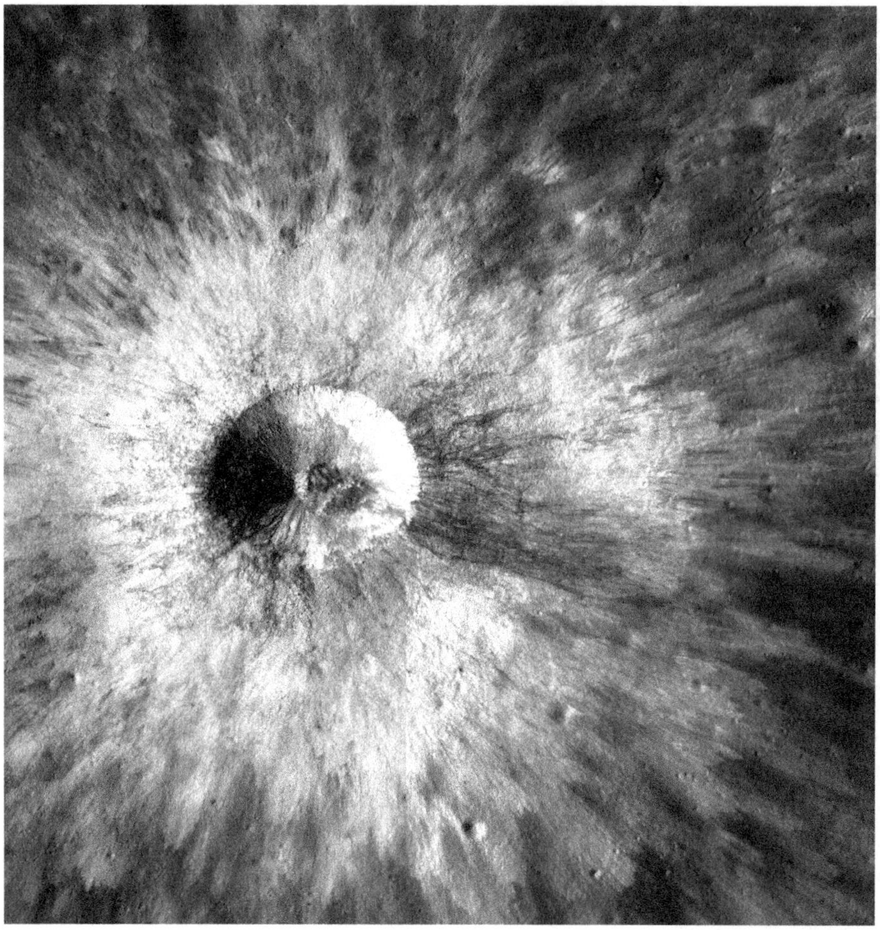

Figure 2.18 A very young, unnamed lunar crater (~1.8 km diameter). This small crater (3.10°N, 281.06°E) has a simple bowlshape and lacks either terraces or a central peak. Notice the bright ejecta rays that surround the crater rim. Darker material is interpreted to be impact melt. LROC image M1295910050LR, NASA/GSFC/Arizona State University.

2.3.2.2 Ejecta Rays

Lunar rays are filamentous, high-albedo deposits occurring radial or subradial to impact craters (Figure 2.18). Rays are primarily composed of locally derived material, but rays of small craters may be primary ejecta deposited on the surface (Elliott et al., 2018). For the crater Copernicus (Figure 1.10), the ray system extends radially for more than 500 km (>5 crater diameters) (Pieters et al., 1985). Using Earth-based spectral and radar data, as well as optical maturity maps derived from *Clementine* images, Hawke et al. (2004) found that lunar rays are bright because of the compositional contrast with the surrounding terrain, the presence of immature material, or some combination of the two. V-shaped features commonly occur on the outer part of the ejecta blanket in association with the rays, and these consist of fragmental debris deposited during the emplacement of the ejecta (Guest and Murray, 1971).

Figure 2.19 These lunar secondary craters, located at 5.50°S, 255.70°E, formed on Mare Orientale Basin ejecta. The largest crater in this image is ~90 m in diameter. The unnamed parent crater is ~2.2 km in diameter and lies ~18 km to the bottom of this scene and is east of the crater Grachev. Note the V-shaped ejecta deposits that point toward the impact site. LROC NAC image M1143115078RE, NASA/GSFC/Arizona State University.

2.3.2.3 Secondary Craters

Secondary craters form when material from a primary crater is ejected and then impacts the surrounding terrain. Such craters can often be identified by their distinctive features, including their occurrence in long chains of craters that are

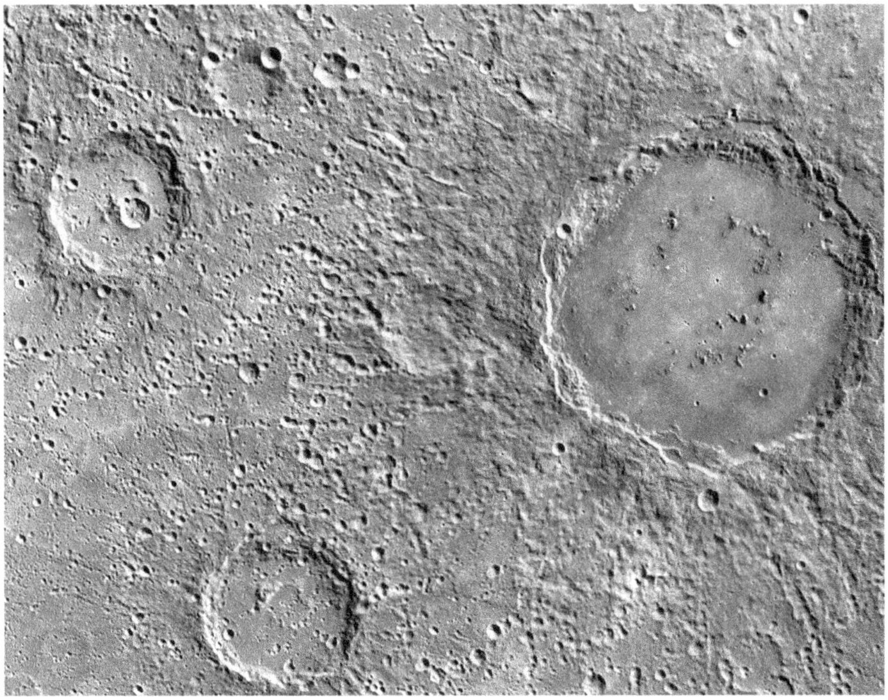

Figure 2.20 Rustaveli Basin on Mercury (at right in this view) has a diameter of 200 km. Notice how most of the ejecta and secondary craters occur close to the rim crest, in contrast to the material ejected from Copernicus crater (Figure 1.10). Despite the well-preserved ejecta blanket, the interior has been flooded by more recent material to leave visible only the tops of a peak ring. The basin is centered at 52.41°N, 82.34°E. JPL Photojournal PIA19283.

approximately radial to the parent crater, the V-shaped ejecta deposits. These V-shaped features point back to the primary crater, or they possess asymmetrical shapes with the rim farther away from the primary crater being less well-formed than the closer rim (Figures 1.10 and 2.19). Because these V-shaped features form from ballistically ejected material, the distribution of secondary craters is controlled by the gravity of the body, with secondary craters located much closer to the primary crater rim on Mercury (Figure 2.20) than those formed on the Moon.

2.4 CRATERS ON VOLATILE-RICH BODIES – GANYMEDE, 1 CERES, AND CHARON

The outer Solar System offers a wide diversity of parameters that may influence the final appearance of an impact crater. Ganymede is perhaps the best example of a Jovian satellite to study the effects of a target made of water ice because there are many craters of different sizes, but not so many craters that their ejecta blankets overlap each other (as is the case for Callisto). The myriads of craters on Callisto, or the few preserved craters on Europa, are also informative as they show differences in their relaxation rate due to differences in the internal temperature distributions within the moons (Moore et al., 1998, 2001; Dombard and McKinnon, 2006). 1 Ceres also has an icy crust and a lower gravity compared to Ganymede, and these attributes will influence cratering on the minor planet (Hiesinger et al., 2016). In the outer limits of the Solar System, Pluto and its moon Charon offer yet another end-member situation due to the colder temperatures at this distance from the Sun (Robbins et al., 2017, 2021).

2.4.1 Interior Features

Impact craters may also be affected by viscous relaxation due to the high temperature of the surface rocks. This could be particularly true for craters formed on the icy satellites, where the lower strength of the substrate can result in the formation of subdued features called "palimpsests." Numerous craters on the moons Ganymede (Figure 2.21), Rhea, Dione, and Tethys have central pits comparable to those within Martian craters (Alzate and Barlow, 2011; Barlow et al., 2017), although central pits tend to be larger than those on Mars, probably due to gravitational scaling (Alzate and Barlow, 2011). Attempts to understand the formation of such features on volatile-rich as well as volatile-poor bodies (Elder et al., 2012; Bland and Bray, 2024) indicate that a complicated interplay of crustal volatiles, target strength, surface gravity, heat flow, and impactor energy along with both uplift and collapse are involved in central pit formation. Flows within some craters formed within icy targets have also been observed, such as the interior of Occator crater on 1 Ceres (Figure 2.22). Here the crater is very young, and so the flows coming off the inner wall may be the equivalent of the impact melt flows that are preserved within the younger lunar craters (Figure 2.14). Occator crater also has a central pit and evidence of hydrothermal alteration (Scully et al., 2020).

Because an icy target has less structural strength than rock, even at the lower temperatures associated with the outer Solar System, deformation of the rim crest is more complex

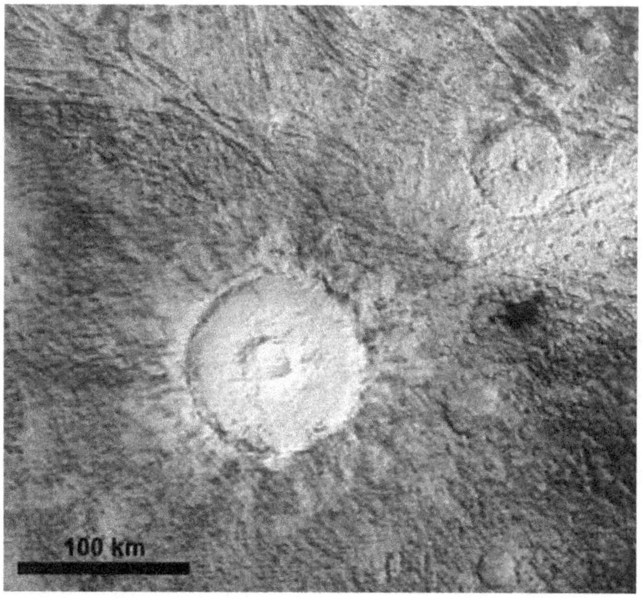

Figure 2.21 Central pits are seen within these two craters on Ganymede and show that such features can form within icy targets. The largest is Osiris crater (38.00°S, 193.69°E), which is 107 km in diameter and contains a 36-km diameter floor pit. In the center of the pit is an upraised dome measuring 19 km in diameter. Domes inside central pits are common on icy bodies and likely represent diapiric activity. Note that there is no evidence of domes inside central pits on Mars, Mercury, or the Moon. NASA Voyager 2 mosaic obtained July 9, 1979.

Figure 2.23 Kokopelli crater (at lower right) on 1 Ceres is ~34 km in diameter is located at 18.30°N, 124.50°E and displays a complicated series of terraces on the inner crater wall. The number of these terraces may indicate that the substrate is much weaker than rock and so slumping has been much easier than on the rocky planets. Illumination is from the right. JPL Photomosaic image PIA21915.

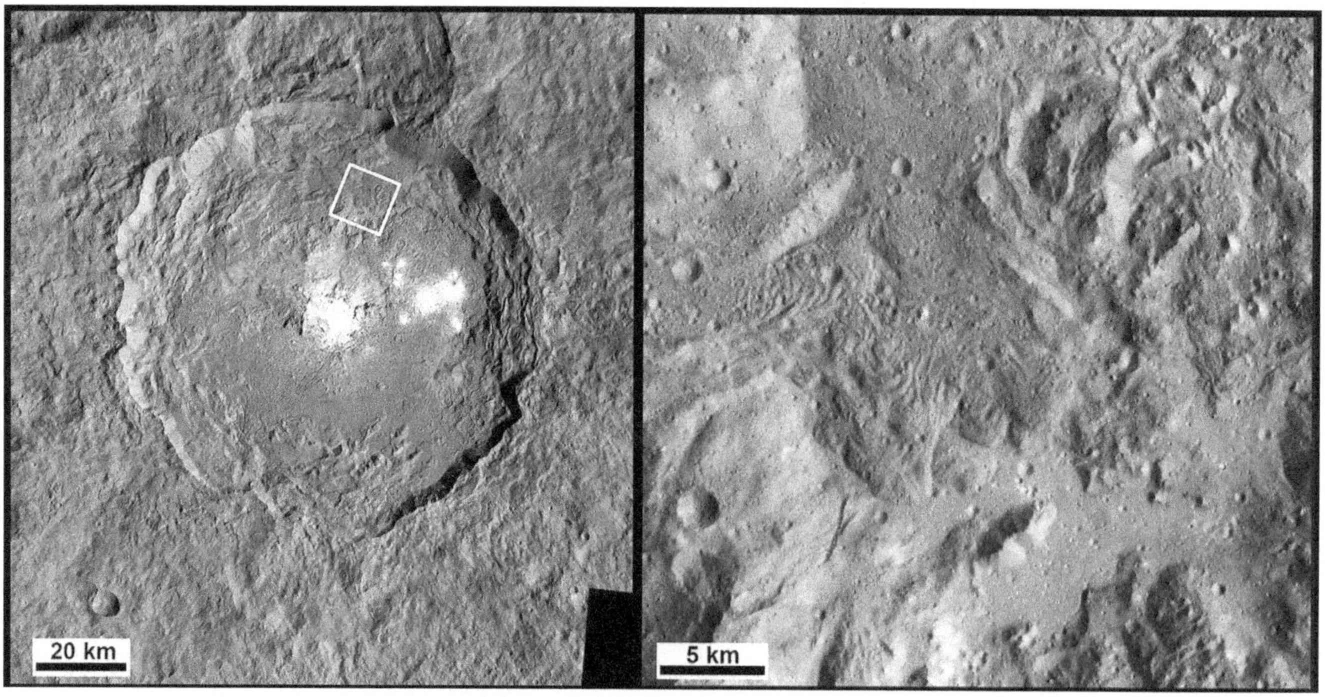

Figure 2.22 Flows can be identified within some craters on the dwarf planet 1 Ceres. Here are two views of the crater Occator (92 km diameter, 19.83°N, 239.34°E), which contains the brightest area on 1 Ceres. At left, the white box denotes the area shown at right. At right, a lobe of material (dust-rich ice?) flowed from the wall (top) toward the crater floor at the bottom of the image. These cryogenic flow features appear to comprise low viscosity material and could either be related to impact melt or an exogenic triggering of cryovolcanic processes (Krohn et al., 2016). Nathues et al. (2020) favored multiple episodes of resurfacing several millions of years after crater formation and the dissipation of the impact-generated heat. JPL Photojournal images PIA20350 and PIA22642.

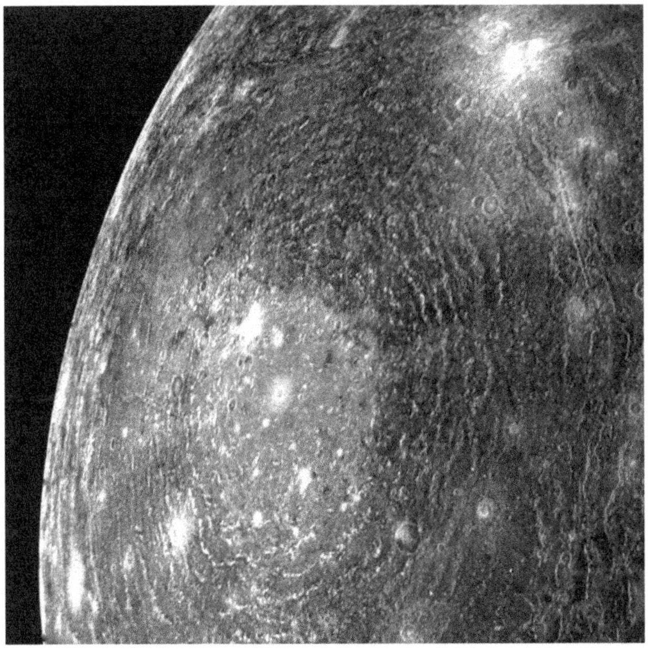

Figure 2.24 Large basins on icy moons have multiple rings and visco-elastically relaxed topography. This is Valhalla Basin on Callisto (15.87°N, 304.75°E), a moon of Jupiter. It has a bright central region, called a palimpsest ~360 km across, an inner ridge and trough zone, as well as striking concentric rings extending up to ~1,900 km from the crater. The outer boundary is not well defined. JPL Photojournal image PIA02277.

than on rocky bodies. For example, the crater Kokopelli on the minor planet 1 Ceres (Figure 2.23) displays a complicated series of terraces on the inner wall. Craters formed on icy worlds also have more rings than those seen on volatile-poor bodies (Figure 2.24). Other interior features that are quite common on asteroid 4 Vesta (Figure 2.25) include pitted material that superficially resembles features we will see on Mars (see Chapter 4). The pitted terrain within the interiors of several craters on 1 Ceres (Sizemore et al., 2017) and 4 Vesta (Denevi et al., 2012) has been linked to the presence of subsurface volatiles.

2.4.2 Exterior Features

The outer Solar System, where ice is the dominant target material, is a good place to search for evidence of ejecta fluidization comparable to that seen on Mars because volatiles must have been incorporated into the ejecta blanket. Examples of annuli of ejecta surround craters on the Jovian moon Ganymede were identified by Boyce et al. (2010). Not only do these craters display one (Gula crater) or two (Achelous crater) ejecta layers (Figure 2.26), but there is also evidence of ejecta moving across the surface. Ganymede is not the only icy moon where layered ejecta have been observed with characteristic morphology like craters on Mars. Charon, a moon of Pluto, also possesses craters with annuli of material surrounding the crater rim (Figure 2.27), and these bear a close similarity to many craters on Mars (Robbins et al., 2018b).

Figure 2.25 Marcia crater on asteroid 4 Vesta (9.56°N, 339.00°E) is ~68 × 58 km in diameter and is one of the youngest craters on the asteroid. Marcia crater contains a large concentration of pits on the floor that resemble pits within certain Martian craters. In this image the illumination is from the right. JPL Photojournal image PIA16182.

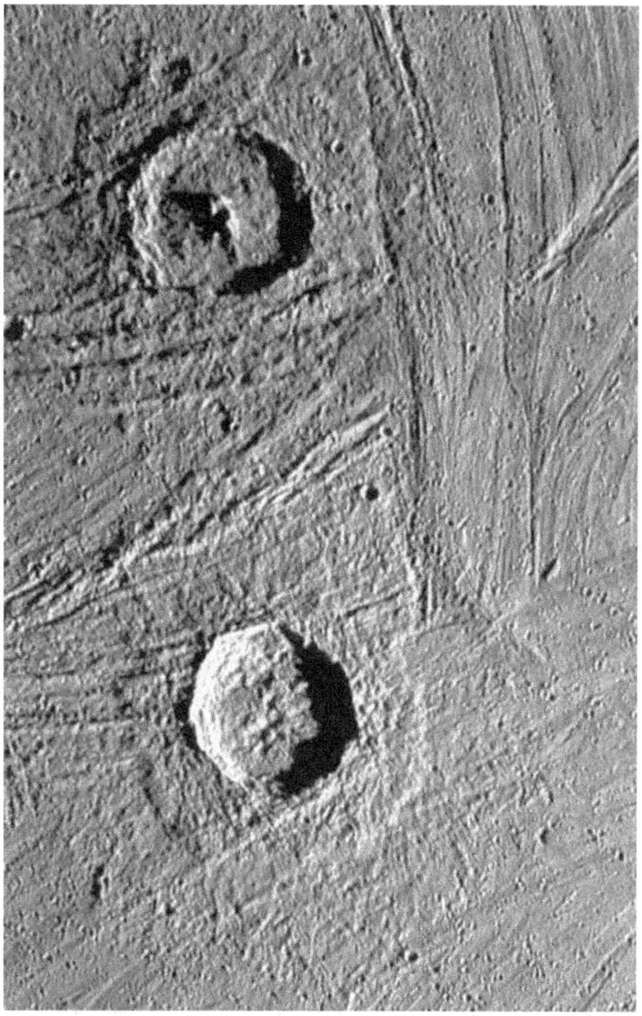

Figure 2.26 Mars is not the only body in the Solar System where layered ejecta craters have formed. Good examples can also be found on Ganymede, a moon of Jupiter (Boyce et al., 2010). The Galileo spacecraft imaged Gula (top) 38 km in diameter (64.39°N, 347.34°E) and Achelous (~40 km) (62.22°N, 347.86°E) craters under low illumination conditions. Illumination is from the right. JPL Photojournal image PIA01660. Other planetary bodies where similar craters with abrupt termini have been identified include Europa, Dione, Tethys, and Charon (Robbins et al., 2018b).

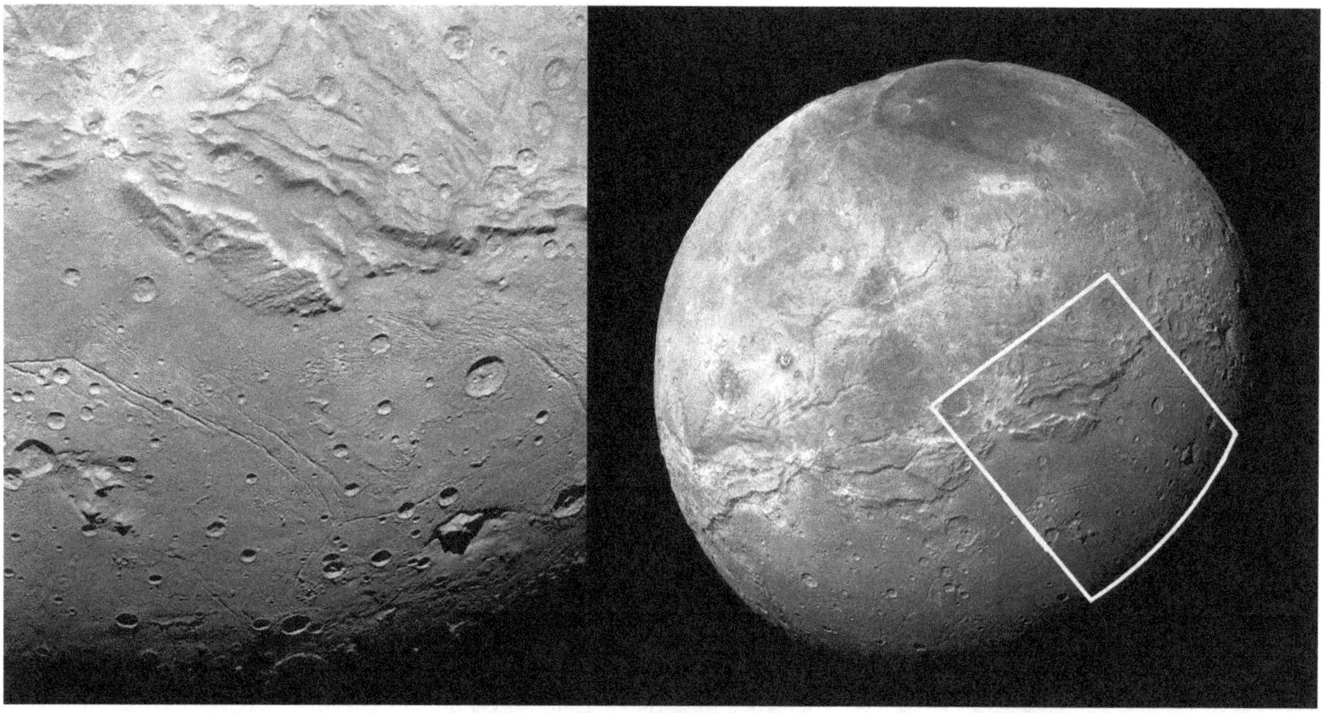

Figure 2.27 At low illumination angles, Charon (a moon of Pluto) displays numerous craters with ejecta blankets that superficially resemble the ejecta layers from fresh craters on Mars. The image at right shows the location of the image at left. JPL Photojournal image PIA19967.

2.5 CRATERS ON BODIES WITH AN ATMOSPHERE – VENUS AND TITAN

The surface of Venus currently has an atmospheric pressure ~75 times that of Earth, while Titan (the largest moon of Saturn) has a surface pressure ~60% greater than on Earth. On Venus, crustal resurfacing means that we only see relatively recent (perhaps formed within the past 1 Ga) craters (Phillips et al., 1992). Fluvial erosion, mass wasting, and burial by dunes have similarly reduced the number of craters observed on Titan (Wood et al., 2010). Nevertheless, the thicker the atmosphere, the greater the deceleration of ballistic ejecta would result in deposits closer to the parent crater rim. A projectile would be slowed by passage through a thicker atmosphere so that it would strike the ground with less energy. Passage through the atmosphere could also break up the projectile creating crater clusters on the ground. Particularly for Venus with its dense atmosphere, the smallest objects can burn up so that there are no small (<~1.5–2.0 km diameter) craters (Phillips et al., 1991, 1992). Other environmental issues, such as the super-hot (>400°C) surface temperature of Venus, will also change the proportion of impact melt produced from an impact of a given size because the target rock will be closer to its liquidus temperature compared to a cold target. A target composed of coherent hot silicate rock would have a different influence on the final morphology of an ejecta blanket compared to a target composed of water ice or other volatiles. Impact craters on Venus may also be affected by viscous relaxation of the crater interior due to the high temperature of the surface rocks (Grimm and Solomon, 1988).

On Titan, only five impact craters have been confidently found, although an additional 44 probable craters have been identified on ~22% of the surface, as imaged by the *Cassini* radar (Wood et al., 2010). At least some of these 44 craters possess central peaks, but they are relatively rare, perhaps due to fluvial erosion (Neish et al., 2016). Terraced inner walls appear to be a common feature and the depth/diameter relationship suggests a modification process with a rate that is constant with time, such as eolian infilling (Forsberg-Taylor et al., 2004; Neish et al., 2013).

2.5.1 Interior Features

On Venus, where surface temperatures are more than 400°C, the proportion of impact melt produced by an impact event might be expected to be appreciably greater. Indeed, *Magellan* radar backscatter images reveal numerous craters with possible impact melt. As in the case of Danilova crater (Figure 2.28), it is uncertain whether the radar-dark materials on the floors are lava flows, impact melt, or eolian deposits. The floor materials of Danilova crater are particularly intriguing because they embay the central peak complex and the terraces on the inner wall. If this material is impact melt, then one would expect the surface to be smooth at the 12.6 cm wavelength of the *Magellan* radar. This contrasts with the melt seen in lunar craters such as Jackson crater (Figure 2.14) where there are cooling cracks and random blocks on the floor that would serve as strong (bright) radar reflectors. While it is possible that the hotter surface temperatures allow melt sheets to cool more uniformly (and smoothly) on Venus than on the Moon, an eolian origin for the dark floor appears more reasonable.

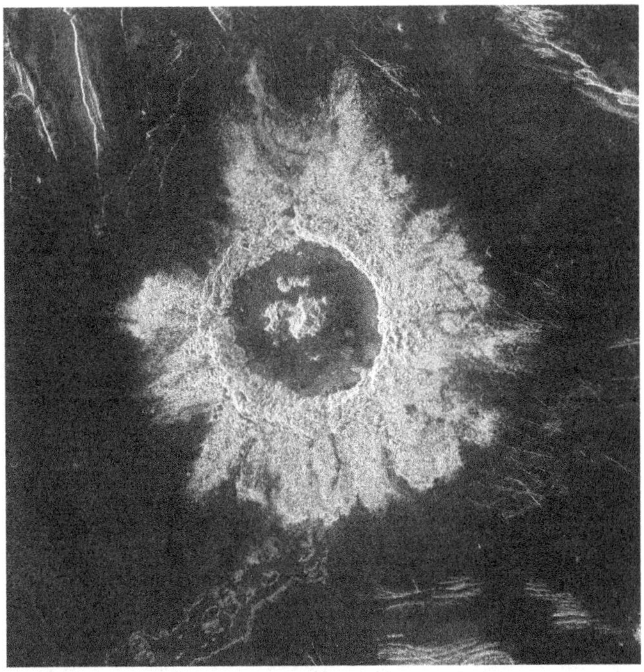

Figure 2.28 Danilova Crater (49 km diameter), in the Lavinia region of Venus, has a complicated central peak surrounded by radar-dark materials. The radar-dark floor may be covered with relatively smooth impact melt, lava flows, or more recent eolian in-fill. The radar-bright ejecta surrounding the crater indicate that the surface is rough at the scale of the radar wavelength (12 cm). Magellan radar image centered at 26.30°S, 337.15°E. JPL Photojournal image PIA00086.

2.5.2 Exterior Features

A key attribute of many craters on Mars is the layered ejecta surrounding the crater (Barlow et al., 2000). On Venus (Figure 2.29), lobes of material superficially similar to ejecta layers around craters on Mars can also be found. But in these instances, it seems likely that the target rocks were sufficiently hot prior to the impact event that the ejecta were at a much higher temperature than on Mars and thus could flow almost as a molten lava flow (Chadwick and Schaber, 1993). Key differences between the ejecta flows on Mercury and Venus compared to Mars are that these flows lack distal ramparts and that there may be large "plates" of material rafted along the direction of flow. In addition, it is rare for the ejecta to be symmetric around the parent crater as the hot ejecta tended to flow in the downslope direction and extend for many crater radii from the rim, as is clearly the case for the Venusian crater Addams (Figure 2.30).

Titan displays but a few impact craters and there are only rare examples of preserved ejecta blankets. The craters Sinlap and Soi (Figure 2.31) show some evidence of ejecta blankets (Le Mouelic et al., 2008; Schoenfeld et al., 2023). Probably due to the rapid rate of erosion on Titan and the low spatial resolution of the image data sets, there is little information to be gained about the cratering process on Titan (Le Mouelic et al., 2008).

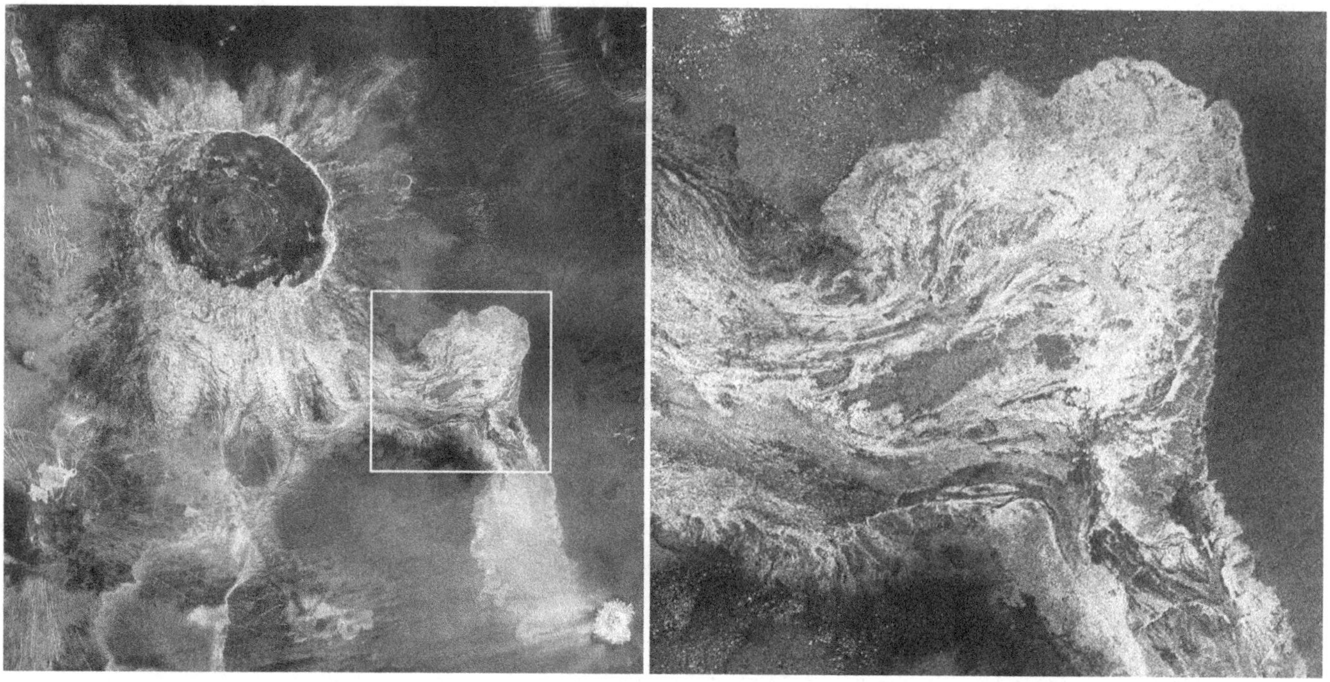

Figure 2.29 Isabella crater (175 km diameter) on Venus (29.70°S, 204.12°E) shows extensive infilling by later material, with only the southern inner wall still visible. The northern part of the ejecta blanket also appears to be partially buried. The most remarkable feature is the extensive flow lobe on the SE exterior of the crater, with a complex flow pattern visible in the ejecta lobe (Miyamoto and Sasaki, 2010). White box in the image at left indicates the location of full-resolution view at right. JPL Photojournal image PIA00480.

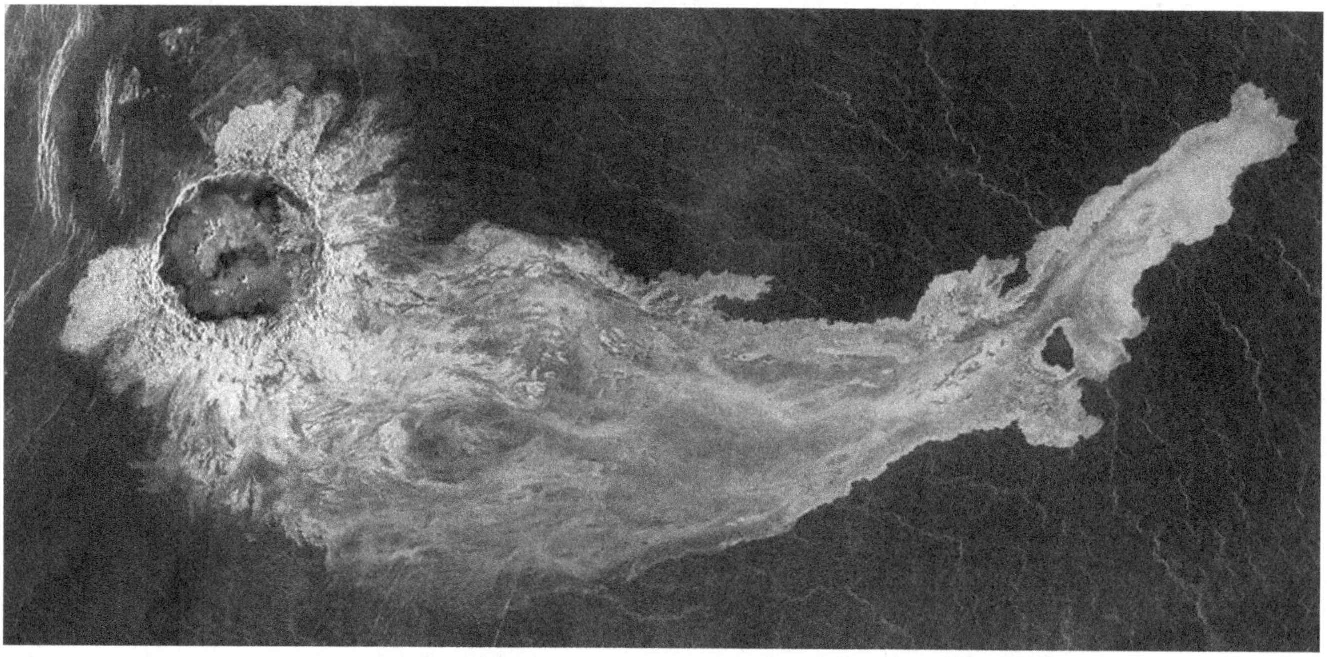

Figure 2.30 The ejecta flow lobe is even more prominent for this crater on Venus than the example given in Figure 2.29. This is Addams crater (56.09°S, 98.81°E), which is 87 km in diameter. The direction of flow was from left to right in this image and displays several channels as well as lobate margins. It is not obvious why such a large lobe of melt was produced in only one azimuth direction or why the crater rim appears unaffected by this large flow of material. NASA Magellan image C2-MIDR 60S093;202, framelets 21 and 29.

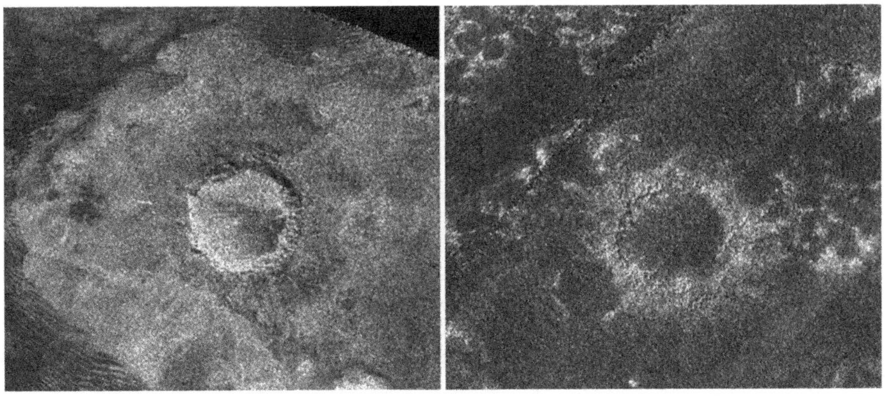

Figure 2.31 There are few fresh impact craters on Titan. Imaging the surface of Titan relies upon the radar system carried by the *Cassini* spacecraft. These images are therefore radar backscatter images. At left is the 80 km diameter relatively fresh crater Sinlap (11.42°N, 343.92°E) with some preserved ejecta surrounding the crater rim (Wood et al., 2010). Hints that the inner wall of the crater may display layering are visible. At right, the extremely degraded Soi crater (24.37°N, 218.99°E) is of comparable size to Sinlap crater and appears to have a flat floor. The ejecta blanket has most likely been lost due to subsequent burial. JPL Photojournal image PIA16638.

2.6 SUMMARY

Through this brief tour through the Solar System, we have seen that there are certain attributes of impact craters that are common to most bodies. As the crater diameter increases, the occurrence of central structures and the deformation of the crater rim become more common and complex. Ejecta blankets are associated with craters of all diameters, although flow rather than ballistically emplaced appears likely on Venus and (potentially) on some icy worlds such as Ganymede and Charon. How these craters differ from those on Mars may be quite subtle or quite pronounced. To best recognize these differences, quantitative measurements need to be made of attributes such as crater depth, rim height, and the radial extent of the ejecta. Thus, in Chapter 3, we delve into how these parameters may vary from place to place on Mars.

Mixed into the planet-specific influences of gravity, the presence/absence of an atmosphere, and the composition of the target, there will be additional changing conditions such as crater size and, hence, depth of the original crater cavity. Mars craters therefore become a test case for a whole series of predictions about what a crater should look like the day after it formed. Of course, for Mars we also expect there to be rapid changes to fresh craters, with the secondary craters and crater rays most affected by the erosional and/or burial potential of the thin Mars atmosphere. Longer term, the interior will experience infilling, and the rim will be degraded.

In Chapter 3 we will see that Mars has a thin atmosphere (~1% as dense as Earth) and a diverse geologic history with multiple rock types and abundant evidence for water and ice on the surface at some time during its history. In Chapters 4 to 8, we will focus exclusively on Martian impact craters. But with this preliminary tour through much of the Solar System, it is valid to now conduct a thought experiment on what we might see on Mars and what might be the spatial and temporal variations to expect in crater morphology.

3

Knowing Mars

3.1 INTRODUCTION

In Chapter 1, we reviewed some of the basic mechanics of crater formation. In Chapter 2, our brief tour through the Solar System gave us some clues into what features are common or rare at impact craters, as well as how local conditions (surface gravity, target rock type, and presence/absence of an atmosphere) may produce different landforms from similar energy impact events. Before we delve into the details of the geomorphology of impact craters on Mars, be it the interior structures, the diverse ejecta blankets, or the ways in which the original landforms have been modified, we must set the stage for placing these craters within the Martian environment. Specifically, it is not valid to assume that all craters of a similar size have been created (and/or preserved) as equals. Both spatial and temporal differences most likely exist due to the crater's position on the planet (e.g., the latitude, elevation, and target rocks) as well as the geologic epoch in which it formed. For example, earlier in Martian history, it is likely that there was more abundant water or ice close to the surface at the time the crater formed compared to recent events.

But first we need to consider the types of information available to the planetary geomorphologist to investigate craters on Mars. This information primarily takes the form of images, as well as topographic and compositional data, collected from Mars orbit by a variety of spacecraft. We then review aspects of the chronology of Mars, from the earliest geologic epoch (the Noachian) until the most recent (the Amazonian), and how the rocks formed during these time periods are distributed across the planet. What we can observe on Mars today is not the way in which the planet has appeared throughout its history. Orbital variations (the "obliquity cycle") would have changed the distribution of surface ice such that the polar caps would have spread over larger (or smaller) areas of the planet compared to what exists today. Obviously, we can only observe the present existence of ice where fresh impact craters have formed during the spacecraft era, but such obliquity changes would have dramatically modified the distribution of ice near the surface and hence influenced the cratering process.

We also consider here the "human-scale morphology" of craters. Several robotic rovers have now explored eroded Martian craters, and they can provide us with their unique field-based observations that can then help with the interpretation of landscapes seen only from orbit. Finally, we ask the question "How can we identify the youngest or least modified craters on Mars?." Several attributes of craters are introduced that collectively can be used to guide this search.

3.2 DATA TYPES

Most of the data used in this book are in image form. Typically, we employ some of the highest resolution images to discuss each feature; image resolution is expressed as the size of an object resolvable by three image elements ("pixels"). Thus, a camera with a pixel size of ~50 cm means that an object ~150 cm across can be resolved on the ground. Scattered throughout this book, we make copious use of images from the Thermal Emission Imaging System (THEMIS) flown on the *Mars Odyssey* spacecraft (Christensen et al., 2004), with both visible wavelength (VIS) and thermal infrared (TIR) data presented. THEMIS TIR data collected both during the daytime and at nighttime are used here. Also used extensively are images from the Context Camera (CTX) (Malin et al., 2007) and the High Resolution Imaging Science Experiment (HiRISE) (McEwen et al., 2007), both of which are instruments flown on the *Mars Reconnaissance Orbiter* (MRO). THEMIS TIR data have a pixel size of 100 m, while images in the visible portion of the spectrum have a pixel size of 18 m. CTX pixels are ~6 m in size and HiRISE pixels are ~25 cm in size. Examples of the field of view and improvement in the spatial resolution of imaging systems from *Mariner IV* in 1965 to that obtained by the MRO (2009 to present) are shown in Figure 3.1.

Baseline topographic information for Mars has been derived from the Mars Orbiter Laser Altimeter (MOLA), which produced a dataset that has been interpolated to a global grid with a pixel size of 463 m at the equator and an elevation precision of ~30 cm (Smith et al., 2001) (Figure 3.2). We will follow the lead of numerous investigators and make copious use of MOLA data for the interpretation of the geometry of craters, particularly the depth/diameter ratio (Garvin and Frawley, 1998; Garvin et al., 2000), while gridded MOLA data have been used to infer subsurface properties from fresh crater geometries (Stewart and Valiant, 2006). THEMIS daytime and nighttime TIR data can determine how quickly a surface changes temperature, which in turn provides an estimate of the average grain size through an attribute called "thermal inertia" (Mellon et al., 2000). Thermal inertia has been used to investigate the variation of particle size across an ejecta blanket (Baratoux et al., 2005; Hill and Christensen, 2017) to help infer what process(es) are responsible for ejecta emplacement.

Both CTX and HiRISE can acquire images on different orbits of the spacecraft that overlap with sufficient accuracy that stereo pairs of the same landform can be produced, from which digital elevation models (DEMs) can be generated at a

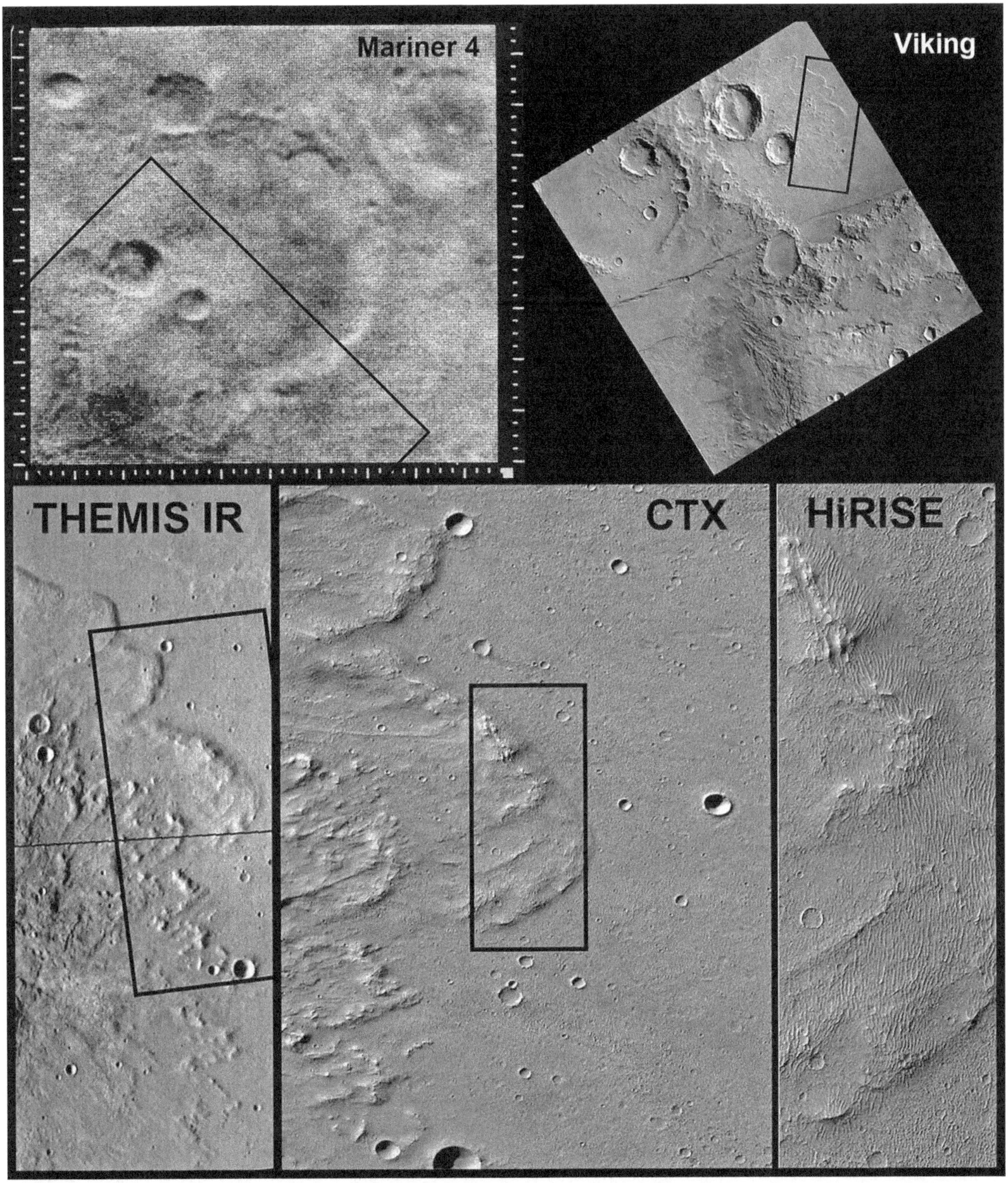

Figure 3.1 See the improvement? Image resolution has strikingly improved from July 15, 1965, when Mariner IV imaged Mariner crater (170 km diameter, 34.63°S, 195.73°E) to the March 30, 2016, HiRISE image of one of the ejecta lobes from a smaller crater (18 km diameter) within Mariner. Boxes show the area of coverage in the next image. But how much has our knowledge advanced since, say, the Viking Orbiter days in the late 1970s and early 1980s? Top left: Image #11 from Mariner IV. Top right: Viking Orbiter frame 562A29. The largest fresh crater is ~32 km in diameter. Bottom left: THEMIS image I675500 reveals that there are multiple layers to the crater's ejecta blanket. Image width is ~28 km. Bottom middle: CTX J01_045357_1449, image width ~23 km. Bottom right: HiRISE image ESP_045357_1450, scene width ~4 km. The vertical ripple pattern in this image is due to a series of eolian dunes that are unrelated to the cratering process.

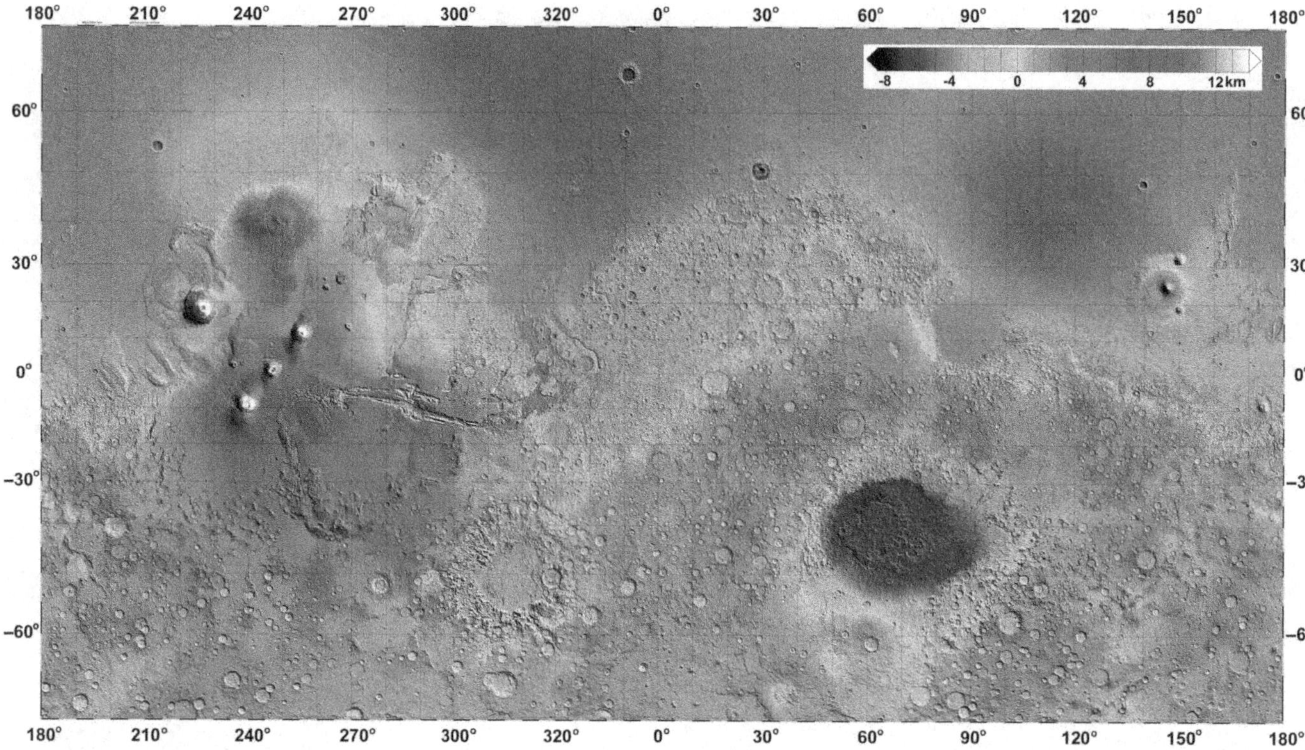

Figure 3.2 MOLA topography of Mars. The color bar at the top right shows the range of elevations from minus 8 to plus 12 km relative to the datum (Smith et al., 2001). Prominent here is the hemispheric difference between the low northern hemisphere and the high southern hemisphere. Hellas Basin is the deep depression at lower right. The Tharsis volcanoes form the high point (in white) to the left of center. The summits of the large volcanoes rise to elevations of ~20 km above the datum. Note: A black and white version of this figure will appear in some formats. For the color version, please refer to the plate section.

resolution of 4 times the pixel size (Figure 3.3). These DEMs have been used to investigate the geometry of a crater cavity and rim crest (Mouginis-Mark et al., 2018). Many other cameras have also imaged craters on Mars, including the High Resolution Stereo Camera (HRSC) (Jaumann et al., 2015), Mars Orbiter Camera (MOC) (Malin et al., 1998), the Mars Color Camera (MCC), and Viking Orbiter Cameras (Arvidson et al., 1980; Davis and Soderblom, 1984). It is worth noting that other data sets have generally been superseded by observations from subsequent missions. Examples of the global data sets used in this book are shown in Figure 3.4.

Important composition data for craters have also been obtained from cameras operating at multiple wavelengths of light, such as the Compact Reconnaissance Imaging Spectrometer (CRISM) on the MRO (Murchie et al., 2007) and the Observatoire pour la Mineralogie, l'Eau, les Glaces et l'Activite (OMEGA), which was flown on *Mars Express* and collects data between 0.5 and 5.2 microns. These instruments have been used to investigate the composition of rocks seen in the interior of craters with the potential identification of ancient rocks in the interiors of craters (Bell, 2008; Michalski and Niles, 2010; Sun and Milliken, 2015). The Gamma Ray Spectrometer (GRS) experiment on *Mars Odyssey* had a spatial resolution of ~300 km (Boynton et al., 1992). GRS measured the abundance and distribution of many elements of the periodic table, including hydrogen, silicon, iron, potassium, thorium, and chlorine. Important for crater studies, GRS data have revealed areas with present-day near-surface ice, which can be correlated with the distribution of different crater morphologies (Barlow and Perez, 2003). The Thermal Emission Spectrometer (TES) on *Mars Global Surveyor* collected high-quality spectral information (Christensen et al., 1992, 1998; Bandfield et al., 2000). TES comprised an infrared (IR) (5.8–50 micron) interferometric spectrometer, as well as broadband thermal (5.1–150 micron) and visible/near-IR (0.3–2.9 micron) radiometers. Along with many other investigations, these data have been used to search for the source area of the Martian meteorites by attempting to match the orbital data with meteorite spectra (Hamilton et al., 2003). TES data have also proven to be particularly helpful in the identification of areas of Mars that have a thick layer of eolian dust (Ruff and Christensen, 2002). This dust layer has importance when searching for the youngest craters (see Section 3.7) because some of the finest structures may be obscured by a few tens of centimeters to a few meters of dust.

Other information about the physical characteristics of the Martian surface has been derived from active (radar) sensors. Earth-based radar measurements have been made for the Equatorial regions since the early 1970s (Pettengill et al., 1971). Radar-derived topographic data were used to measure the depths of large craters prior to the availability of MOLA data (Roth et al., 1989). Sounding radars can peer beneath the surface; the Shallow Radar (SHARAD) on MRO has a wavelength of ~15 m (20 megahertz), with a horizontal

Figure 3.3 Topographic data for a spectacular small (~900 m diameter) crater within the northern plains (38.15°N, 191.58°E), generated from stereo HiRISE images. The elevation range is from –4,242 to –3,875 m relative to the datum and clearly shows the asymmetric distribution of the ejecta. Left: HiRISE image DTEEC_025735_2185_025801_2185. Right: A monoscopic view of the same crater, showing the prominent partially eroded continuous ejecta blanket. HiRISE image ESP_025735_2185. Note: A black and white version of this figure will appear in some formats. For the color version, please refer to the plate section.

resolution of between 0.3 and 3.0 km and a vertical resolution of ~15 m (Seu et al., 2007). SHARAD data can be used to investigate the ice content of the ejecta blankets of pedestal craters (Nunes et al., 2011). For the polar regions, SHARAD can penetrate the ice cap to reveal numerous impact craters on a surface that appears to be Hesperian in age (Putzig et al., 2018). The *Mars Express* spacecraft also carried a radar system. The Mars Advance Radar for Subsurface and Ionospheric Sounding (MARSIS) radar operated at four frequencies between 1.3 MHz and 5.5 MHz and has been able to detect geologic features such as old impact basins with a depth of burial of ~2.0–2.5 km (Watters et al., 2006).

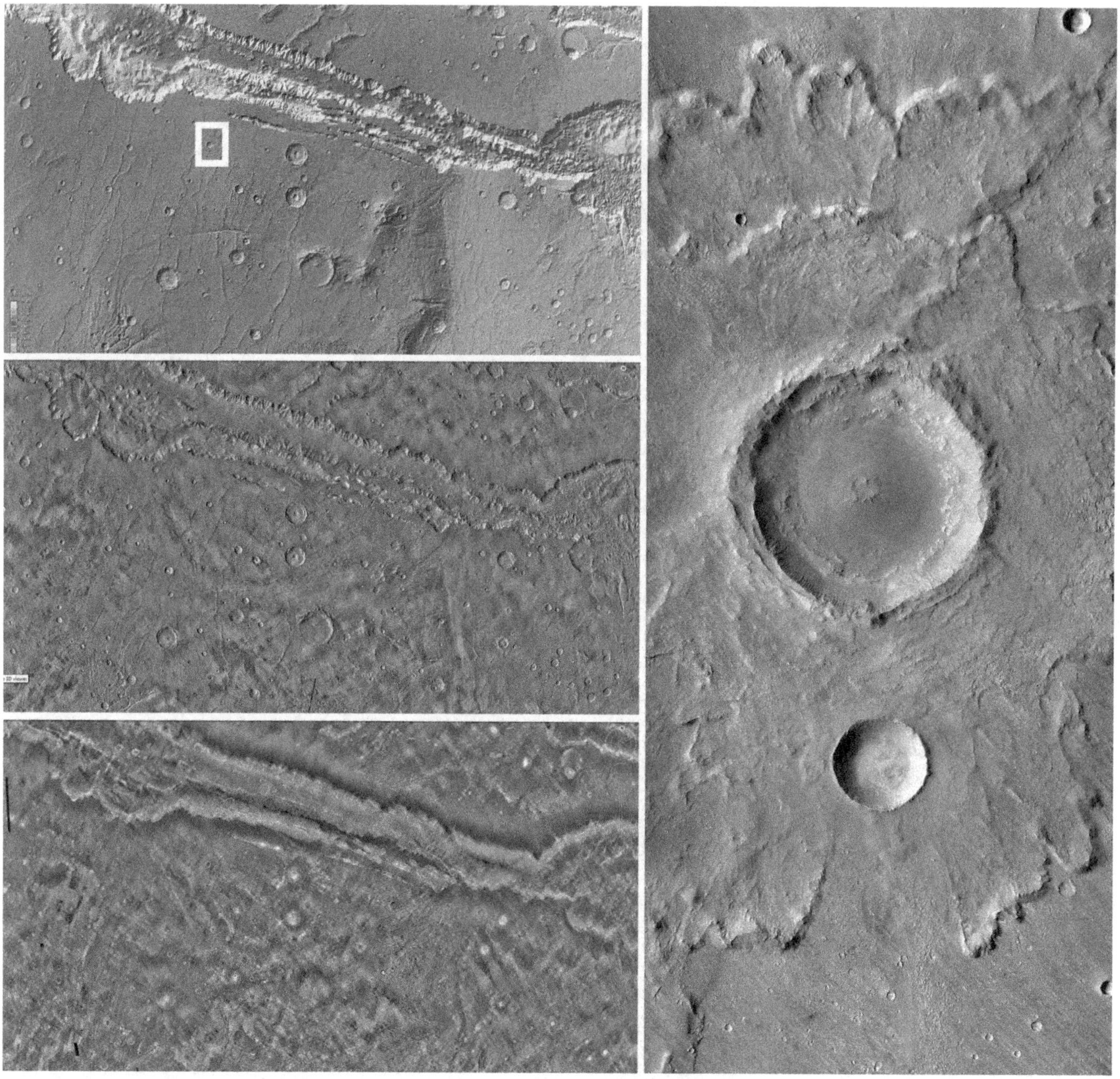

Figure 3.4 Comparison of four global data sets used extensively in this book. The top left image is a colorized rendition of the MOLA topography (color scale is the same as Figure 3.2). White box outlines coverage at right. Middle left is a THEMIS daytime IR image, and bottom left is THEMIS nighttime TIR image. Each image displays part of Thaumasia Planum and the eastern part of Valles Marineris (12.80°S–24.53°S, 282.48°E–309.30°E). At right is an example of a CTX image (frame G05_020312_1642). CTX provides near-global coverage at ~6 m/pixel resolution. The larger crater is 15.6 km in diameter and is located at 15.62°S, 292.41°E. Note: A black and white version of this figure will appear in some formats. For the color version, please refer to the plate section.

3.3 AGES OF MARTIAN ROCKS

The oldest geological period on Mars is called the Noachian, which has an uncertain age but probably corresponds to the lunar Pre-Nectarian to Early Imbrium periods of 4,100 to 3,700 Ma ago. The Noachian includes the interval known as the Late Heavy Bombardment (Werner and Tanaka, 2011; Tanaka et al., 2014). The next youngest period is the Hesperian Period, which is an intermediate and transitional period in Martian history. The beginning of this period followed the end of the Late Heavy Bombardment and probably corresponds to the lunar Late Imbrium period, around 3,700 Ma. During the Hesperian, Mars changed from the wetter and perhaps warmer world of the Noachian (Andrews-Hanna and Lewis, 2011; Palumbo et al., 2020) to the dry, cold, and dusty planet we see today. The end of the Hesperian Period is uncertain and could range anywhere from 3,200 Ma to 2,000 Ma. The most recent period on Mars is the Amazonian. The Amazonian is thought to have begun around 3,000 Ma, although it could have started much more recently. The period is sometimes subdivided into the Early, Middle, and Late Amazonian (which continues to the present day). The Amazonian period has been dominated by impact crater formation and eolian processes, although recent volcanism may have occurred in the Tharsis region and within Cerberus Fossae (Plescia, 2003; Jaeger et al., 2007; Voigt et al., 2023). Noachian-aged targets are most likely highly fragmented and so may not have structural strength compared to Amazonian-aged units. Thus, we would expect to see differences in the interior of fresh craters formed in the older rocks (specifically, a lack of terraces due to the lower mechanical strength of the rocks) and those formed in younger rocks. Of course, most of the oldest, most heavily eroded, craters on Mars are found within materials formed during the Noachian.

3.4 PHYSIOGRAPHY OF MARS

Many review papers and books on the physiography of Mars have been published, as well as the publication of an excellent 1:25,000,000-scale geologic map that uses the latest satellite data sets (Tanaka et al., 2014). The reader is referred to the exceptional book by Carr (2006) and to the review paper by Carr and Head (2010) plus the references therein for much of the background material. In addition, some of the most pressing puzzles of Martian geology were reviewed by Head (2007).

Here we first focus on the type and age of the rock within which an impact crater formed, and the elevation at which it formed. Geographic location, whether the crater formed at low or high latitude, is also important. These attributes are all expected to influence the final geometry of the crater cavity as well as the morphology of the ejecta blanket (Mouginis-Mark, 1979a; Barlow and Bradley, 1990; Barlow and Perez, 2003). Recognize too that impact craters have formed throughout the history of Mars, so that undoubtedly there will be a temporal component to the way in which a projectile of any size was influenced by the planet's surface conditions. At some periods in Martian history, liquid water or ice was most likely close to the surface (Squyres, 1989), while the atmosphere will have changed in density due to orbital changes of the planet (Ward, 1973; Baker et al., 1991).

At the most basic level, Mars exhibits a hemispheric dichotomy manifested in the crustal structure, topography of the landscape, and the age of the surface (Scott and Carr, 1978; Tanaka et al., 2014). This difference is most strikingly seen in the global map of topography (Figure 3.2) derived from MOLA (Smith et al., 2001). The subdivision between the two hemispheres follows an approximate great circle tilted ~30° to the equator. South of this boundary, the terrain is generally at higher-than-average elevation and is heavily cratered indicating an old relative age. Major impact basins (e.g., Hellas and Argyre) over 1,000 km in diameter are located within the southern highlands and each may have hurled ejecta hundreds to a few thousand kilometers across the surrounding landscape (Figure 3.5). This basin ejecta would have formed the target material for younger craters that subsequently formed in this area. Younger craters smaller than a few tens of kilometers in diameter that formed within these basin ejecta may thus have distributed their energy in a different way than similar-sized craters formed within sediments or young volcanic terrain. Parts of the highlands also preserve ancient magnetic anomalies, which no doubt date from the earliest period of crustal formation and preservation (Acuna et al., 1999). Retention of these magnetic materials may indicate that some parts of the Southern Highlands comprise ancient volcanic rocks.

To the north of the hemispheric boundary, there are younger units at generally lower elevations (Figure 3.2). These terrains are less heavily cratered and are therefore younger than the units to the south. Circling the globe at mid- to high-northern latitudes are vast, generally flat, smooth plains that numerous investigators have speculated are the desiccated seas (Baker et al., 1991; Head et al., 1999). Sedimentary deposits in these regions are likely to be tens to hundreds of meters thick and may be responsible for spatial variations in the depth/diameter ratio of craters due to the thickness of this sedimentary veneer (Boyce et al., 2005). These sediments may well have buried ancient crust. Ancient, buried impact basins as well as numerous quasi-circular depressions have been identified within Utopia and Isidis Basins (Frey et al., 2002). Several different types of impact craters were formed within these sediments including pedestal craters, which may provide clues to the recent (Amazonian) climate history of Mars (Kadish and Head, 2011).

In addition to these low-lying sedimentary regions, the northern hemisphere includes the two major volcanic regions on the planet that have several giant shield volcanoes rising to >15 km above Mars datum. The Tharsis volcanic region is centered near 10°N, 240°E, and includes the largest of the Martian volcanoes, Olympus Mons (Figure 3.5). Together with the other giant shield volcanoes (Arsia, Pavonis, and Ascraeus Montes), the elevation differences across different flanks of the volcano afford the opportunity to study potential elevation effects on the cratering process while keeping the country rock the same (Mouginis-Mark, 2015). The second volcanic province is Elysium, which includes the steep-sided Elysium Mons volcano centered at 25°N, 145°E. Lava flows from both the Tharsis and Elysium volcanic centers can extend >200 km from the summit area. The youngest volcanic flows are believed to be in the Cerberus Fossae region, and here flows may be only a few tens of millions of years old (Vaucher et al., 2009; Jaeger et al., 2010; Voigt et al., 2023).

Valles Marineris is a tectonic rift that undoubtedly formed via multiple geologic processes that included dike emplacement, regional faulting, and erosion (Schultz, 1998). Many parts of Valles Marineris have most likely been expanded through fluvial erosion, and there is good evidence for ponding and catastrophic flooding within the central portion of this ~4,000 km long canyon system (Harrison and Chapman, 2008). Valles Marineris opens to the southeast of

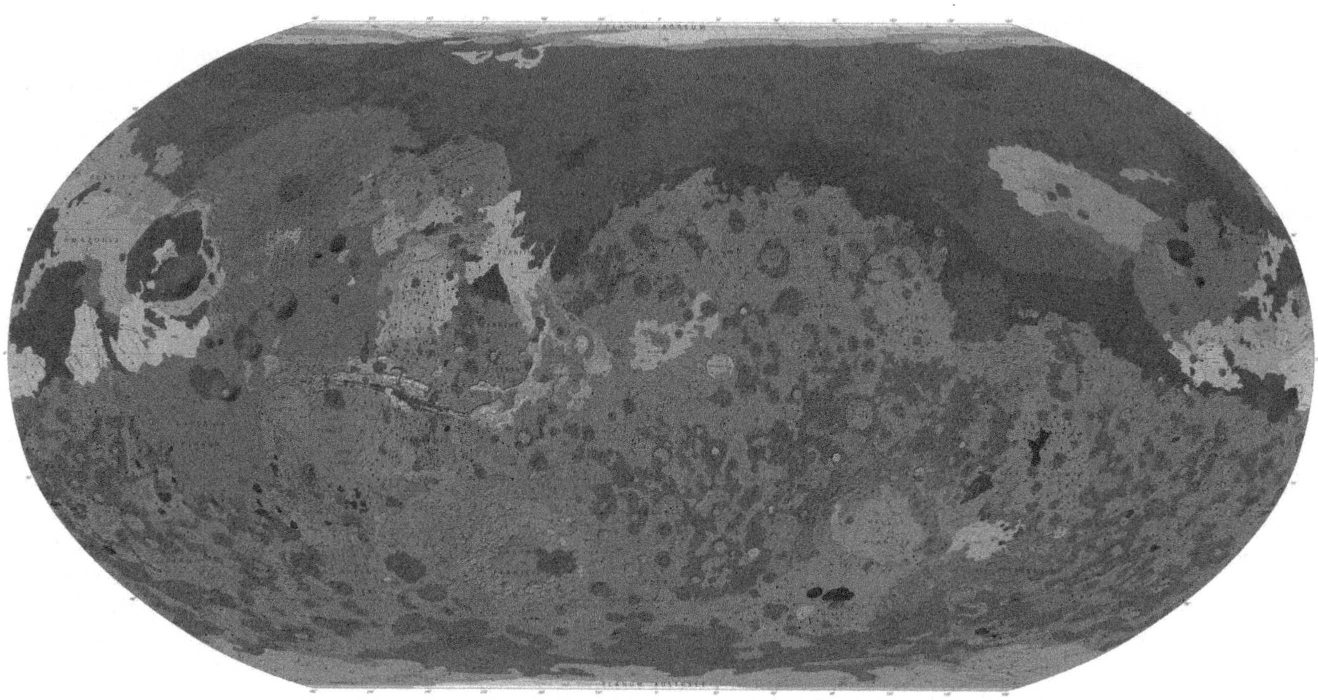

Figure 3.5 Tanaka et al. (2014) geologic map of Mars created at a scale of 1:20M using the Robinson pseudocylindrical projection. Of particular importance for the study of impact craters is the subdivision between the old Noachian-aged highland to the south (brown units) and the relatively young northern plains (green unit). Young volcanics in the Tharsis and Elysium regions are shown in orange and red. The younger craters are shown as yellow dots. Compare this map to the topographic data displayed in Figure 3.2. Note: A black and white version of this figure will appear in some formats. For the color version, please refer to the plate section.

Tharsis and ends in channels and flow features reaching into the northern plains. Outflow channels such as Kasei, Maja, and Ares Valles fed fluid-derived sediments across Chryse Planitia (Baker and Kochel, 1979; Warner et al., 2009), and other "mega-flood channels" originated from tectonic fractures elsewhere on Mars (Burr et al., 2009). At both poles, geologically young sequences of layers can be seen, and these consist of water ice at the North Pole and carbon dioxide ice at the South Pole (Howard et al., 1982; Fishbaugh and Head, 2001).

When studying Martian impact craters, it is important to recognize that craters of the same size could have formed under significantly different conditions depending upon the distribution of volatiles and atmospheric conditions. This is well illustrated by Figure 3.6, which shows two comparable-sized craters in western Utopia Planitia with different interior morphologies and very different ejecta deposits. Such changing conditions could have been the result of obliquity variations; changes of more than 40° may have taken place (Ward, 1973) with the result that water-ice-rich materials related to ice ages may occur in both hemispheres from mid-latitudes to the poles (Head et al., 2003). Variations in the onset diameter for the layered ejecta deposits as a function of location on the planet may well reflect the spatial distribution of volatiles at the time of crater formation (Barlow et al., 2001).

Relative ages of different units can be derived from superposition, cross-cutting relationships, and the numbers of impact craters formed on exposed surfaces. Modeling surface ages requires an estimate of the production rate of craters over time (Hartmann, 2005; Werner and Tanaka, 2011). These age determinations rely in part on the known age of lunar samples and the number distribution of craters in an area. As such the production functions may be biased toward a chronology that predicts ages that are either too old or too young, but for most geologic units on Mars, we just do not know their absolute age. Firm data points on the age of Martian rocks do nevertheless come from the radiometric ages of the SNC meteorites from Mars that have been tossed to Earth by impact events. Meteorite ages cluster between 165 and 1.35 Ga (Nyquist et al., 2001), and these ages may be useful to calibrate the crater production models of Hartmann (2005) and Neukum et al. (2010). Recently, Herd et al. (2024) proposed a novel approach involving models of shock conditions within certain SNC meteorite groups, as well as geomorphic criteria, to identify the freshest craters (see Section 3.7) in order to link each rock to a specific place of origin. Crucially for the study of young craters, these meteorites not only provide the crystallization age of the target rock but also the "ejection age" of the sample. This ejection age corresponds to the formation age of the crater. If proven to be valid, such an approach would be highly valuable as the ejection and crystallization ages of the meteorite groups represent rigorous constraints on the ages of the parent craters as well as the crystallization ages of the terrain in which each crater formed. While almost all the SNCs have ages corresponding to the Amazonian, a single meteorite, ALH84001 (Mittlefehldt, 1994), has a much older age, possibly ~4.1 Ga (Lapen et al., 2010). This rock most likely represents part

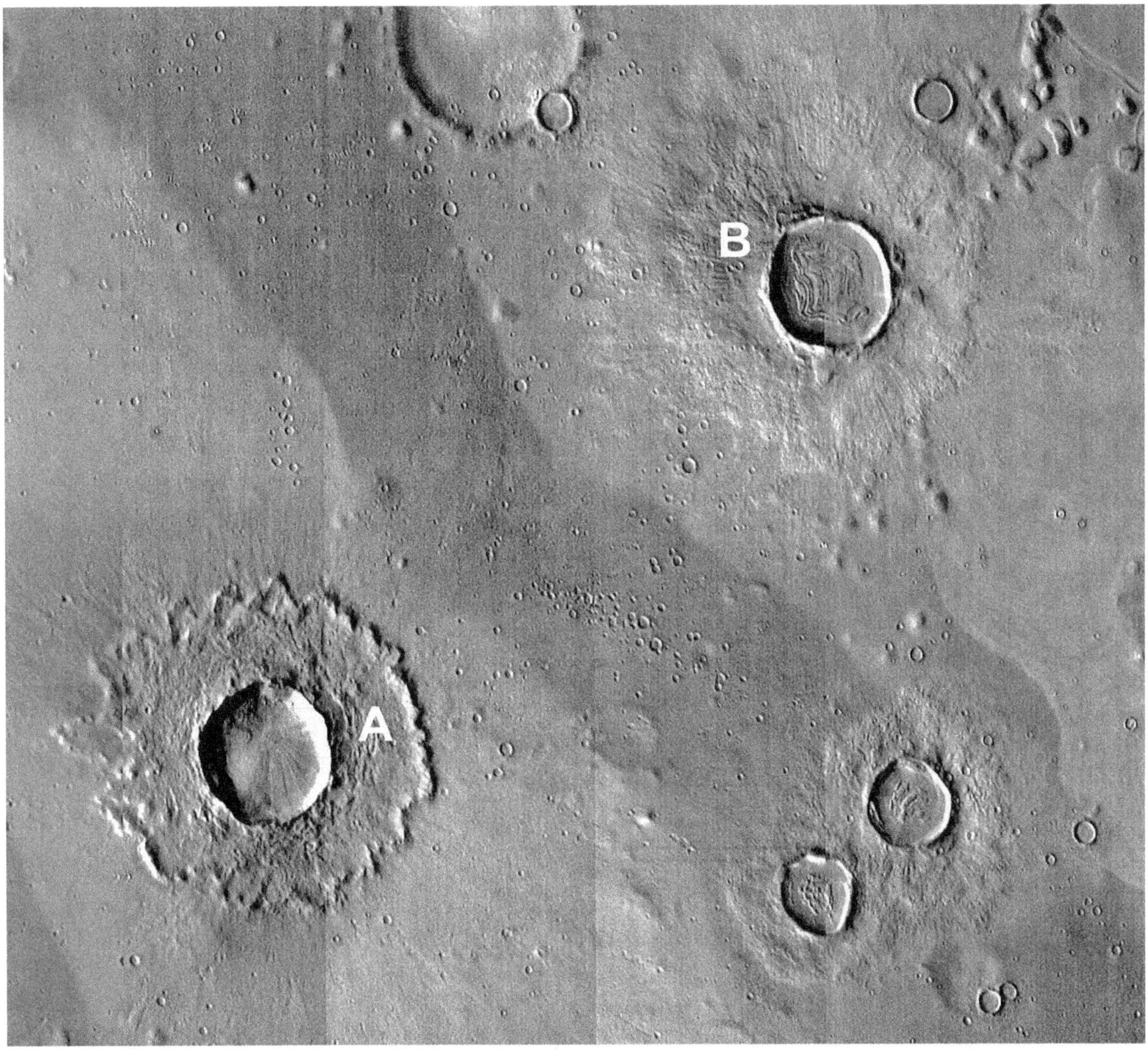

Figure 3.6 Temporal variations in atmospheric, surface, or subsurface properties most likely explain the differences in morphology of the two similar-sized craters ("A" and "B"), although it is also possible that there is a buried geologic boundary running between these two craters. Crater "A" has a well-formed distal rampart surrounding the ejecta blanket, whereas crater "B" has two concentric ejecta layers, a set of radial grooves, and only a faint distal margin. Both craters are ~11.5 km in diameter and lie within western Utopia Planitia (36.34°N, 81.03°E). The dark band crossing this image represents a cooler (higher thermal inertia) unit. Mosaic of THEMIS daytime TIR images I03068002, I09871022, I22201012, I36850018, and I45449034.

of the Noachian highlands. No source crater has yet been proposed for this ancient rock.

The atmosphere of Mars is currently very thin (~6 mb at the datum) but could have played an important role in the formation of the ejecta blankets of craters (Schultz and Gault, 1979). The major atmospheric gas (carbon dioxide) is in dynamic equilibrium with its solid phase, resulting in significant changes in abundance and pressure due to temperature changes because of the obliquity and the long-term rate of atmospheric loss. It is estimated that a total of ~34 m of global equivalent water layer lies in the Martian polar-layered deposits and shallow ground ice (Carr and Head, 2015). Carr and Head (2015) considered the potential changes with time for the near-surface water inventory, with a high of ~62 m at the end of the Hesperian. Kite et al. (2014) proposed that increases in the paleo-pressure of the atmosphere could be detected from the distribution of ancient craters. Analysis of gases trapped within the SNC meteorites delivered to Earth suggests that the evolution of the atmospheric mass as well as the near-surface temperature and pressure can all be identified (Moyano-Cambero et al., 2013).

Various regions of Mars show differences in their erosional histories that may affect the final form of impact craters. For example, the Arabia region (centered at approximately 25°N, 30°E) shows ample evidence for an extensive, several-hundred-meter-thick deposit of horizontally layered material subsequently subjected to erosion (Moore, 1990; Fassett and Head, 2007). This mantle appears to have influenced the degradation and preservation state of craters in the 1–5 km diameter range (Barlow, 1995). An extensive dust mantle of variable thickness has also been identified on Mars (Ruff and Christensen, 2002). This dust mantle assists in the identification of impact events that have taken place during the past few years, where tonal differences between the new ejecta and the existing dust deposits make the new crater identification relatively easy (Daubar et al., 2014).

The two polar caps of Mars present challenging environments for the analysis of crater geometry due to the rapid deposition of ice and, most likely, a more rapid rate of erosion. Specifically, Garvin et al. (2000) reviewed the MOLA altimetric data for craters near the North Pole and found that the amount of cavity infill may be as high as ~80%. South polar deposits exhibit relatively young crater retention ages due to recent resurfacing (Pathare et al., 2005). Large polar craters such as the ~36 km diameter Louth crater (Brown et al., 2008) include water ice mounds.

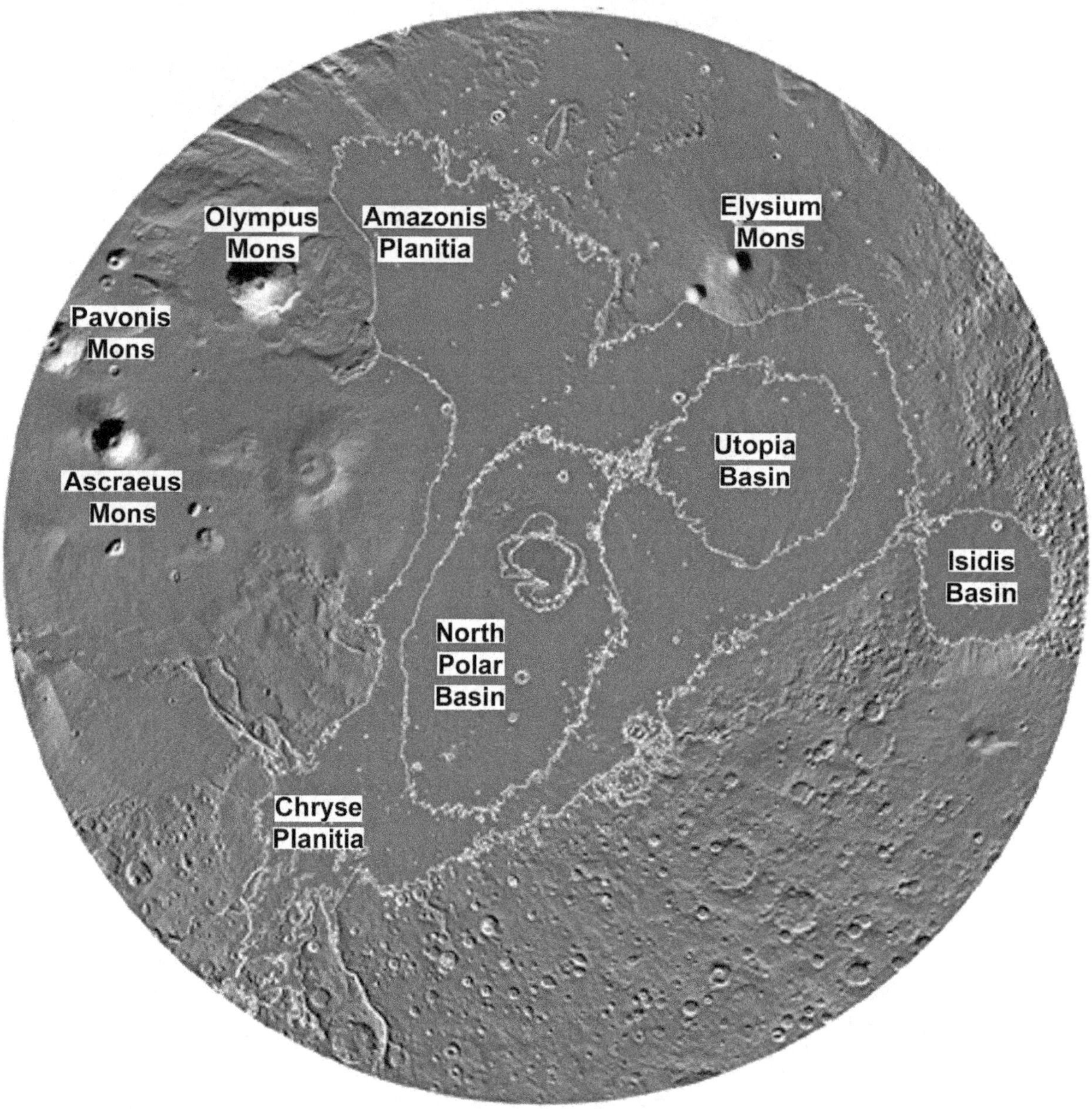

Figure 3.7 Prominent basins and major landmarks in the northern hemisphere of Mars are outlined and labeled in this polar stereographic projection. Two contours are shown, an outer one at an elevation of –3,500 m and an inner one at –4,350 m. These contours may approximate the spatial extent of paleo-oceans that once occupied the northern plains. The base image is a shaded relief image derived from MOLA data. Reprinted from Carr and Head (2003) "Oceans on Mars: An assessment of the observational evidence and possible fate," *Journal Geophysical Research: Planets*, vol. 108. Their Figure 2, with permission from John Wiley and Sons.

3.5 MAJOR VARIATIONS WITH TIME

3.5.1 Potential Northern Ocean

A variety of unusual geomorphological features in the northern plains led Baker et al. (1991) to propose that catastrophic flooding, triggered by extensive volcanism within the Tharsis region, led to repeated ocean formation. Such an idea was further elaborated by Head and Carr (2003) (Figure 3.7). Episodic flooding of the northern plains (Fairen et al., 2003) could in turn have led to a thick sedimentary deposit in the northern plains that influenced the geometry of impact craters (Boyce et al., 2005). Such deposits may potentially be linked to unique ejecta morphologies of Type-1 double-layered ejecta (DLE) craters (see Chapter 6). On rare occasions, impacts into a transient ocean (Ormo et al., 2004) may have taken place so that tsunami deposits are possible (Rodriguez et al., 2016). Indeed, Costard et al. (2019) have suggested that the Lomonosov crater (~120 km diameter) formed during the existence of an ocean in the Late Hesperian/Early Amazonian.

3.5.2 Obliquity Changes

Today the obliquity of Mars is 25.2°, which is quite like Earth's value of 23.5°. Seasons on Mars are therefore like those on Earth, but due to Mars' greater distance from the Sun, they last longer and are colder. Mars experiences dynamic changes in obliquity with geologic time (Figure 3.8) due to the secular precession of both the spin axis combined with periodic oscillations of the inclination of the Martian orbital plane (Figure 3.9) (Forget et al., 2017). First identified by Ward (1973), the consequence is that the polar caps have not always been the size that they currently are. At times of high obliquity, the polar caps were most likely much more extensive than they are today. Hence, craters that we study today may have formed in targets that were at one time under ice sheets but that are now ice-free. The geomorphic consequences of impacts into layered icy material were modeled by Senft and Stewart (2008); they discovered that dramatic changes in crater cavity excavation due to ice within the target could produce shallower craters, icy flow-related features, and crater lakes. The detailed morphology of craters originally

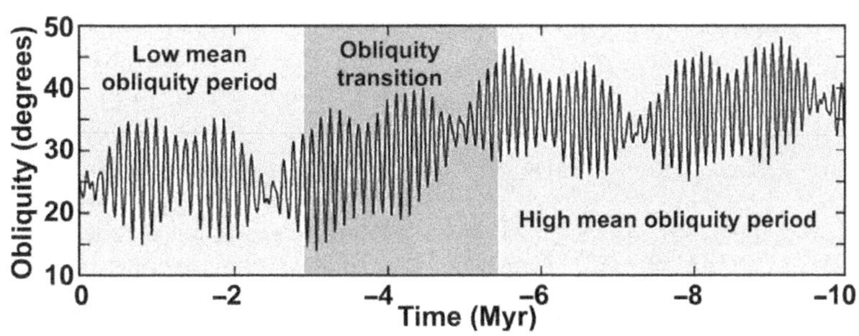

Figure 3.8 Obliquity variations of Mars over the past 10 million years, showing that present-day Mars is experiencing a lower mean obliquity than 5–10 million years ago. Such obliquity changes inevitably must have caused the redistribution of ice across the surface of Mars, which in turn would have affected any craters that formed at that time. Reprinted from Levrard et al. (2004). "Recent ice rich deposits formed at high latitudes on Mars by sublimation of unstable equatorial ice during low obliquity," *Nature*, vol. 431, pp. 1072–1075, Figure 1, with permission from Springer Nature Customer Service Center.

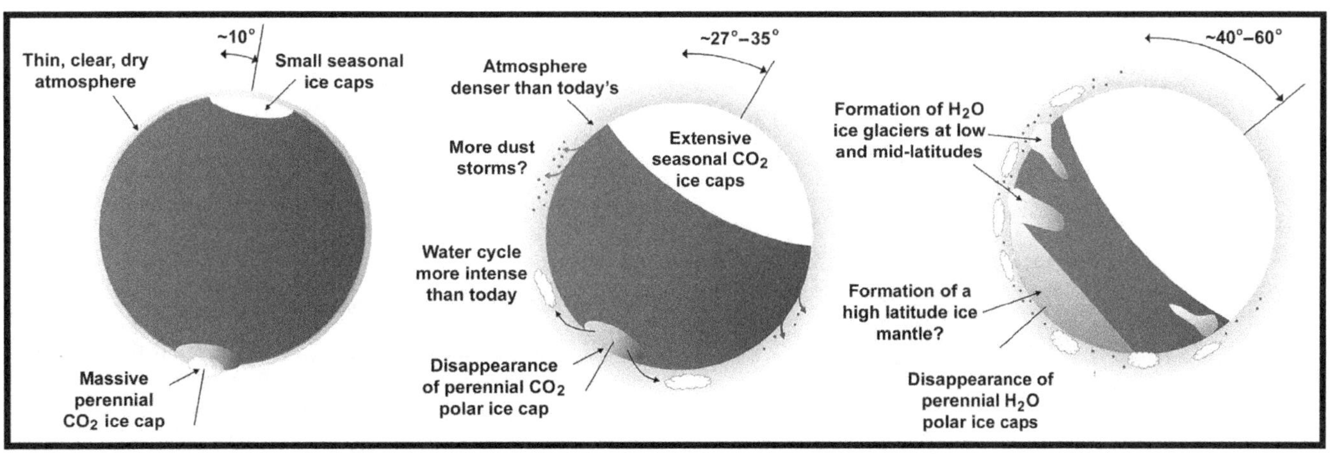

Figure 3.9 Distribution of the polar ice caps at various obliquities. Left, at an obliquity of 10°. Middle at ~27°–35°, and at right at 40°–60°. Notice that a very extensive seasonal ice cap is predicted to form in the northern hemisphere at high obliquity and may well have covered terrain within which new craters formed, no doubt affecting the final crater morphology and ejecta characteristics. Redrafted from Forget et al. (2017), "Recent climate variations," in *The Atmosphere and Climate of Mars*, with permission from Cambridge University Press.

formed in surface ice but now in an ice-free region has not been investigated in detail, although some crater attributes such as "excess ejecta" have been linked to ice-rich paleodeposits (Kadish and Head, 2011).

What might the possible obliquity variations have been over the past 250 Ma (Laskar et al., 2004) or even further back in time when the impact cratering rate was much higher? It seems likely that ice would have dramatically moved around the planet during obliquity excursions (Levrard et al., 2004) and thus complicates the geomorphic analysis of many landforms. For instance, Schultz and Lutz-Garihan (1982) speculated that the Medusae Fossae Formation (MFF), which lies near the Equator, could be a polar deposit from when the spin axis was on its side, and that craters superimposed on this deposit typically have anomalously large secondary craters potentially indicating an easily compressed substrate. But more recent interpretations of the MFF (Ojha and Mittelholz, 2023) support a volcanic origin that does not depend upon obliquity variations. Nevertheless, sounding radar may indicate an ice-rich deposit with a higher percentage of dust and sand than polar layered deposits (Watters et al., 2007). Other observations by Schultz and Lutz-Garihan (1982) suggested that oblique impact craters that were not aligned with the current equator of Mars may be lost Martian moons that had spiraled into the surface at a time when the orientation of the planet was different from that which we see today.

Figure 3.10 This unnamed 28 km diameter crater (located at 30.05°S, 37.38°E) displays gullies on both its northern and southern inner walls, with the northern (polar-facing) gullies much less degraded than the southern (Equator-facing) gullies. Such a situation is unusual as gullies typically are only observed on the poleward-facing wall. The crater lies within a slightly larger (38 km) crater. Mosaic of CTX images D09_030686_1501, P16_007162_1497, and P19_008309_1497.

There is good evidence that climate variations may have conspired to modify individual craters (Goldspiel and Squyres, 2011). While there is consensus that fluvial gullies are preferentially located on the poleward-facing inner wall of a crater (Christensen, 2003; Dickson and Head, 2009), some craters show varying degrees of erosion at different azimuths around the perimeter of the rim. This is suggestive of different episodes of gully formation and, hence, different climatic conditions. Figure 3.10 illustrates such an example.

The identification of youthful near-surface ice created during the last phase of high orbital obliquity (i.e., less than 100,000 years ago) can be inferred from landforms that are indicative of viscous creep (Mustard et al., 2001). Craddock and Howard (2002) made the case for a warm, wet early Mars based upon their analysis of degraded craters in the southern highlands. During degradation, the diameter of a crater is expected to increase due to back-wasting, and an unusual amount of enlargement of craters ~50 km in diameter or larger required the existence of crater fill to explain this discrepancy. This fill has subsequently been removed as the climate changed. Further signs of climate change in the southern highlands come from Type-1 DLE craters (see Section 6.2 for a description of this type of crater morphology). While the exact conditions necessary for the formation of this type of crater have not been resolved, most Type-1 DLE craters formed in the northern plains where the target material is believed to be fine sediments associated with the hypothesized northern ocean. Such conditions are not thought to exist in the southern hemisphere, but local sedimentary traps (such as partially infilled preexisting craters) could be an explanation.

Figure 3.11 presents an intriguing puzzle as to the physical properties of the rim and the inner ejecta layer. Here a morphologically fresh crater has experienced a second impact on the southwest rim, destroying part of the rim. Prominent grooves and pressure ridges can be seen on both the inner and outer ejecta layers. Surprisingly, the ejecta blanket from this second impact is almost totally absent. The most plausible explanation appears to be that at this southern latitude (45°S), there was a mantle of material covering the larger crater that has subsequently been removed from the entire crater. Such an interpretation seems reasonable because some of the grooved ejecta from the large crater can be seen exposed in "windows" of erosion in a mantle deposit on the southern side of the smaller crater (Figure 3.11(C)). Instead of ejecta covering part of the inner ejecta layer of the larger crater, the main consequence of the second impact has been to provide a pathway into the larger crater floor for debris flows.

3.5.3 Distribution of Near-Surface Ice

If ice in the target had a significant effect on the morphology of an impact crater, it is reasonable to ask what the latitudinal distribution of this ice might be. Investigators have used numerical models (Fanale et al., 1986) and data from the *Mars Odyssey* Neutron Spectrometer (Feldman et al., 2011; Pathare et al., 2018) to infer this near-surface distribution of ice. One of the most immediate measures of the distribution of ice at the present day comes from observations of craters

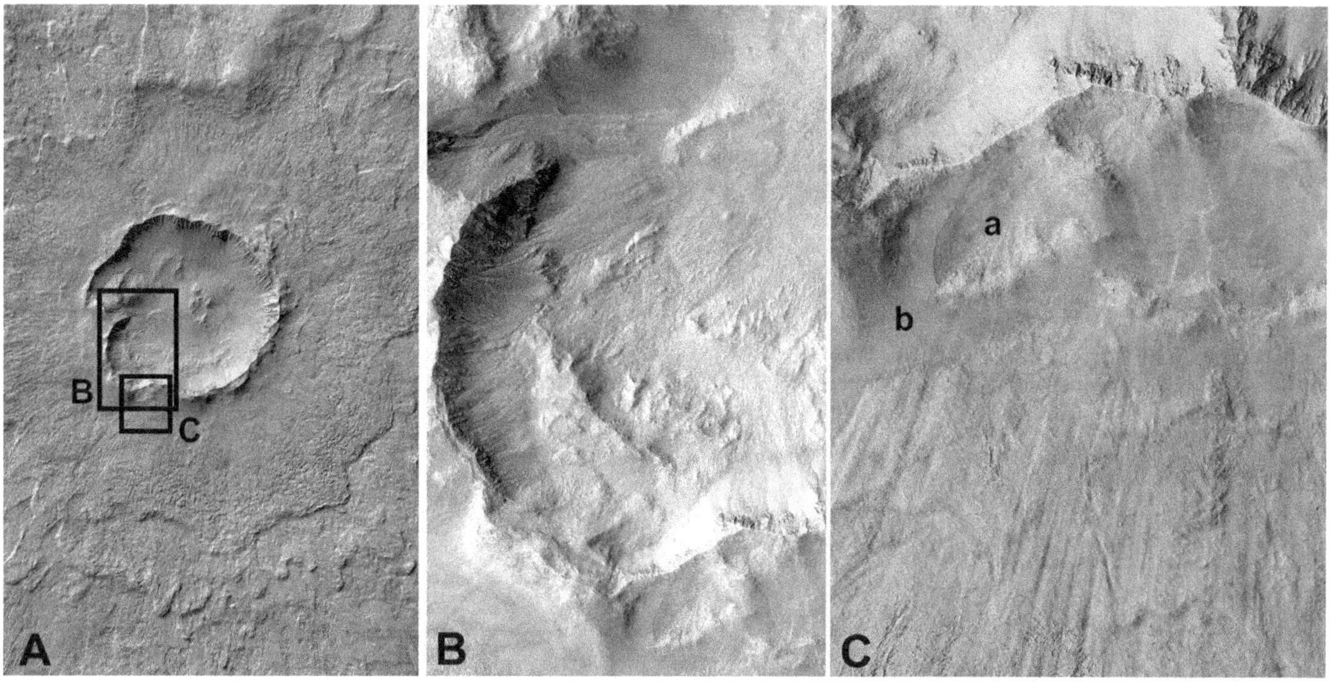

Figure 3.11 (A) Overview of an 11.3 km Type-1 DLE crater located at 45.39°S, 25.66°E. This crater may be a good example of a now-removed mantle of material that existed between the formation of the larger crater and the smaller ~4.4 km crater on the SW rim. The boxes denote the areas of coverage of the higher-resolution images. CTX image G15_024080_1343. (B) The smaller crater on the SW rim was ~4.4 km in diameter, although the eastern half has now been totally removed by erosion. No ejecta from this smaller crater can be found on the ejecta layer or on the floor of the larger crater, strongly suggesting that a mantle of material (now removed) existed when the small crater was formed. (C) Details of the southern rim of the smaller crater. Windows through a mantle ("a") occur on the outer wall. Grooves from the inner ejecta layer of the older crater can be seen to emerge from beneath this mantle ("b"). Both images are parts of HiRISE frame ESP_039970_1340.

Figure 3.12 A recently formed 20 m diameter crater on Mars. Repeat coverage with satellite images confirms that this crater formed during the past decade. Crater located at 44.20°N, 164.20°E. The white ejecta is water ice, which must be within a few meters of the surface at the present time. HiRISE image ESP_016954_2245. Note: A black and white version of this figure will appear in some formats. For the color version, please refer to the plate section.

in the diameter range of 4–24 m that have formed during the spacecraft era (Figure 3.12) and have excavated ice within a few meters of the surface (Dundas et al., 2014). Dundas et al. (2021) found that the lowest latitude at which ice currently exists is near 39°N, which is at the edge of a region where impact exposures of ice are common between 40° and 50°N. While this near-surface ice is unlikely to influence the morphology of larger craters, as they excavate materials from much greater depths, craters smaller than ~50 m in diameter may provide good insight into the present-day distribution of near-surface ice.

3.6 HUMAN-SCALE MORPHOLOGY OF MARS

We have some information about how Martian craters appear from the ground by virtue of the rovers *Opportunity, Spirit, Curiosity*, and *Perseverance*. Data from the *Opportunity* rover provided detailed information on the rim structure of Endeavour crater (Squyres et al., 2012; Crumpler et al., 2015) where erosion has brought segments of the crater rim to the surface (Figure 3.13). Several meters of the ejecta blanket were identified with *in situ* mapping conducted from the rover data. While not a specific target of the *Curiosity* mission, images of Slangpos crater have revealed layering in the inner wall of the crater, as well as asymmetries of the crater rim crest not visible from orbital images (Figures 3.14 and 3.15). Viewed from ~30 km, it has not been possible

Figure 3.13 A scene from the panoramic camera on the Opportunity rover on sol 3,663 of that mission (May 14, 2014). This view is looking east from a point on the western rim of Endeavor crater (2.25°S, 354.78°E) called "Pillinger Point." The eastern rim of Endeavor crater is on the distant horizon, about 22 km away. NASA JPL photojournal image PIA18394.

Figure 3.14 At left we see Slangpos crater (4.8 km in diameter, 4.73°S, 136.81°E), as imaged by the CTX (D09_030735_1746). At right, the crater is imaged on October 25, 2017, from the ground by the *Curiosity* rover as it was climbing the "Vera Rubin Ridge" heading toward Mt. Sharp (which is the slope in the lower left foreground). The rim of Gale crater lies beyond Slangpos and is ~40 km away from the rover at the time this image was obtained. Part of JPL Photojournal image PIA22210. "A" to "D" mark common places in the two images. Slangpos crater is ~5.3 km in diameter. North is to the right.

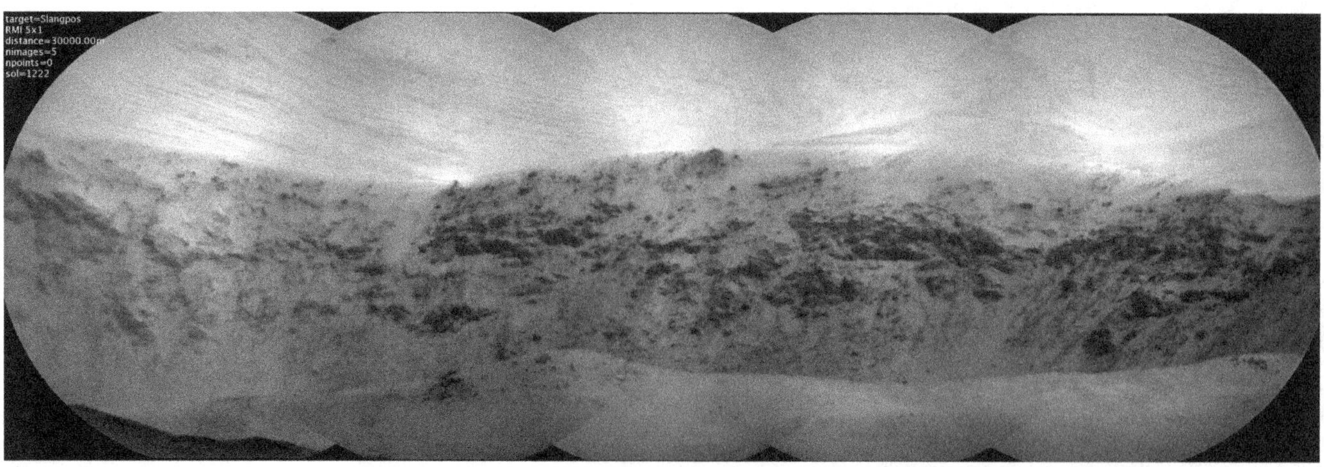

Figure 3.15 View of the western inner wall of Slangpos crater, as imaged from ~30 km by the *Curiosity* rover's ChemCam Remote Micro-Imager. This part of the rim shows numerous layers near the rim crest and is close to point "B" in Figure 3.14b. Frame CRO_505973557PRC_F0521162CCAM02221L1.

to resolve details of the ejecta layer surrounding Slangpos, although it appears to be relatively thin. Images from the ChemCam instrument reveal layering in the inner crater wall. The *Opportunity* rover conducted surface investigations, including copious imaging, of Endeavour and Victoria craters (Squyres et al., 2009; Grant et al., 2016), and the *Curiosity* rover is exploring the interior of Gale crater (Grotzinger et al., 2015). As of mid-2024, the *Perseverance* Rover, after investigating a deltaic deposit on the western floor of the crater, is climbing out of Jezero crater to investigate some of the rim units and, potentially later in the mission, explore outcrops of ejecta from the Isidis impact event.

3.7 IDENTIFYING THE YOUNGEST CRATERS ON MARS

Throughout the rest of this book, we will make extensive use of examples of morphologically "fresh" craters to illustrate different attributes. It will be important to connect the diverse attributes to help identify (potentially) the youngest and/or closest-to-pristine craters. It is often that the youngest craters provide the best clues to the cratering process and the condition of the target material at the time of crater formation, which in turn will guide the interpretation of older craters where significant erosion has taken place.

The youngest craters on Mars are, of course, those formed during the spacecraft era, and these can be unambiguously identified by their absence in early satellite images and their presence in images acquired later (Dauber et al., 2016). Dauber et al. (2022) reported the identification of 1,203 new craters in the diameter range of 58 m to as small as <1 m with date constraints from orbital images. However, these spacecraft-era craters are all smaller than 1 km, and so do not provide an adequate insight into the freshness of craters larger than a few kilometers in diameter.

"Young" does not necessarily mean "best preserved." For example, dust in the Tharsis region will often rapidly mask some of the finer-scale features of a crater, such as a ray system. Crater modification may also be influenced by the diameter of the crater, with some features (e.g., impact melt) being more rapidly buried (perhaps even during the terminal phases of crater formation) if the crater is large and thus experiences terrace formation and wall collapse near the terminal phase of crater formation. Here our quest is to find the landforms that might indicate the chronologically youngest craters. Tornabene et al. (2015, 2016) initiated a discussion of which attributes should be used to identify young craters on Mars. We adopt their strategy and use six different attributes that might collectively indicate a very young crater. While no single attribute can unambiguously support the idea that any one example is a very young crater, the existence of three, four, or five of the following attributes strongly suggests that the crater is young.

3.7.1 Superposed Craters

In the absence of dated samples collected *in situ* on Mars, the crater-size frequency distribution of mapped terrains, determined by crater counts, is currently the only quantitative way to constrain the relative age of landforms, be they impact craters, lava flows, or other geologic features (e.g., Neukum and Hiller, 1981; Hartmann, 2005; Werner and Tanaka, 2011). Determining the ages of very young surfaces relies on using populations of small, often sub-km diameter impact craters due to their relative abundance compared to craters of larger size. Warner et al. (2015) suggested that because of variability in cratering over an area of 100 km^2 as well as the susceptibility of smaller craters to resurfacing, crater counts derived from small area samples are subject to major uncertainties. These smaller craters are less reliable as their size-frequency distribution is more susceptible to alteration via erosion and/or burial (Williams et al., 2018). In a detailed study of some of the apparently youngest craters on Mars using crater size/frequency measurements, Lagain et al. (2021) reported a model age for Tooting crater, which shows several of the other attributes in this list, of 7.7 ± 3 Ma. This is comparable to the age of ~9 Ma derived from independent crater counts described by Hartmann et al. (2010). But if Herd et al. (2024) are correct in the identification of Tooting crater as the source for the Tissint meteorite (which has an ejection age of 1.1 ± 0.1 Ma), this implies that the crater counting method overestimates the age of the crater by almost 8 Ma.

3.7.2 Pitted Floor Material

Tornabene et al. (2012) identified a densely pitted crater-related unit within 205 Martian craters from CTX and HiRISE data (see also Section 4.3.1). These craters range in diameter from ~1 to 150 km and are found at elevations between −5.5 and +5.2 km. The pits are polygonal to quasi-circular depressions that often occur in dense clusters and range in size from ~10 m to as large as 3 km. They are intrinsically different from the pits seen within certain craters on Mercury (Figure 2.15) because on Mars they have raised interconnected rims and are found within materials that appear to be unconsolidated. The pits on Mars typically have rims that lack well-defined ejecta deposits and present no clear overlapping stratigraphic relationships. Interestingly, the size of the pits appears to scale with both the host crater's diameter and the size of the host deposit (i.e., larger pits are found within the largest deposits). Such material can occur on the crater floor, on flat surfaces of terrace blocks, or within flat depressions on the exterior ejecta deposits. Boyce et al. (2012) proposed a model wherein the pits formed by the explosive degassing of water from within the host material, which they assumed to be water-bearing, impact melt-rich breccia at the time of deposition. Because the pits formed in very friable material, and have low relief, it is inferred that they would be rapidly eroded or buried; thus, where pits are seen, the crater is not heavily modified and is most likely quite young.

3.7.3 Depth/Diameter Ratio

Several investigations have shown that the most morphologically fresh Martian craters are relatively deep compared to most other craters (Tornabene et al., 2018a), presumably because of the lack of infilling post-formation. Interestingly, there is a particularly good correlation between the deepest craters for their size (i.e., those with a large depth/diameter

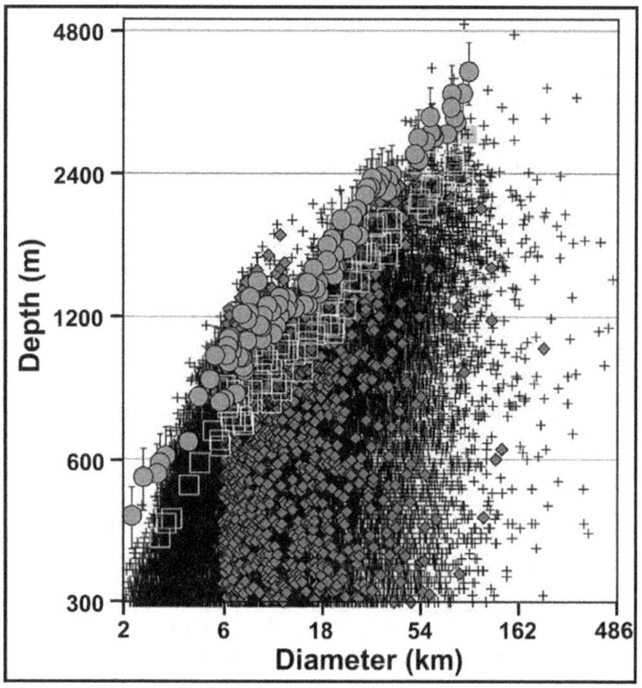

Figure 3.16 Log-log scatter plot of depth versus diameter for pitted material-bearing craters >2 km in diameter (large gray circles), 2,268 craters from Boyce et al. (2005) (small dark gray diamonds), and 71,871 measurements (crosses) from the database of Robbins and Hynek (2012). All the pitted craters lie toward the proportionally deepest side of the plot for a given crater diameter. Compare this plot with a vast number of craters to that shown in Figure 1.12 where there were only a few (204) lunar craters with a wide range of degradation states included in the data set. Reprinted from Tornabene et al. (2018a). "A depth versus diameter scaling relationship for the best-preserved melt-bearing complex craters on Mars," *Icarus*, vol. 299, pp. 68–83, Figure 4b, with permission from Elsevier.

ratio) and the occurrence of pitted material deposits in the interior of the crater (Figure 3.16). Another aspect of a large depth/diameter ratio may be the occurrence of unusually large (both in height and width) central peaks (Mouginis-Mark and Garbeil, 2007). This appears to be reasonable because as a crater degrades, thereby reducing the depth/diameter ratio, it is highly likely that either the walls will erode to produce crater fill, or eolian processes may partially fill the crater floor. In both these situations, the reworked material will start to bury the lower reaches of the central peak.

3.7.4 Impact Melt

While impact melt flows are preserved on many fresh lunar craters (e.g., Bray et al., 2010; Stopar et al., 2014), very few Mars craters preserve clear evidence of impact melt (see Section 4.4.3). This may either be the consequence of volatiles in the target preventing the formation of melt (Kieffer and Simonds, 1980) or the mechanical destruction of the melt sheet soon after crater cavity formation. The preservation of impact melt may be inversely correlated to the formation of the pitted terrain on the crater floor (Boyce et al., 2012), or to the stability of rim material that did not collapse to form a terrace block. If the calculations of Kieffer and Simonds (1980) are correct, the volatiles needed to produce the pitted terrain should also act to destroy impact melt. Tooting crater (Mouginis-Mark and Boyce, 2012) is the clearest example of impact melt preserved on part of the rim that did not collapse to form a terrace block (see Section 4.4.3). Pangboche crater is another example where impact melt has been identified, this time within the crater interior. This may be because the crater formed at the highest elevation for an impact crater on Mars and was therefore least influenced by volatiles in the target rocks (Mouginis-Mark, 2015).

3.7.5 Ejecta Rays

Filamentary ray systems are common for young craters on the Moon. Rays are filamentous, high-albedo deposits occurring radial or subradial to the parent crater. Lunar crater rays are bright compared to the surrounding terrain because of a compositional contrast with the surrounding terrain, the presence of immature material, or some combination of the two (Hawke et al., 2004). Because the deposits are thin, recognizing rayed craters on Mars is likely to be more difficult because eolian processes will rapidly mask their occurrence; thus, a rayed crater on Mars is most likely relatively very young (see Section 7.9). McEwen et al. (2005) provided one of the earliest detailed studies of a rayed crater on Mars, investigating the crater Zunil in Cerberus Planum. A system of rays was mapped as far as 1,600 km from the crater. McEwen et al. (2005) identified ~10^7 secondary craters 10–200 m in diameter, many of which were concentrated in a radial pattern associated with material morphologically similar to lunar crater rays. Dauber et al. (2022) reported that rays from very recent impact events are more likely to form from impacts into more consolidated material and may be enhanced by lower atmospheric pressure at higher elevations. Harris et al. (2023) developed an updated catalog of rayed Martian craters, as identified in daytime and nighttime THEMIS TIR data. A total of 118 rayed craters larger than 3 km in diameter were identified by this method. Rayed craters appear to occur on all terrains and were found at a wide range of elevations (from −5.0 km to +4.8 km). Model ages for 30 of these craters suggest ages <20 Ma. Hartmann et al. (2010) found that the youngest rayed craters' floors at decameter scale do not display "simple" excavation/deposition geology that can be attributed to the impact process alone.

3.7.6 Thermal Inertia

A high thermal inertia for a feature on Mars is indicative of blocky materials (Palluconi and Kieffer, 1981; Edwards et al., 2009). On Mars, impact craters can have very different thermal inertia values on their rims (Beddington et al., 2021) and their ejecta blankets. Hoover et al. (2021) suggested that it might be possible to identify layered ejecta that have thermal properties consistent with volatile fluidization. From 50 craters in the diameter range of 5–70 km that they investigated, there is no overwhelming support for ejecta fluidization (or an atmospheric entrainment mechanism). Nevertheless, Beddington et al. (2021) concluded that rim thermal inertia could be used

Table 3.1 *Morphologic characteristics of some of the best candidates for large (>1 km diameter) fresh craters on Mars.*

Crater	Latitude	Longitude	Diameter (km)	Pitted	d/D	Rays	Rim TI	Ejecta TI	Sec.	Melt
Gratteri	17.71 S	160.18 W	7.3	No	0.120	Yes	Yes	No	Yes	No
Tomini	16.27 N	234.2 W	7.4	No	0.142	Yes	Yes	Yes	Yes	No
Pangboche	17.22 N	133.62 W	10.2	No	0.093	No	Yes	No	Yes	Yes
Zunil	7.80 N	193.9 W	10.4	Yes	0.110	Yes	Yes	Yes	Yes	No
Canala	24.33 N	80.09 W	11.0	Yes	0.093	No	Yes	No	Yes	No
Steinheim	54.51 N	169.43 W	11.6	No	0.071	Yes	Yes	No	Yes	No
Corinto	16.93 N	218.39 W	13.5	Yes	0.073	No	Yes	No	Yes	No
Jaisalmer	33.49 N	84.14 E	15.0	Yes	0.063	No	Yes	No	No	No
Tooting	23.10 N	154.4 W	28.0	Yes	0.067	No	Yes	Yes	Yes	Yes

The presence/absence of pitted material ("Pitted"), rays, secondary craters, and impact melt was determined from inspection of CTX and HiRISE images. High rim thermal inertia ("Rim TI") and ejecta thermal inertia ("ejecta TI") were determined from daytime and nighttime THEMIS TIR data. The depth/diameter ratio ("d/D") was measured from individual MOLA profiles across each crater using JMARS software. The number of measurements of the rim varied from 36 for the Jaisalmer crater to 10 for the Canala crater. Maximum depth was taken to be the average of the five lowest points on the crater floor. Because Pangboche crater formed on the steep flanks of Olympus Mons, the rim height is more difficult to measure, and thus an average depth of 954 m was used. Interestingly, only Corinto and Tooting craters are included in the analysis of Herd et al. (2024) as candidate source craters for the SNC meteorites.

to estimate impact crater degradation states and thus relative ages in areas of Mars with low dust cover and similar target properties. The best-preserved craters have floors that are characterized as lower thermal inertia areas enclosed by higher inertia wall units (Piatek et al., 2018). Ejecta blankets may also have a high thermal inertia but do not exhibit any radial pattern in their thermal characteristics. This is consistent with the idea that the floor units contain porous fill and/or talus deposits and slumps from the steep inner walls, and that the ejecta layers may contain blocks. Crater walls should exhibit both the widest range and highest overall thermal inertia values and are most likely associated with blocky deposits and potentially exposed bedrock. Lower values are consistent with finer deposits lying on top of wall blocks, such as the mass wasting/talus deposits.

Given these six attributes, we can list some of the freshest and youngest craters on Mars. Table 3.1 provides a checklist for the nine craters most likely to be fresh craters as identified by Tornabene et al. (2006, 2018a). We note that Tooting crater is the only crater in this list for which both fluidized ejecta and impact melt can confidently be identified on the rim. Furthermore, Tooting and Corinto are the only craters in Table 3.1 that have been identified as the probable sources for certain SNC meteorites (Herd et al., 2024), and these rocks have ejection ages (i.e., the age of the impact event) of 1.1 and 2.9 Ma, respectively. Chakpar (0.6 ± 0.1), Domini (1.2 ± 0.2 Ma), and Kotka (11.3 ± 0.6 Ma) are the other very young craters that may have ejected certain meteorites identified by Herd et al. (2024). Pangboche crater also displays impact melt but has none of the other characteristics of fluidized ejecta layers, most likely due to the high elevation on the upper slopes of Olympus Mons (Mouginis-Mark, 2015).

3.8 SUMMARY

Having seen how Mars may conspire against our ability to interpret craters, it is now time to take a closer look at the diversity of crater morphologies. Keep in mind that what we see today is but a snapshot of the geologic evolution of the planet; craters may have formed in very different climatic and surface conditions than what we see today. Similarly, the geology of Mars is sufficiently complex that the two-dimensional view that we have from satellite data may fool us into thinking that there are not striking differences in the stratigraphy of any location on Mars that would then play a role in the excavation of a crater cavity. And if the crater cavity were excavated within a complex target stratigraphy, then it is highly likely that this diversity of materials would also be represented in the materials contained within the ejecta blanket.

That said, in Chapter 4 we will explore the diversity of landscape that characterizes the interiors of fresh (i.e., relatively unmodified) Martian impact craters. Subsequent chapters will then review the features of modified crater interiors (Chapter 5) that are commonly seen, as well as fresh (Chapter 6) and modified (Chapter 7) ejecta blankets.

4

Fresh Crater Interiors

4.1 INTRODUCTION

In Chapter 1, we introduced some aspects of the impact cratering process, from the contact of the incoming projectile with the surface, compression of these materials, crater cavity excavation and modification, and finally ejecta emplacement. The diversity of the resultant landforms throughout the Solar System was then illustrated in Chapter 2. The formation of craters on Mars is influenced by the external environment, such as the type of rocks within which the crater formed and the distribution of ice or water within those rocks. So, in Chapter 3, we explored some of the geologic and climatic vagaries of Mars as they might modify a "generic" crater. It became clear that the rocks on Mars have a variety of ages and origins. There is also a great variation in elevation across the planet with a consequence that atmospheric pressure could be different for similar energy impact events. In addition, the distribution of ice over time may have been drastically influenced by changes in the orbital inclination of the planet and thus an ice-free area today may once have been buried by a more extensive polar cap. In this chapter we delve into greater detail of the morphology of individual craters. We review what the freshest, and hence the most likely youngest, craters look like. In Chapter 5, we will show some of the typical ways that the crater interior may have been modified by erosional and depositional processes.

Some of the earliest comparisons of crater morphology-size relationships were conducted by Cordell et al. (1974), Arvidson (1974), and Smith (1976). Other early investigations were more focused on the interior characteristics (Wood et al., 1978) and the geometric properties of the craters (Pike, 1980a). To this list of interior features must be added rare examples of features preserved in very young craters (e.g., impact melt) as well as landforms created by climate-induced processes (e.g., glacial features). Martian impact craters display a wide range of interior morphologies, ranging from those created during or shortly after crater formation to those that were produced by subsequent geologic processes. Barlow (2015) defined nine basic interior morphology classifications for a total of 7,669 craters: central peaks (N = 1013), peak rings (N = 9), floor deposits (N = 5,173), flat floors (N = 43), chaotic texture (N = 105), wall terraces (N = 428), symmetric floor pits (N = 525), asymmetric floor pits (N = 38), and summit pits (N = 335). Table 4.1 lists the characteristics of these nine morphologies for craters in the northern hemisphere (Barlow, 2015). These interior features are dependent upon the diameter of the parent crater. In addition, crater depths, rim heights, and volumes of fresh impact craters are directly related to the crater diameter. Variations from the "pristine" relationships (see Section 3.7) reveal the effects of target properties at the time of impact and subsequent modification through erosional and depositional processes.

Table 4.1 *Size range of the interior features of Martian craters.*

Morphology	Diameter range (km)	Median diameter (km)	Dp/Dc
Central peak	5.0–139.8	10.9	0.05–0.76
Peak ring	18.5–328.3	18.5	0.17–0.71
Floor deposits	5.0–438.4	9.1	N/A
Flat floor	6.3–88.4	11.2	N/A
Chaotic texture	5.5–279.9	37.6	N/A
Wall terraces	5.5–103.1	17.6	N/A
Symmetric floor pit	5.0–89.1	14.8	N/A
Asymmetric floor pit	9.0–103.3	25.7	N/A
Summit pit	5.5–125.4	15.7	0.13–0.63

Dp = central peak basal diameter
Dc = crater diameter
Source: Reprinted from Barlow (2015), "Characteristics of impact craters in the northern hemisphere of Mars," *Geological Society Special Paper* 518, pp. 31–63. Used with permission from the Geological Society of America.

4.2 FLOOR FEATURES

4.2.1 Bowl-Shaped Craters

Simple bowl-shaped craters dominate the fresh crater population on Mars (Cintala et al., 1976). The typical morphology of a fresh bowl-shaped crater is well illustrated in Figure 4.1, where boulders of ejected material can be seen both on the crater floor and the ejecta blanket. No obvious signs of wall collapse or a flat floor can be found. In an extensive analysis of ~385,000 craters larger than 1 km in diameter, where "D" is the crater diameter, Robbins and Hynek (2012) demonstrated the utility of craters $1 < D < 3$ km as a proxy for estimating the relative regional ages of the surface.

4.2.2 Flat-Floored Craters

Only when the crater diameter approaches 2–3 km does the interior of a Martian crater start to possess a flat floor (Figure 4.2).

Figure 4.1 A very fresh bowl-shaped crater at 4.74°N, 46.19°E. This crater is ~1,050 m in diameter. Note the extensive boulder field on the outer rim of the crater as well as numerous boulders and small landslides on the crater interior. HiRISE image ESP_027890_1850.

Figure 4.2 Telz crater is a flat floor crater located at 21.17°N, 111.11°E, and is 2.6 km in diameter. Note that this crater lacks the boulders on the rim compared to the example shown in Figure 4.1, suggesting that Telz crater is the oldest of the pair. HiRISE image ESP_018512_2015.

This flat floor most likely originates from material sliding down the inner walls and collecting in the middle of the crater. As with simple bowl-shaped craters, there is no sign of structural failure of the inner walls.

4.2.3 Central Peaks

Craters with central peaks comprise ~15% of all craters in the northern hemisphere with complex interior morphologies (Barlow, 2015). Many central peaked craters occur in the highland regions, but they are also found in the younger plains units. Figure 4.3 illustrates that there are two different morphologies to Martian central peaks. Typically, multi-layered ejecta (MLE) craters (see Chapter 6 for a complete definition of crater types) have peaks with a craggy appearance and are quite often asymmetric. In contrast, many Type-1 double-layered ejecta (DLE) craters have central peaks with a smoother appearance and are typically more symmetric in outline. Central peaks transition to peak ring structures as crater diameter increases (Pike, 1980a; Bray et al., 2012). In some instances, central peaks reflect the structural fabric of the target material with blocks within the peak complex having prominent lineations (Pan et al., 2015).

As we saw in Chapter 1, central peaks are formed from uplifted material that originally was at depth beneath the preimpact surface (see Section 1.4). Thus, craters with central peaks offer the possibility of exploring subsurface crustal compositions. Using thermal infrared (TIR) and near-IR spectral data, Pan et al. (2015) conducted a global study of craters between 10 and 200 km in diameter (N = 866) and identified 26 central peaks that showed distinctive spectral properties compared to the surrounding terrain. Most of these craters were found in western Noachis Terra, Tyrrhena Terra, within the northern rim of Hellas Basin, and a few on the northern plains. At THEMIS TIR spatial resolution (100 m/pixel), spectrally distinct central peaks (SDCPs) are generally restricted to latitudes equatorward of 45°. The lack of SDCPs at higher latitudes suggests that a latitude-dependent degradation process, such as periglacial activity, may have obscured the altered peak units such that they are now indistinguishable from surrounding materials.

Observed central peak heights and basal diameters naturally depend upon the amount of infilling of the parent crater. Remarkably, there has been little work done on the height of Martian central peaks as a function of parent crater diameter, presumably because central peaks are one of the first features to be buried as the crater is infilled with sediment or lava. Except for craters such as Gale and Nicholson, both of which have unusually large central mounds (Bennett and Bell, 2016), there is a tendency for peaks to become buried by subsequent crater infilling. This raises the interesting possibility that the ratio of peak height to crater diameter may be a surrogate for estimating the rate of infilling and/or the approximate age of the crater, but this potential has not been exploited in the literature. Such an investigation is made more difficult because both peak height and crater diameter are difficult to measure accurately with Mars Orbiter Laser Altimeter (MOLA) because it is sometimes not easy to find a laser altimetric profile in the right location (see Figure 1.14). In the case

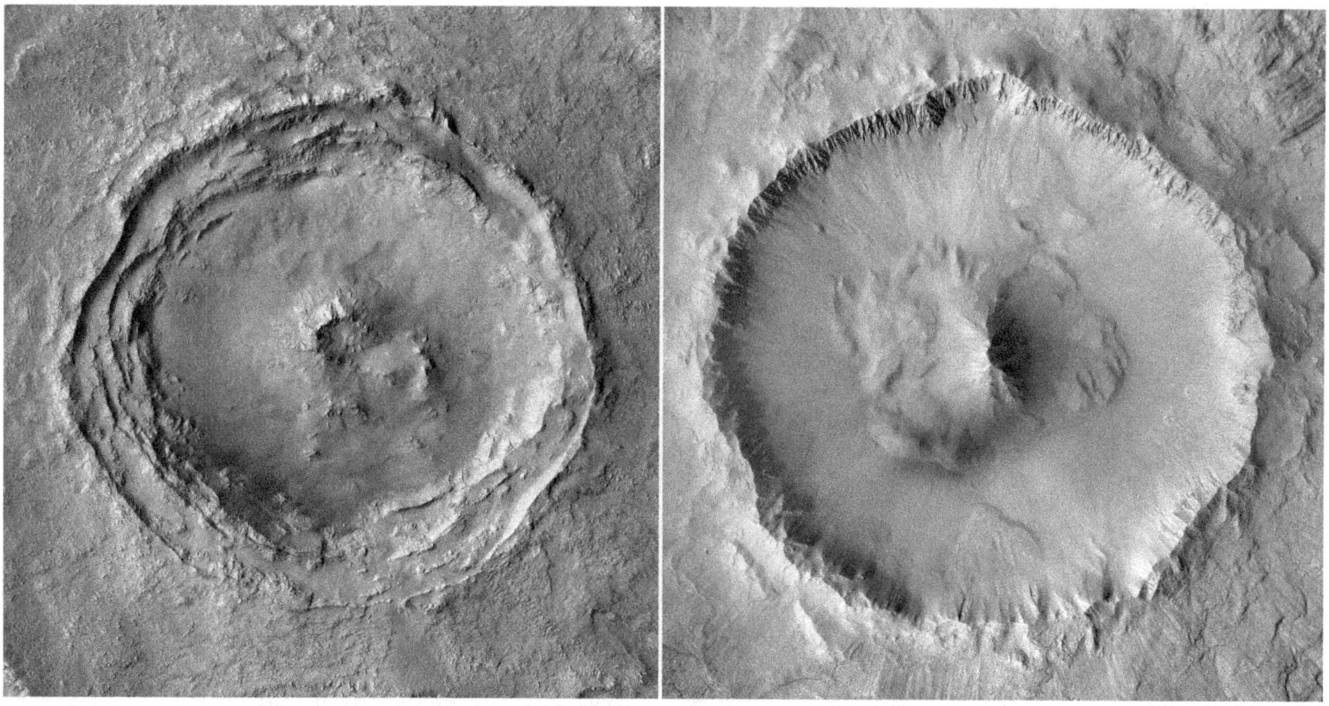

Figure 4.3 Here we see different morphologies of central peaks. On the left is a craggy type of peak, most often seen in MLE craters. Santa Fe crater is 20.3 km in diameter and is located at 19.39°N, 312.03°E. CTX B02_010382_1994. At right is Arandas crater (42.43°N, 343.88°E), which is ~24.4 km in diameter and has a large "smoother" central peak with broad talus deposits present on the steeper slopes. This type of central peak is often the type seen within Type-1 DLE craters. CTX image D04_028802_2227.

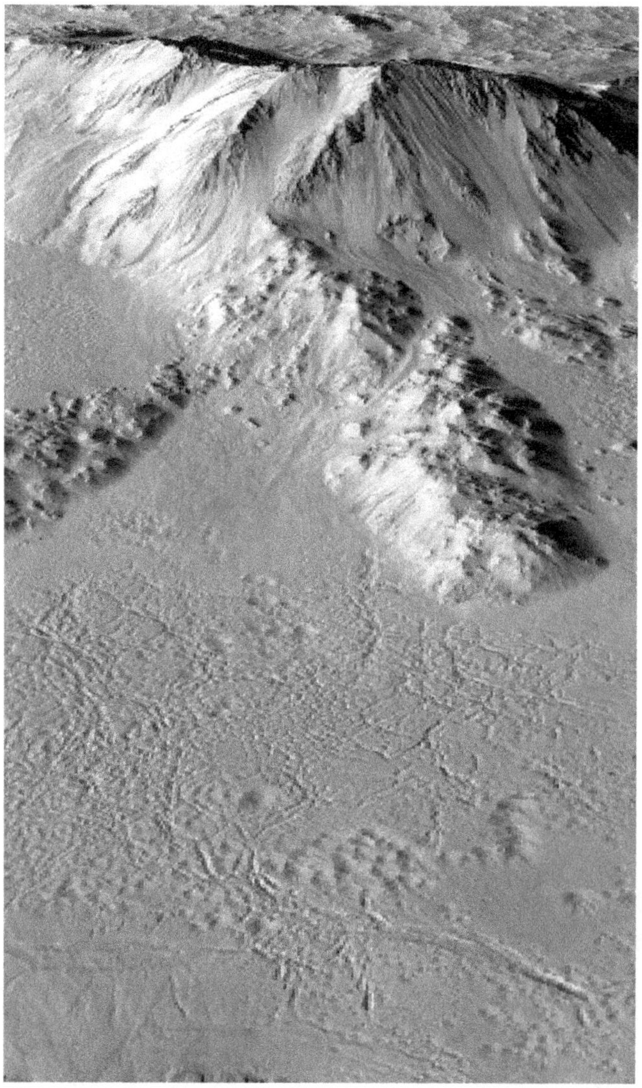

Figure 4.4 Oblique computer-generated view looking north from above the south rim of Tooting crater toward the central peak complex. The elevation difference between the summit of the central peak and the intersection of the peak and floor material is ~1,110 m, making this one of the highest central peaks on Mars relative to the crater diameter. This view illustrates the hummocky terrain surrounding the central massive. Is this material exposed segments of the crater floor, or a broader uplifted part of the central peak? Typically, in other craters, this material may well be buried by recent fill-material so it is not clear what the situation is here. At the bottom of the image is fractured terrain (cooled impact melt?) and pitted terrain (degassed floor material?). Also note the talus that drapes over many parts of the peak suggests that the peak may be relatively friable. Produced from a stereo pair of HiRISE images, frame numbers PSP_001538_2035 and PSP_002158_2035.

of the 28 km diameter Tooting crater (Figure 4.4), the peak rises ~1,110 m above the crater floor (Mouginis-Mark and Garbeil, 2007). The central peak diameter (~9.8 × 8.3 km) is also appreciably larger than the 6.0 km predicted by the relationship developed by Hale (1983) for 1,672 Martian craters and lies at the +1-sigma value for craters with the diameter of Tooting. The examples cataloged by Bennett and Bell (2016) are exceptions to this trend and are typically located near the boundaries of previously identified large-scale sedimentary deposits in Arabia Terra.

4.2.4 Peak Rings

Unlike the Moon, where the Orientale Basin serves as a good example of a relatively fresh basin >500 km in diameter, Mars does not have an unmodified basin. The Isidis (~1,500 km), Chryse (~1,600 km), Argyre (~1,800 km), and Hellas (~2,300 km) Basins have all been extensively infilled by sediments and modified by other processes (Figure 4.5). Schultz and Frey (1990) reviewed the morphological characteristics of these large basins: that is, basins having diameters 300 < D < 1,850 km are morphologically comparable to the lunar Orientale Basin. Argyre-type basins (1,850 < D < 3,600 km) are characterized by a rugged annulus and concentric grabens. The largest, Chryse-type basins (D > 3,600 km) have extremely shallow topographic profiles and numerous concentric structures expressed as scarps, massifs, and channels. Many probable buried impact basins with peak rings have also been identified, either using MOLA topography (Frey et al., 1999) or sounding radar data (Watters et al., 2006).

4.2.5 Central Pits

First described by Wood et al. (1978), central pits and pitted central peaks are quite common on Mars (Barlow, 2006; Alzate and Barlow, 2011; Barlow et al., 2017). These pits may occur as rimless depressions on the crater floor, or they may be pits with raised rims or pits at the summit of a central peak (Figures 4.6–4.8). Alzate and Barlow (2011) conducted a survey of central pit craters (Figure 4.9). The ratio of pit diameter to the diameter of the parent crater is generally consistent among all three pit types (Barlow, 2006). Alzate and Barlow (2011) also recognized that both floor pits and summit pits are found in the same regions and within the same topographic and geologic unit ranges. Floor pit craters exhibit a peak in the 10–15 km diameter range, and summit pit craters peak in frequency between 10 and 20 km diameter (Figure 4.10).

Martian floor pits do not appear to be the result of a central peak collapse, as the median ratio of the peak to crater diameter is about twice as high for central peaks/summit pits than values for floor pits (Barlow et al., 2017). An explosive origin for the central pits within craters was proposed by Williams et al. (2015) who studied floor pits within 654 craters ~10 km in diameter or larger within 60° of the Equator. They based this idea on the presence of raised rims and blocky ejecta detected with THEMIS TIR data, which suggests that coarse-grained materials are distributed proximally around many central pits on the floors of their host craters. A statistical study of central pit craters was conducted by Peel et al. (2019) to identify the most likely mode of formation. Three different mechanisms were considered: (1) explosive release of water vapor; (2) uplift and collapse of rock at the crater

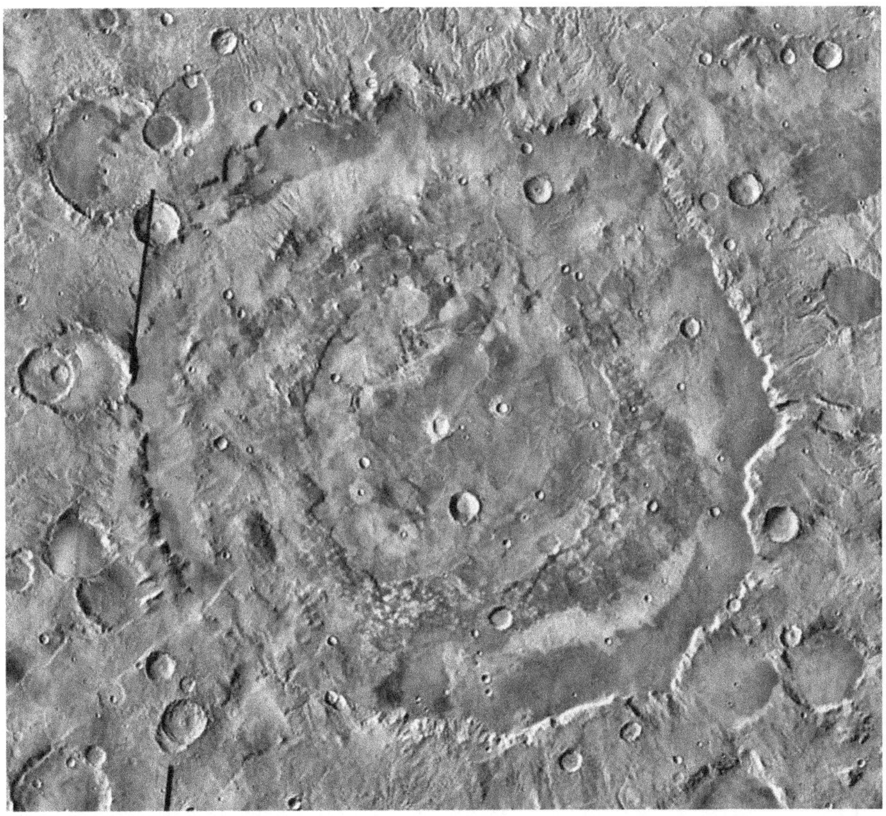

Figure 4.5 There are no fresh basins on Mars, and so examples such as the ones for the Moon and Mercury (see Figure 2.12) do not exist. But Huygens Basin (~470 km diameter at 13.88°S, 55.58°E) is one of the best examples of a basin on Mars. Numerous ridges are present on the central basin floor, and these may have been formed during deformation of the floor associated with the lava infill within the basin interior. THEMIS daytime TIR mosaic.

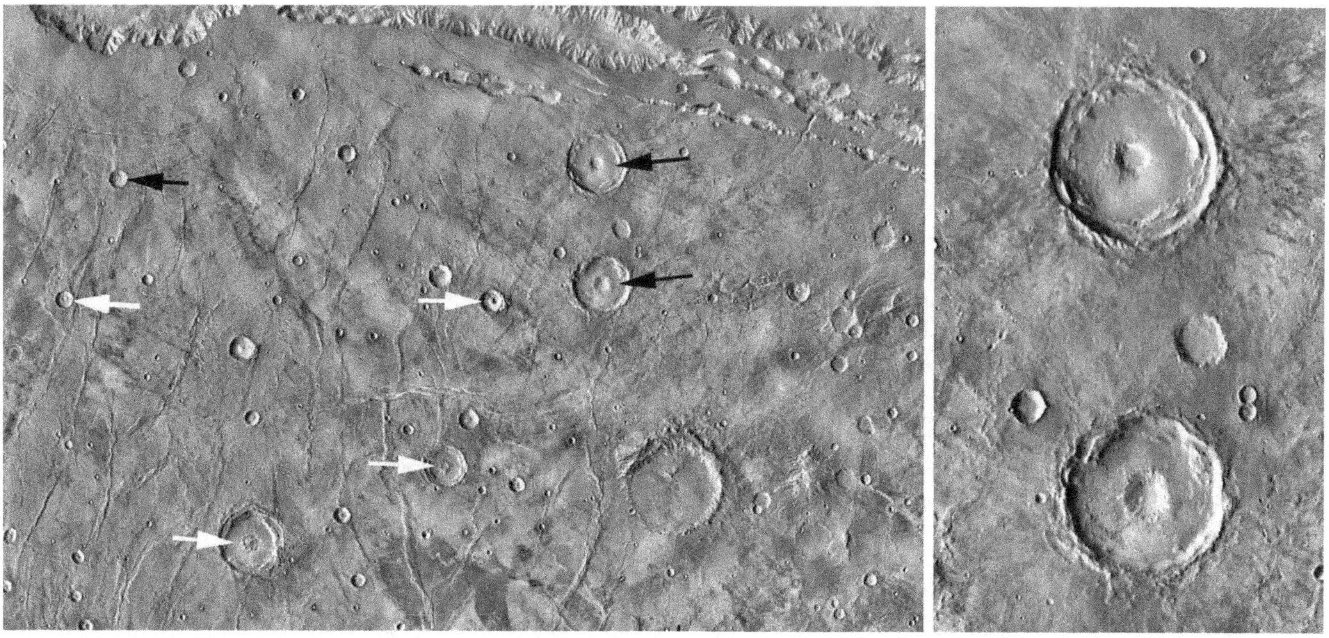

Figure 4.6 Left: Craters with central pits (black arrows) and pitted peaks (white arrows) within Thaumasia Planum. Right: The upper crater is Arima crater (15.73°S, 296.23°E), which is ~53 km in diameter. The occurrence of two large craters so close together each with a central pit raises the interesting possibility that local target properties might promote the formation of central pits in this region. Both images are parts of THEMIS daytime TIR mosaics.

Figure 4.7 Detailed view of a central pit with a raised rim at Mazamba crater (27.48°S 290.24°E). Numerous strata are visible in the walls of the pit while the floor has been covered by later sediments. Mazamba crater is 52.3 km in diameter. CTX image P18_008036_1522.

Figure 4.8 The central structure of Bamberg crater (39.78°N, 356.80°E). Bamberg crater is ~54 km in diameter. Note that the pit has a rim around the perimeter except to the north, where a prominent secondary chain is found (outside this image and beyond the rim crest). The southernmost part of the pit rim has the morphology of a central peak. Beyond the prominent crags and the pit, the central structure forms a broader topographic rise with sedimentary fans surrounding the feature. CTX image B02_010301_2200.

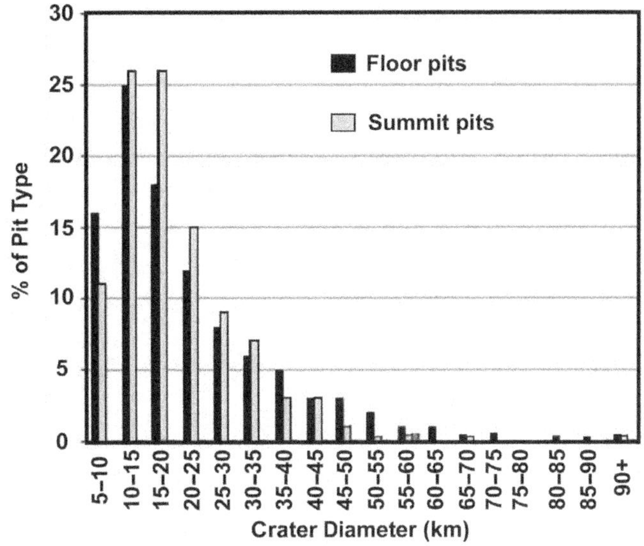

Figure 4.10 Central pit craters on Mars as a function of parent crater diameter. Results are normalized to the total number of pit craters of a specific type (i.e., the total number of floor pit craters and the total number of summit pit craters). Floor pit craters exhibit a peak in the 10–15 km diameter range, and summit pit craters peak in frequency between 10 and 20 km in diameter. Reprinted from Alzate and Barlow (2011), "Central pits on Ganymede," *Icarus,* vol. 211, pp. 1274–1283, Figure 11. With permission from Elsevier.

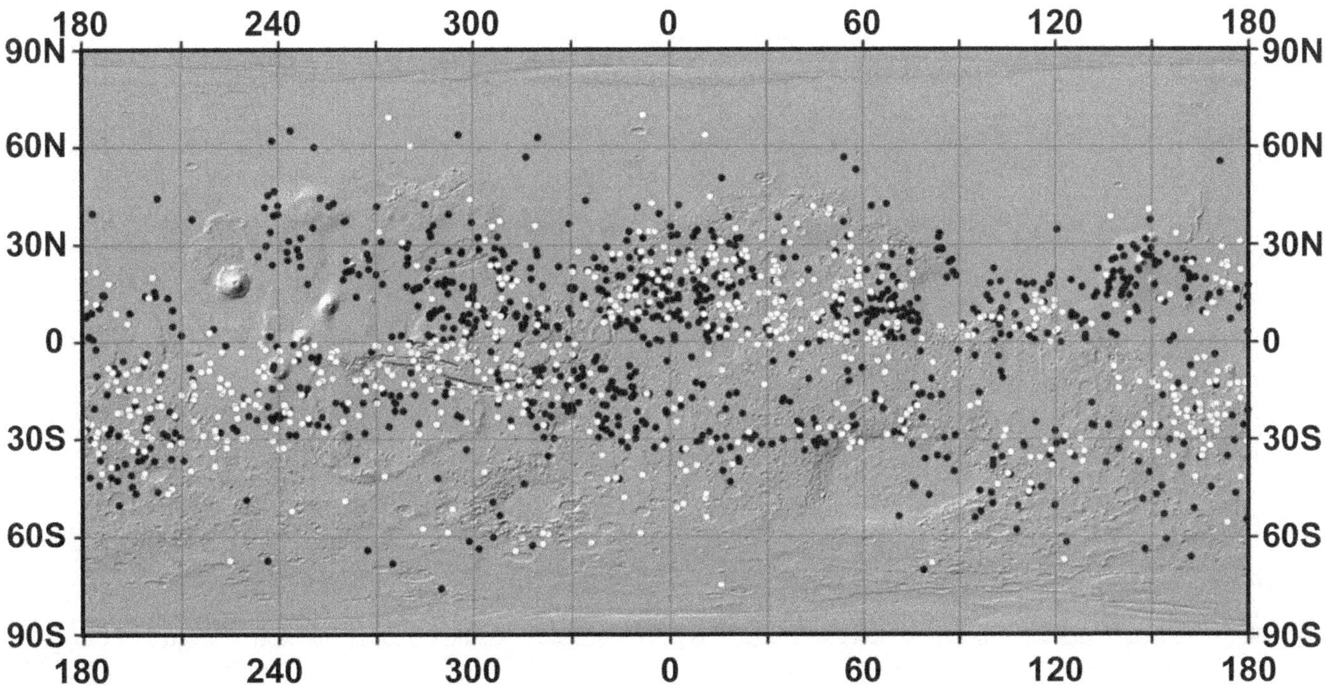

Figure 4.9 Distribution of Martian central pit craters. Both floor pits (black) and summit pits (white) are seen in the same regions and within the same topographic and geologic unit ranges. Reprinted from Alzate and Barlow (2011), "Central pits on Ganymede," *Icarus,* vol. 211, pp. 1274–1283, Figure 10. With permission from Elsevier.

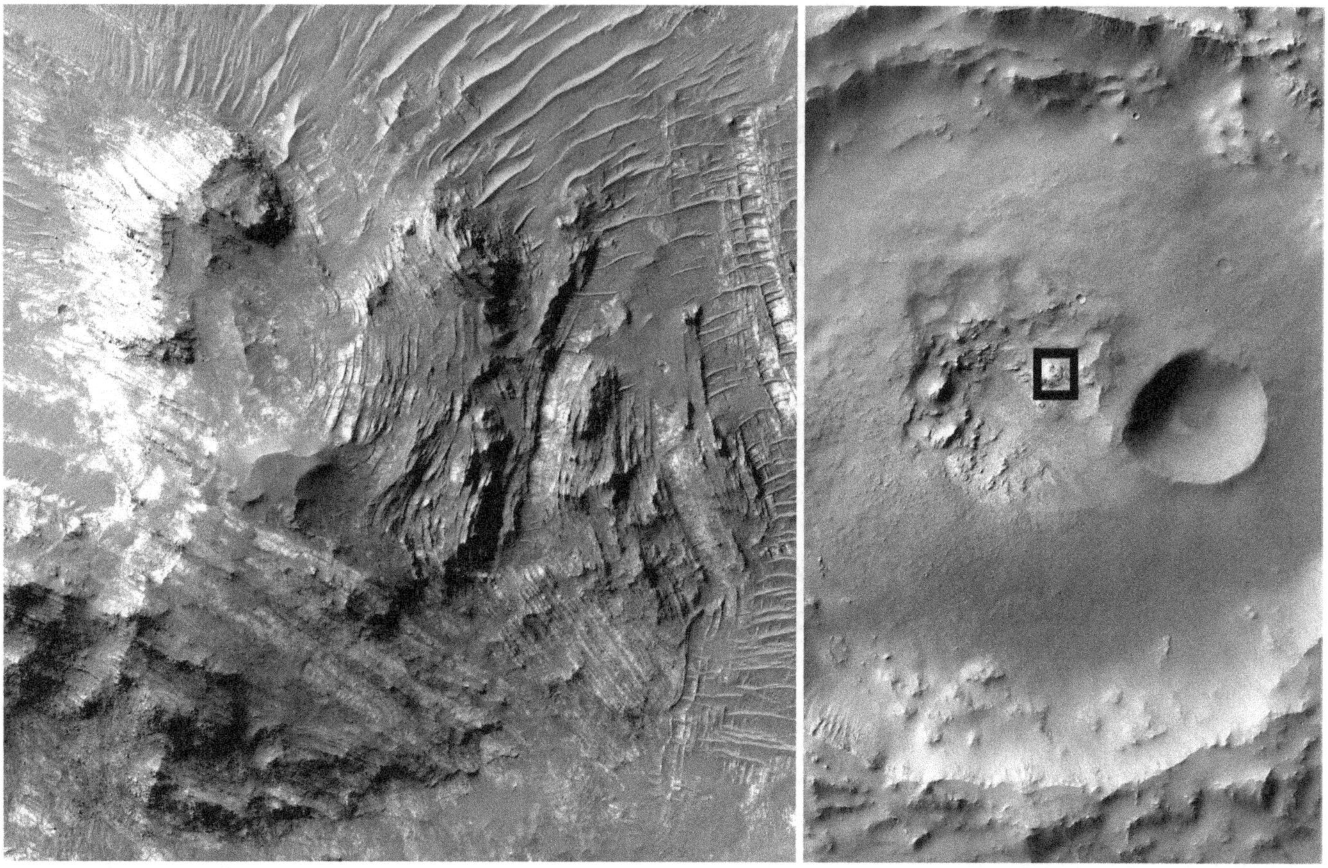

Figure 4.11 Layering is clearly visible within the central pit of Betio crater (23.07°S, 281.30°E), which is 32.4 km in diameter and formed within the ridged plains materials south of Valles Marineris. This complex layering could reflect the structure of the target material, or it could have formed due to shearing caused by the impact. Left: HiRISE image ESP_016805_1565. Right: CTX B18_016805_1567. Box at right denotes coverage of the image at left.

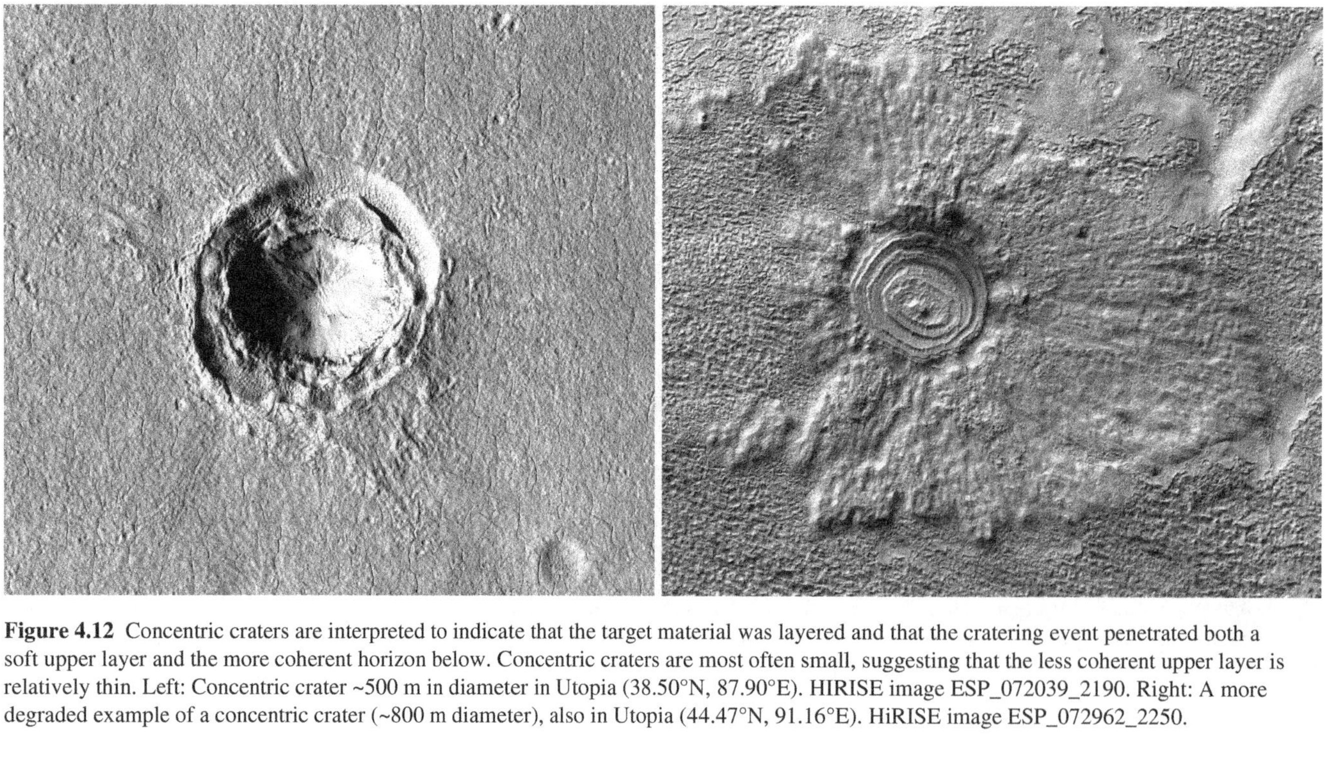

Figure 4.12 Concentric craters are interpreted to indicate that the target material was layered and that the cratering event penetrated both a soft upper layer and the more coherent horizon below. Concentric craters are most often small, suggesting that the less coherent upper layer is relatively thin. Left: Concentric crater ~500 m in diameter in Utopia (38.50°N, 87.90°E). HIRISE image ESP_072039_2190. Right: A more degraded example of a concentric crater (~800 m diameter), also in Utopia (44.47°N, 91.16°E). HiRISE image ESP_072962_2250.

center; and (3) drainage of liquid, produced by the impact, into cracks in the crater floor. While several examples are consistent with each of these models, the results did not point to a single mechanism for central pit formation and thus it seems likely that multiple processes may be responsible.

Central pits can reveal considerable information about the preimpact subsurface. Megabreccia and highly deformed bedrock can be seen at high resolution at craters such as the unnamed crater in Solis Planum (Caudill et al., 2012). Layering in the central structure, such as the central pit of Betio crater (Figure 4.11) and several central peaks, has been shown to be compositionally different from the surrounding terrain (Pan et al., 2015). This observation is consistent with the idea that central peaks represent material uplifted from depth during the cratering process.

4.2.6 Concentric Craters

Concentric craters on any planetary surface form a unique class because the interior contains a concentric second depression. Concentric craters are relatively common at crater diameters of a few kilometers. On the Moon and have been used to deduce the regolith thickness (Quaide and Oberbeck, 1968; Bart et al., 2011; Yue et al., 2019). No comparable study has been conducted for concentric craters on Mars, but the inference is that a similar explanation holds for the numerous concentric craters that can be observed (Figure 4.12). However, certain concentric craters remain enigmatic in origin. Figure 4.13 illustrates one of several concentric craters near the outflow channel Hrad Vallis. These craters have a stepped structure to their interiors and concentric fractures around the rim. Although morphologically quite fresh in appearance, there is no discernible topographic relief of the rim. In addition, the craters exhibit unusual thermal properties with the material surrounding the crater being relatively cold during the day and relatively warm at night (Morris and Mouginis-Mark, 2006). This indicates that the ejecta from the craters are unusually massive compared to the surroundings. Because these craters all formed within a geologic unit inferred to be an inflated lava flow, a more plausible mode of origin for the concentric craters appears to be that they are all inflation pits, comparable to examples found within the McCarty's lava flow in New Mexico (Hamilton et al., 2018). The higher thermal inertia may consequently be explained by fractures in the rim material exposing more coherent rock.

Figure 4.13 A crater with numerous concentric fractures around the rim near the source of the outflow channel Hrad Vallis (34.65°N, 141.75°E). The diameter of the outermost set of concentric fractures is ~2 km. HiRISE image ESP_016045_2150.

4.3 OTHER INTERIOR FEATURES

4.3.1 Pitted Terrain

A pitted unit believed to be impact-generated is found on crater floors (Figure 4.14), as well as pools on interior terraces blocks, and on ejecta blankets near rims of fresh single and multi-layered ejecta of craters (Boyce et al., 2012; Tornabene et al., 2012). Indeed, the unit is one of the main features that are thought to be associated with very young craters (see Section 3.7.2). This type of deposit is common for fresh Martian craters, although similar pitted material has also been observed on 4 Vesta (Denevi et al., 2012) and 1 Ceres (Sizemore et al., 2017). This morphology has no lunar analog. Although there are places in this unit that lack pits, typically pits are found in fields of closely packed groups where individual pits share their rims with the adjacent pits that give the surface of the deposit a foam-like appearance. Previous workers have suggested that this unit is composed either of water-bearing impact melt (or melt breccias) or impact-generated fine-grain sediment. Boyce et al. (2012) and Tornabene et al. (2012) suggested that the foam-like geometry of the closely spaced pits and the shared rims makes it unlikely that they formed by collapse or sublimation. Instead, the pits were most likely caused by diffusion and boiling of the water trapped within the host materials in a manner comparable to the way mud pots form in hydrothermal regions. The pits therefore most likely formed during the final stage of crater formation. Pits may also have a terrestrial analog. Newsom et al. (1986) identified many chimney-like degassing pipes within the suevite deposits of the Ries crater in Germany. These pits apparently formed by high gas pressure in the hot (>500°C) fluidized suevite immediately after deposition. The most likely source of the gas was volatiles for the melted and shocked basement inclusions. Suevite-like deposits are therefore likely to exist on Mars because of the presence of volatile-rich target rocks.

The proposed mechanism requires substantial water (or ice) in the target materials prior to impact to form the pits, so that the global distribution of this pitted material (Figure 4.15) may provide an indication of near-surface volatiles within the younger crater population. Clifford (1993) estimated the porosity of the Martian near-surface rock to average ~20% but could

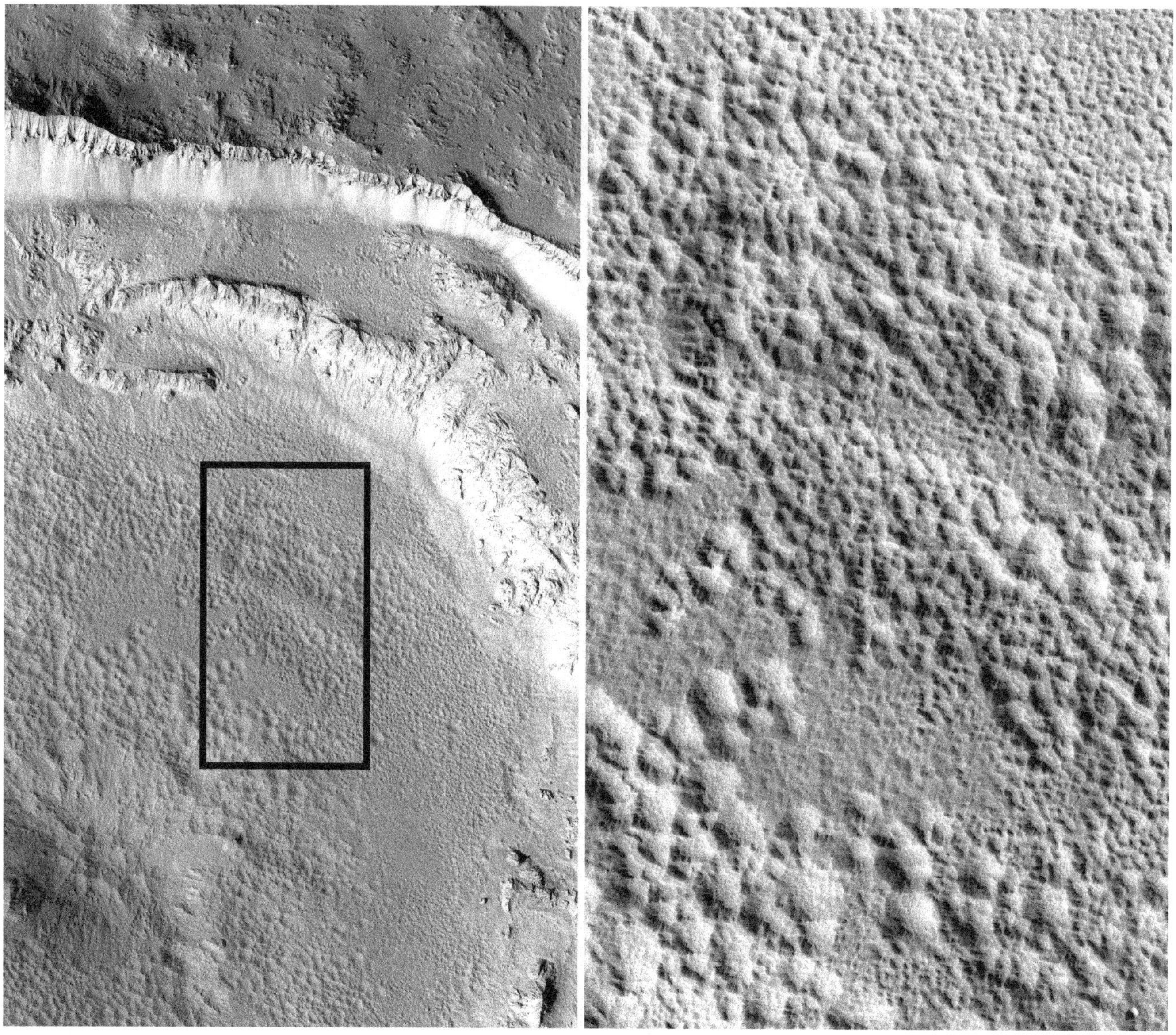

Figure 4.14 Pitted material in Corinto crater (13.3 km in diameter at 16.95°N, 141.70°E). The image on the left shows the location of the higher resolution image on the right. Both images are part of HiRISE image PSP_003611_1970.

be as much as 50%. Furthermore, based on the estimated cryosphere thickness, Clifford (1993) used these values to calculate that an equivalent of between 374 and 460 m of water could be stored evenly in the subsurface over the whole planet. This suggests that there was easily enough water supplied from the target rock to saturate the pitted materials. Additional water also could have been supplied from basement material of craters smaller than about 25 km in diameter, while larger craters would have excavated deeper than the predicted bottom of the aquifer (Clifford, 1993).

4.3.2 "Swirl" Texture

On the Moon, swirl floor texture is most common in craters in the size range 20–30 km but can also occur at larger crater diameters (Smith and Hartnell, 1978). This crater morphology appears to be transitional between simple bowl-shaped craters and larger craters where discrete terraces have formed (Schultz, 1976). In the case of Mars, there are few clear examples of this type of interior morphology, but one of the best examples is Karzok crater, which lies near the summit of Olympus Mons volcano (Figure 4.16). Why there are so few Martian craters with this morphology is unclear, but quite possibly the absence of volatiles in the near-surface layers of the target may have promoted the formation of terraces at a smaller diameter. The swirl texture within the Karzok crater would be consistent with this idea, as the high elevation would decrease the possibility of volatiles in the target. Unfortunately, the direct correlation between high elevation and the occurrence of the swirl structure may not be valid, as Karzok crater formed on the steep eastern flanks of Olympus Mons volcano and thus the interior morphology may be the

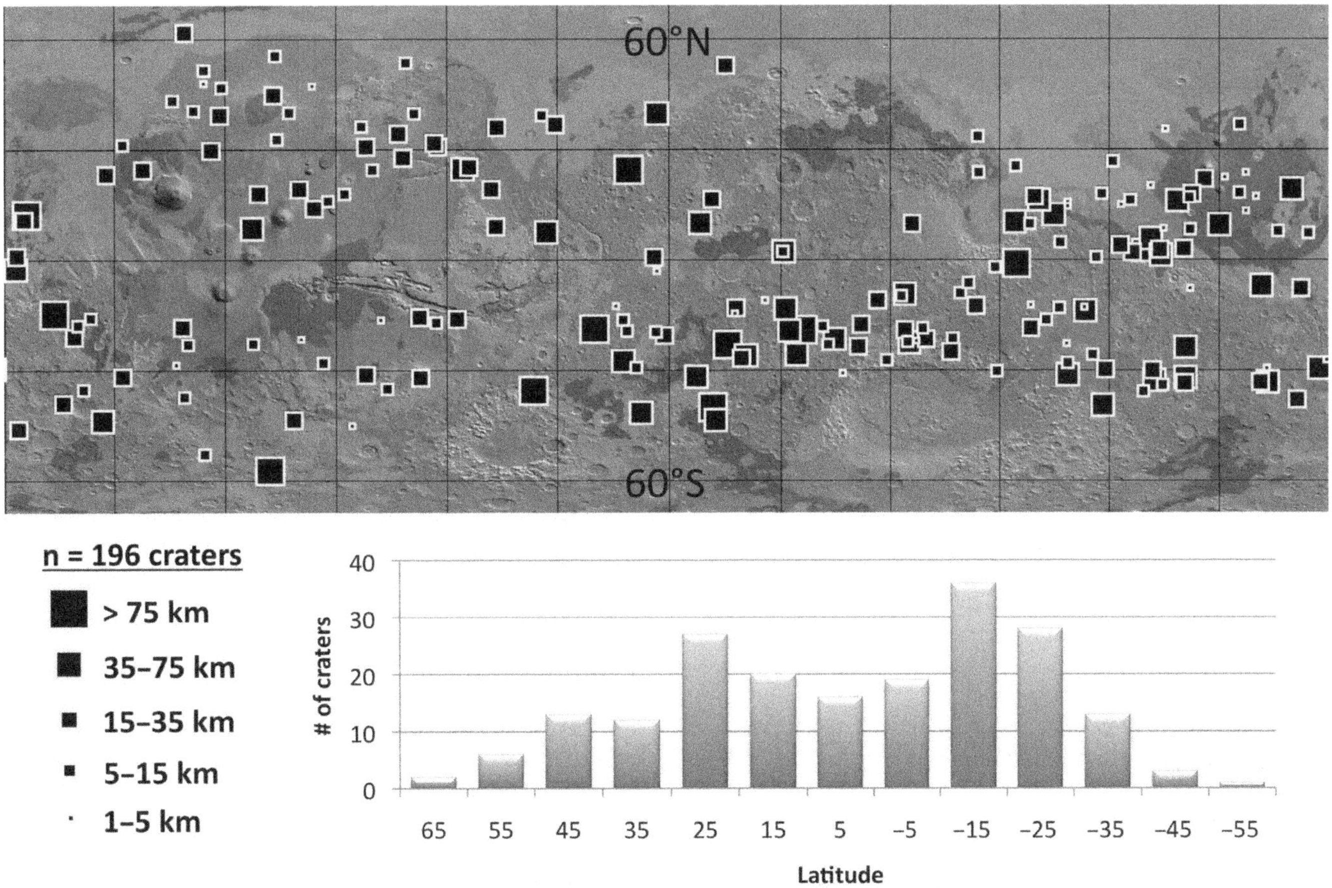

Figure 4.15 Global distribution of all pitted material-bearing craters identified by Tornabene et al. (2012), "Widespread crater-related pitted materials on Mars: Further evidence for the role of target volatiles during the impact process," *Icarus*, vol. 220, pp. 348–368, Figure 2. With permission from Elsevier. Craters range from ~1 to 150 km in diameter and show no obvious correlation with latitude or with target material. The interpretation is, therefore, that these pitted craters are the randomly distributed sample of the most recent large craters to form on Mars. If water is required to exist within the target material in order for the pits to form, then this global distribution implies that water was globally distributed in the relatively recent past. Note: A black and white version of this figure will appear in some formats. For the color version, please refer to the plate section.

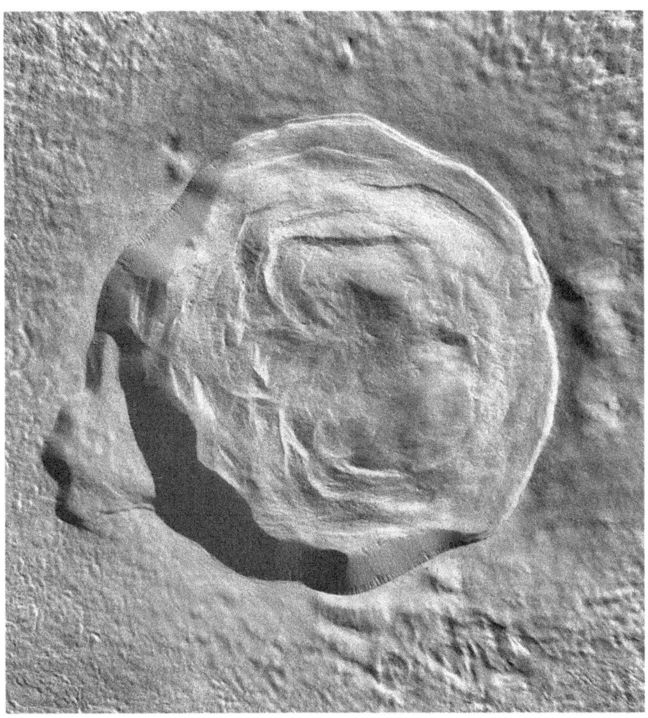

Figure 4.16 Karzok crater (15.6 km, 18.42°N, 228.26°E) shows one of the best examples of swirl texture on the floor of a Martian crater. This texture (rather than a flat floor with pitted material) may have formed because the crater is at high elevation (almost 19 km above datum) or because the crater formed on the relatively steep slope of the eastern flank of Olympus Mons volcano. In this context, the downslope direction is to the right of the image so that the swirl texture generally flows down the topographic slope. CTX image B09_013075_1985.

consequence of large-scale downslope creep of materials across the crater floor. Therefore, such creep might not take place on crater floors that have less of a regional tilt.

4.4 WALL AND RIM FEATURES

4.4.1 Terraces

Early in the analysis of craters on Mars, Cintala et al. (1976) recognized that some impact craters >~1 km in diameter had terraced walls. As with the interpretation of lunar impact craters, these terraces are thought to originate from structural failure and the collapse of the crater cavity toward the end of the crater formation process (Settle and Head, 1979; Pearce and Melosh, 1986; Melosh, 1989). Few detailed studies have been conducted since the early *Viking Orbiter* era (i.e., the late 1970s and early 1980s), except that Mouginis-Mark and Boyce (2012) recognized that terrace blocks within the 28-km diameter Tooting crater (Figure 4.17) were typically wider where the rim appeared to have failed and expanded the crater radius in that azimuth direction.

Wall terraces are commonly observed in craters that also possess central peaks or central pits. Sometimes, as illustrated in Figure 4.18, the collapse of the inner wall can be quite complex, with some layers of the target rock producing scree slopes on the wall while at other depths beneath the original target surface rugged, scree-free slopes can be seen. In larger craters, such as the crater Curie (Figure 4.19), terraces can occupy ~50% of the interior of the crater. Curie crater (114.1 km in diameter, 28.80°N, 355.23°E) may offer important insights into how the cavities of larger (basin-sized) craters appeared soon after a large crater was formed. Curie crater is transitional in size from the impact craters illustrated in this book to the older craters a few hundred kilometers in size such as Gale (155 km, 5.34°S, 137.79°E), Baldet (180 km, 22.98°N, 65.27°E), Herschel (304 km diameter, 14.33°S, 129.73°E), Antoniadi (394 km, 21.61°N, 60.70°E), and Huygens (470 km diameter, 13.69°S, 55.44°E). All these larger craters have a significant amount of fill material that hides the original structure of the inner wall. Although the floor of Curie crater is now buried by sedimentary materials, a relatively large (~14 km diameter) central peak rises ~1,700 m above this floor and the rim crest is ~3,500–3,700 m above the crater floor. The most prominent exterior deposits are lobate ejecta flows to the south of the crater, and several prominent secondary crater chains to the northwest and south that extend up to 120–140 km from the crater rim crest.

Terraces of Curie are markedly different from those observed at smaller craters (see Section 4.4.1). The narrowest wall unit (12 km wide) lies to the northwest (Figure 4.20) and displays some layering near the rim crest. No layering can be identified in the eastern side of the crater and the width of the wall material varies from ~17 to 25 km. Extensive slump material on the southwest side (Figure 4.21) is ~25 km wide with no indication of terrace blocks.

It is instructive to consider what these observations of the interior of Curie crater might mean for the interpretation of other more degraded craters on Mars that exceed ~100 km in diameter. There is a clear diversity of wall units around the rim crest of Curie crater. Some azimuths show terrace blocks and high-standing rim units that are comparable to the wall units of the smaller crater. Other azimuths display no prominent terracing but instead show a broad slump region comprised of "jumbled terrain." Perhaps this jumbled terrain represents a more extensive failure of the cavity wall during its formation? It may also reflect differences in the strength of the target rocks, as craters of this large size most probably possessed excavation cavities that excavated a few to perhaps ~10 km below the surface. As most Mars craters larger than 100 km are found in the ancient highlands of the southern hemisphere (Robbins and Hynek, 2012), the target must have undergone considerable reworking early in Solar System history to produce strong regional differences in stratigraphy that influence subsequent crater morphologies.

Figure 4.17 Computer-generated oblique view of the southern inner wall of Tooting crater (23.21°N, 207.75°E). Notice how the slumped wall blocks preserve the stratigraphy of the rim crest. In this view the rim rises ~2,100 m above the crater floor. DEM derived from CTX images P01_001538_2035 and P03_002158_2034.

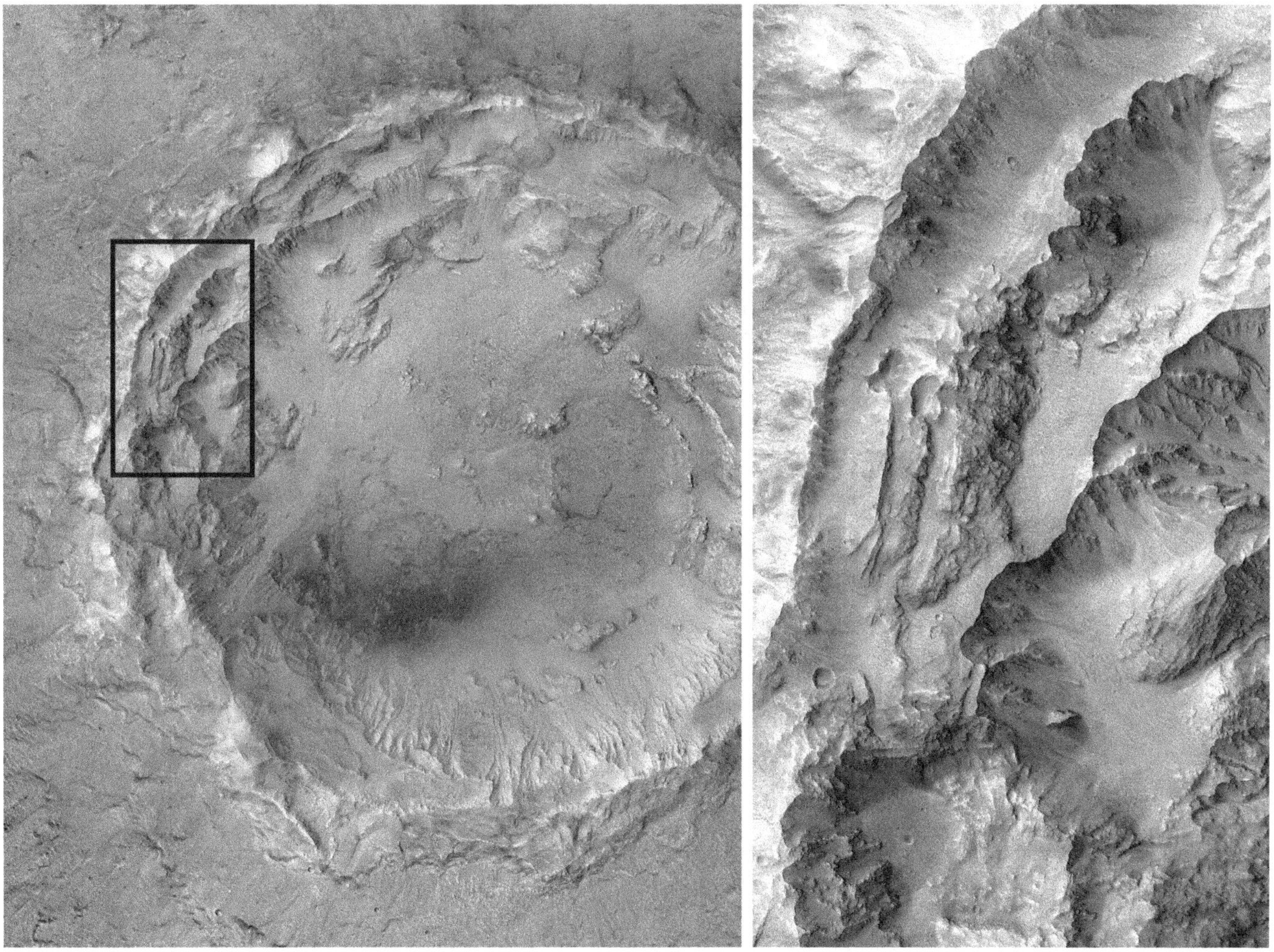

Figure 4.18 The inner wall of Tehachapi crater (27 km in diameter, located at 36.58°N, 329.57°E) displays a complex sequence of wall collapse and terrace formation. The box at left denotes the boundaries of the detail shown at right. The image on the left is CTX frame B18_016737_2168, and at right HiRISE image ESP_017370_2170.

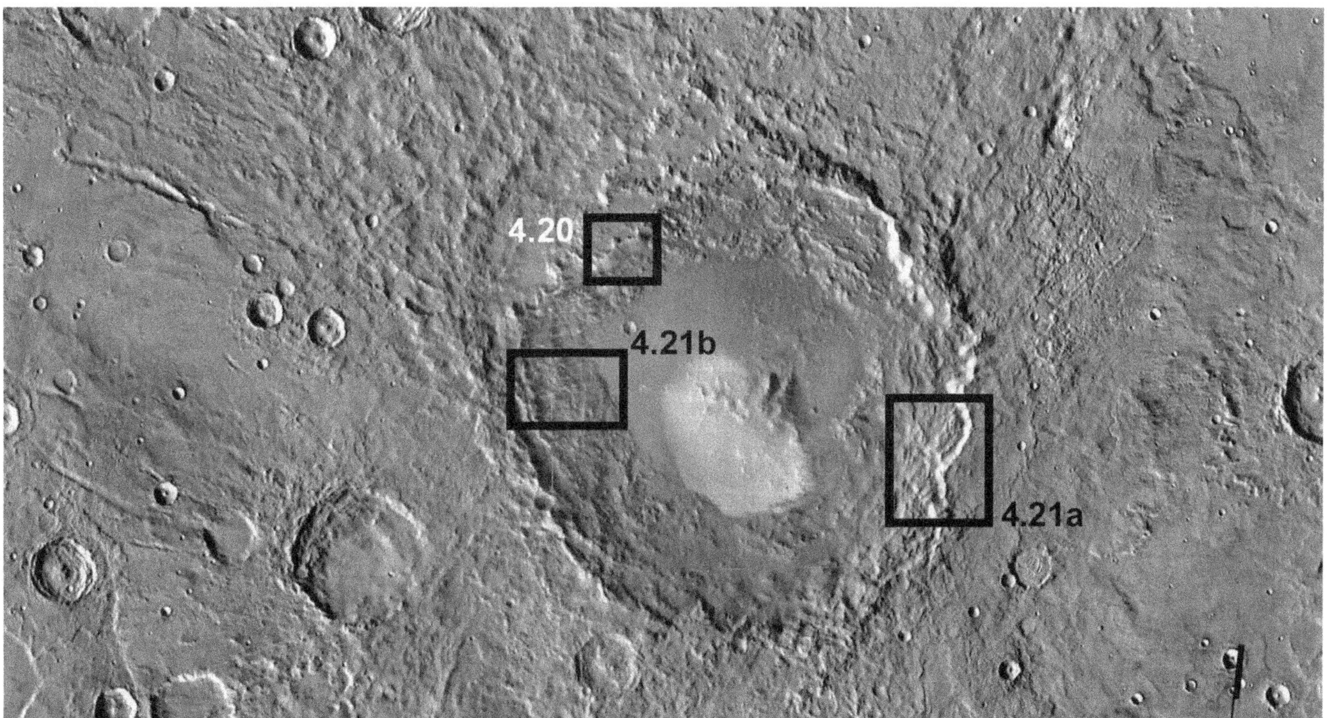

Figure 4.19 Curie crater (114.1 km diameter, 28.83°N, 355.21°E). Massive wall terraces can be seen here. Boxes mark the locations of Figures 4.20 and 4.21. THEMIS daytime TIR mosaic.

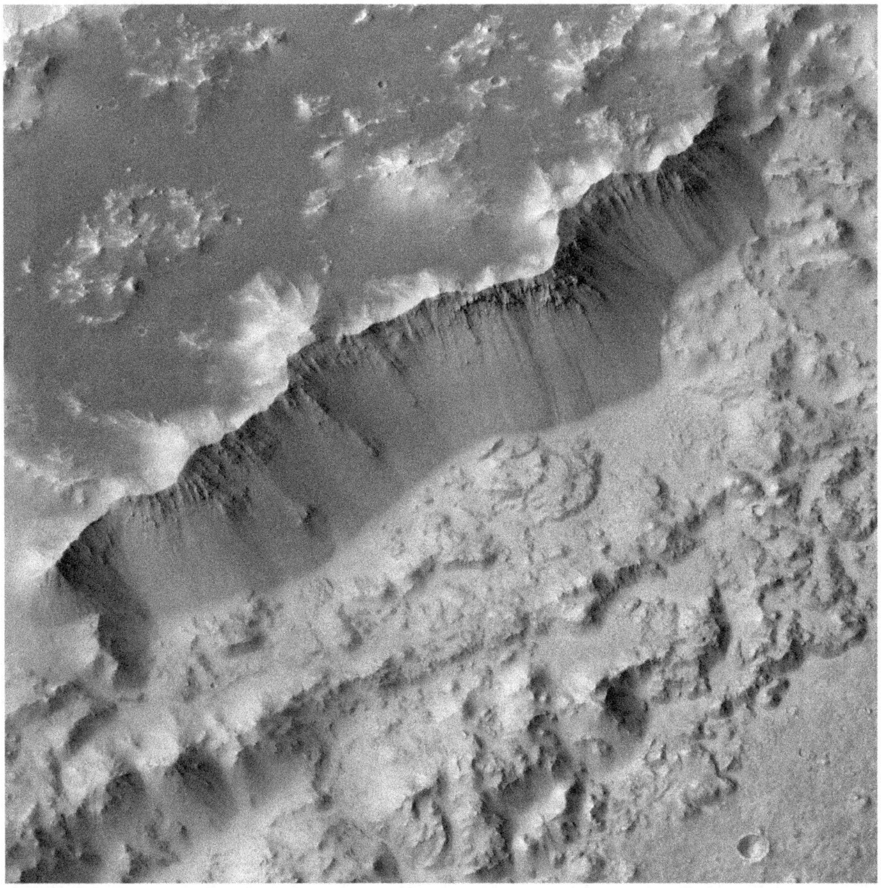

Figure 4.20 Segments of the rim crest of Curie crater, particularly on the northwest rim, have high-standing walls covered with talus. At the base of this wall, there is a relatively narrow set of poorly formed terraces extending to the crater floor (at lower right). See Figure 4.19 for location. CTX image B20_017435_2090.

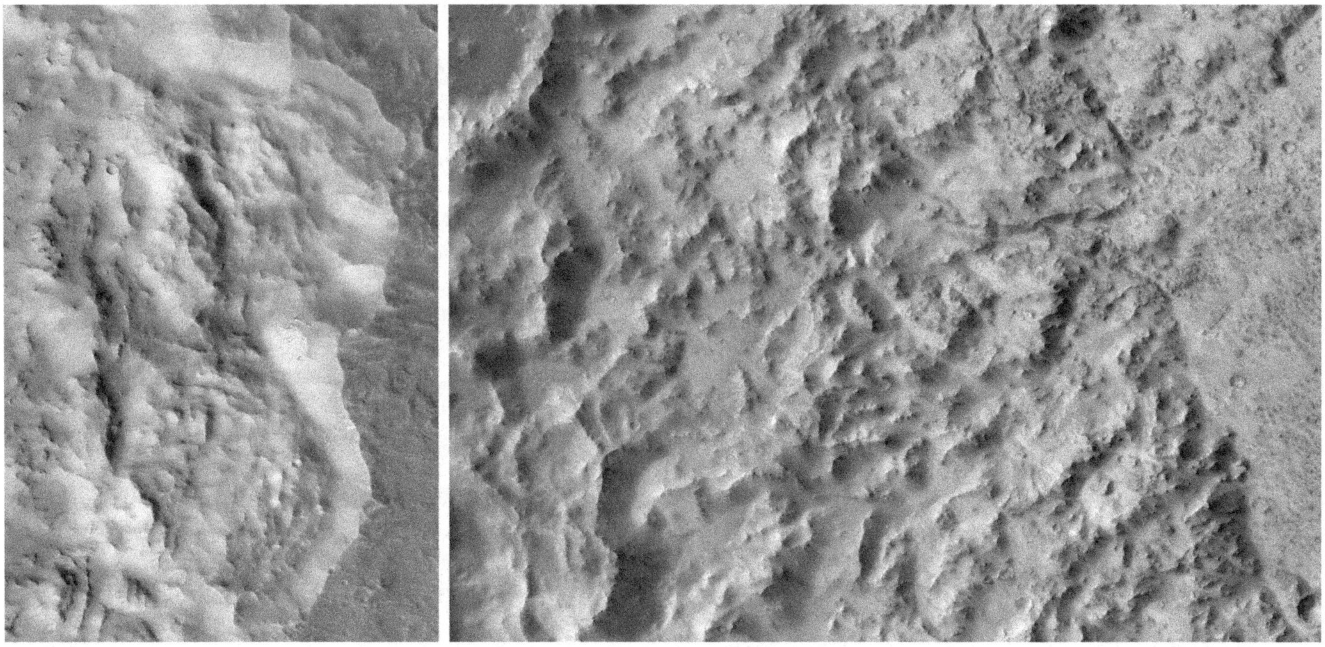

Figure 4.21 Left: Scalloping on the eastern wall of Curie crater with no clear sign of terraces. CTX image P13_006148_2091. Right: Extensive slumping on the western wall almost has the appearance of a giant landslide. See Figure 4.19 for the location of each image. CTX image D20_034854_2075.

For craters such as Gale (155 km), there is the suggestion that the western rim may have been comparable to the northwest rim of Curie crater, while the eastern wall of Gale appears like the eastern wall of Curie, although the subsequent lake fill identified by Grotzinger et al. (2015) has no doubt buried the lower sequence of the rim stratigraphy here. Still larger craters, such as Herschel and Huygens craters, preserve almost no wall units due to subsequent partial

Figure 4.22. It looks quite different from the ground than from orbit! Here we see the western rim of 155.3 km diameter Gale crater (5.34°S, 137.78°E) from the Curiosity rover as it started to ascend Mt. Sharp on October 25, 2017. The base of the rim is ~30 km away, and the peaks on the horizon are ~40 km distant. JPL Photojournal image PIA22209. See also Figure 3.14 for another view of the rim of Gale crater.

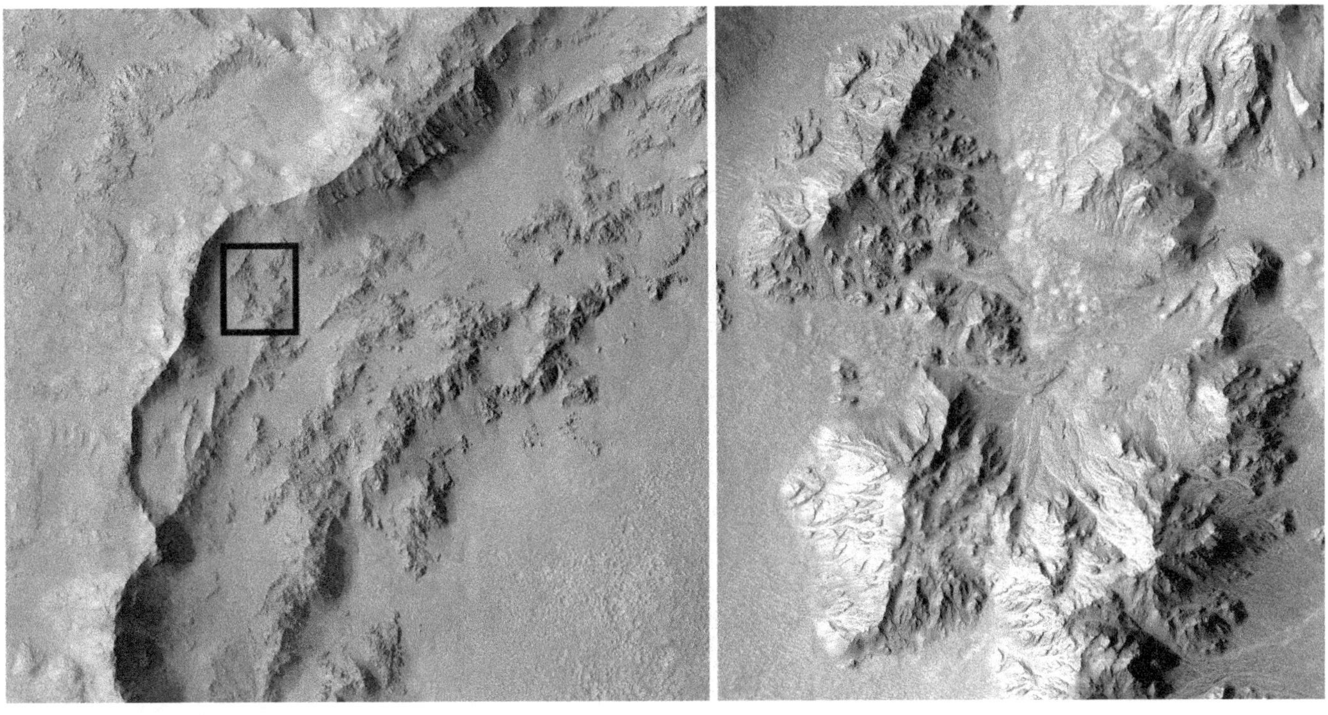

Figure 4.23 The northwest inner wall of Mojave crater (58.5 km diameter, 7.54°N, 326.97°E). Left, showing coverage at right, CTX P03_002167_1877. Right, showing numerous small fans of material coming off the high-standing blocks of the terrace. HiRISE image PSP_002167_1880.

in-filling by mare-like lavas. And yet the southeast wall of Herschel crater shows hints of jumbled terrain comparable to the southwest floor of Curie crater. Such observations may provide some insight into the shallow subsurface structure of the interiors of these large craters as well as provide a constraint on the thickness of the lava in-fill. The concentration of wall terraces in highland terrains most likely indicates that wall collapse was facilitated by the weaker target strength in these regions. A fine example of the erosion of terrace blocks within Gale crater was imaged by the *Curiosity* rover as it ascended Mt. Sharp (Figure 4.22).

Most terrace blocks have a comparable morphology to the rim of the parent crater, but in some cases, such as at crater Mojave (Figure 4.23), which is a relatively young Late Hesperian to Amazonian-age crater, diverse fans of probable fluvial origin are observed on the crater walls. These fans show minimal evidence of fluvial erosion (Williams and Malin, 2008), suggestive of the liberation of water for surface runoff. Such surface runoff has also been proposed for Bamberg crater (Mouginis-Mark, 1979b). Fans are most clearly seen at Mojave crater and have been proposed to have formed by precipitation (Goddard et al., 2014). Such features raise the possibility of several different ways in which precipitation might have taken place: (1) snowfall and snowmelt on the young, hot, crater materials; (2) precipitation from the impact plume; or (3) degassing of volatiles from terrain containing impact melt. Mojave crater is therefore potentially the source of its own precipitation.

4.4.2 Rim Uplift

The elevated rim in simple craters results from the structural uplift of preimpact target rocks and the deposition of a coherent proximal ejecta blanket at the outer edge of the transient cavity. This is best seen on fresh lunar craters (Sharpton, 2014). Clear examples of layered lava flows close to the rim crest of Martian craters have been identified for Tooting crater (Figure 4.24) and Pangboche crater (Mouginis-Mark, 2015). For ten study craters in the diameter range 8.2–53.9 km, Sturm et al. (2016) found that the mean structural rim uplift at the final crater cavity (i.e., post-terrace collapse) makes up ~81% of the total rim elevation (meaning that the ejecta comprises only 19% of the topographic expression of the rim).

4.4.3 Impact Melt

Morris et al. (2010) identified fractured flows on the southern rim of Tooting crater. The morphology of the flows (Figure 4.25) is comparable to observations of materials interpreted to be cooled sheets and flows of impact melt on lunar craters such as the craters Copernicus (Howard, 1975; Howard

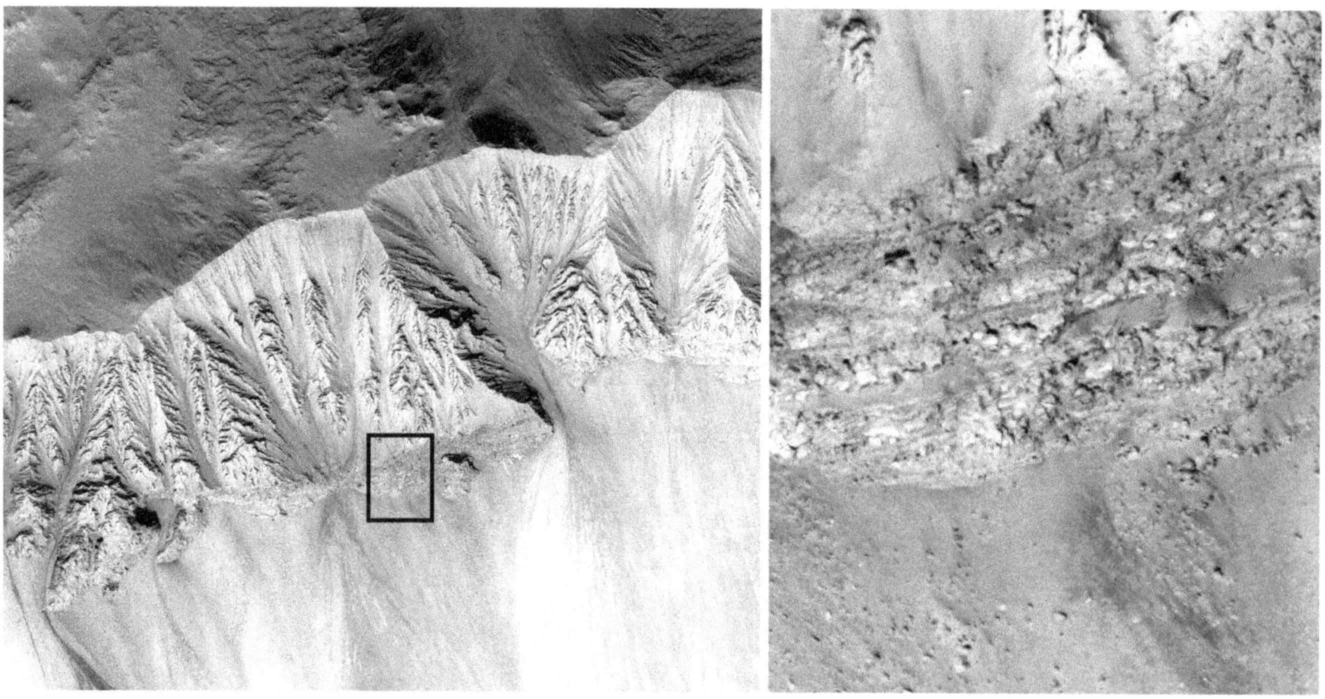

Figure 4.24 Example of rim uplift: Four distinct units can be identified in this vertical section of the eastern wall of Tooting crater, which is ~590 m high between the rim crest and the layers identified at right (black box in image at left denotes area). The exterior ejecta deposits are at the top of the section, which overlies a layer of more dissected material, then a series of coherent layers, which are then partially buried by scree materials. The sequence of layers is most likely preimpact lava flows that formed the target material. These flows now lie ~237–258 m above the preimpact surface and represent the structural uplift of the target (Mouginis-Mark and Boyce, 2012). Width of image at right is ~200 m. HiRISE image PSP_007907_2035.

Figure 4.25 Portion of HiRISE image PSP_001538_2035 detailing lobate flows ("f") on the southern rim of Tooting crater. Flow direction was both toward crater cavity at the top of the image and toward the bottom in image center. The crater rim crest is at the top left of this image. Width of image is ~1,700 m. Compare this morphology to that seen at fresh lunar craters (Figure 2.13).

and Wilshire, 1975), Necho (Hawke and Head, 1977), and King (Heather and Dunkin, 2003). Perhaps significantly, these flows are only present on the parts of the crater rim that do not appear to have collapsed toward the center of the crater, suggesting that more flow material could have originally been formed but has been lost at other azimuths around the rim.

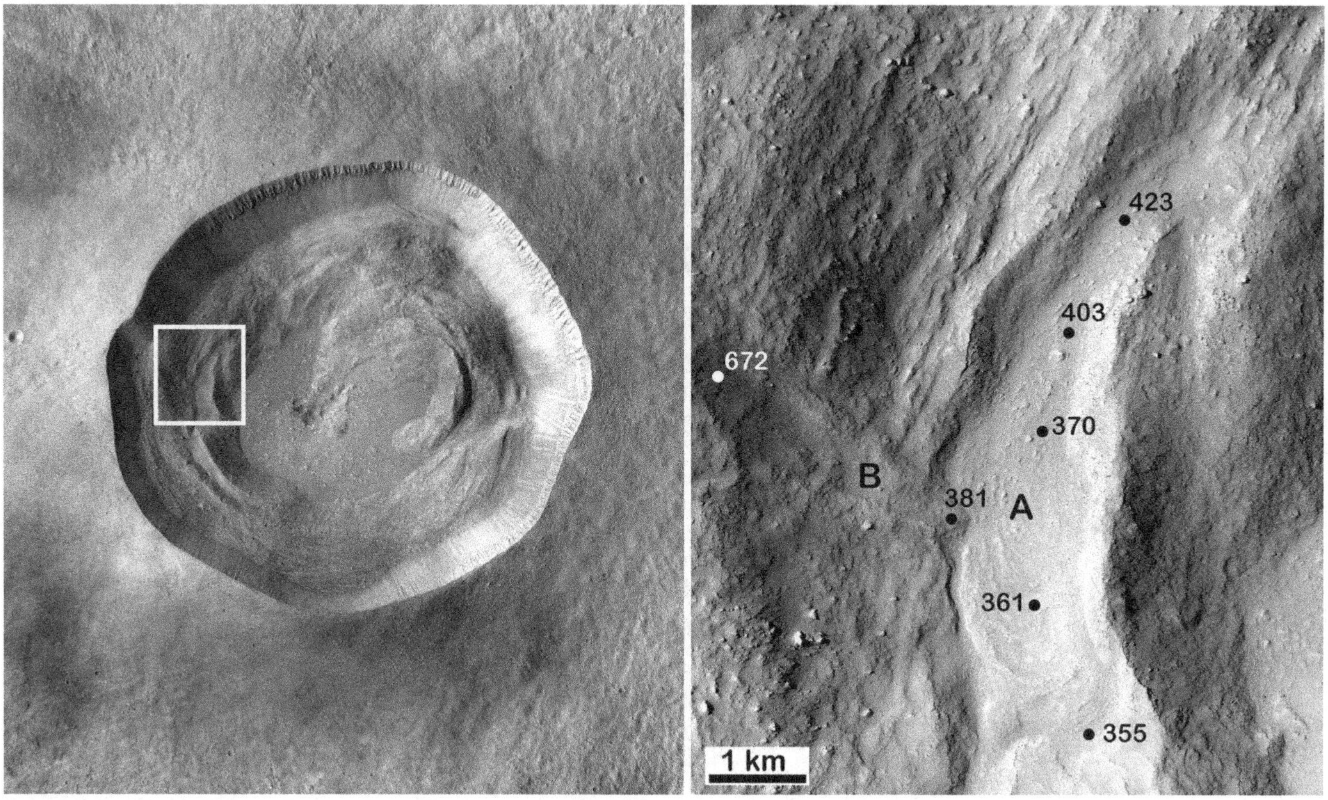

Figure 4.26 Left: Pangboche crater (17.31°N, 226.58°E; 10.2 km diameter) has a complex interior with a central pit and multiple terraces. White box indicates area of coverage of image at right. CTX image P02_001643_1974. Right: Lobate flow ("A") perched on the western inner wall of Pangboche crater. This flow, interpreted to be solidified impact melt, is ~190 m above the crater floor (at lower right). The flow direction was from the top of the image. Also note the subsequent flow ("B") that postdates this larger flow. Black circles give elevations in meters above an elevation of 20 km above datum. Part of HiRISE image ESP_027646_1975. Reproduced from Mouginis-Mark (2015). "Cratering on Mars with almost no atmosphere or volatiles: Pangboche crater," *Meteoritics and Planetary Science,* vol. 50, pp. 51–62, Figure 7. Reproduced with permission from John Wiley and Sons.

Subsequent deformation, perhaps minutes to hours after the original crater formed, may have destroyed this material via collapse of the inner wall (Mouginis-Mark, 2015).

Evidence for extensive impact melt can also be found on the inner walls of Pangboche crater (Figure 4.26). The clearest example of impact melt is perched at an elevation of ~190 m above the crater floor. This flow is about 1 km wide and 6 km long and displays festoon ridges (indicators of flow) that point in the downslope direction. In addition to this flow, a second smaller flow overrides part of the larger flow, indicating that drainage of material was multi-phased. Both flows are like those identified on the Moon; similar features, also interpreted to be impact melt, have been identified in the lunar craters Kepler (Ohman and Kring, 2012) and King (Ashley et al., 2012).

4.5 SUMMARY

This chapter has shown the diversity of interior and rim features of impact craters on Mars. Of necessity, it is easiest to explore this diversity when morphologically fresh craters are studied, and the degree of erosion is low. In general, the craters illustrated in this chapter are predominately Amazonian in age, although some attributes of Hesperian-aged craters are sufficiently well preserved that they too can illustrate well-preserved landforms. But Mars has had an extended geologic history with extensive erosion of landforms characterizing both the oldest, Noachian Period, as well as much of the Hesperian. Thus, in Chapter 5, we will look at some of the most common forms of modified interiors.

5

Modified Crater Interiors

5.1 INTRODUCTION

In Chapter 4 we saw a variety of interior features of fresh Martian craters. It is likely that erosion and deposition will rapidly (on a geologic timescale) modify some or all of these attributes. Subtle features, such as veneers of impact melt on the rim or the pitted terrain on the crater floors, are both likely to disappear long before the basic geometry of the crater can no longer be studied. Central peaks may also be buried by sediment before the terraces are no longer visible, even though the rate of infilling diminishes through time as crater wall heights are reduced and the area of deposition on the crater floor increases (Forsberg-Taylor et al., 2004). Craddock et al. (2018) conducted an analysis of different morphologic parameters of modified craters. These parameters included depth, crater wall slope, crater floor slope, the curvature between the interior wall and the crater floor, and the curvature between the outer wall and the surrounding landscape. They concluded that craters in both the initial and terminal stages of modification have consistent morphology regardless of their geographic location, suggesting that the processes that operated during early Mars (i.e., the Noachian period) were globally ubiquitous. Slightly less degraded craters show geographic variations in morphology potentially related to the spatial differences in the type and intensity of geologic processes, implying that the ancient climate may also have varied across regions (Mangold et al., 2012).

As we look at the range of morphologic features of craters on Mars, it is therefore prudent to see some of the most common ways that crater interiors change over geologic time. Thus, in this chapter we take the younger examples, as illustrated in Chapter 4, and show some of the common ways that craters may be modified. Even craters that are classified as morphologically fresh may have experienced modification. This might take the form of chemical weathering of the floor or deposition of eolian or ice deposits within the crater cavity. While most of this book is focused on relatively small (<100 km diameter) craters, larger craters and impact basins also display a wide range of interior landforms that show the extent of modification. We include here some examples of this basin modification.

5.2 FLOOR FEATURES

5.2.1 Dunes

Eolian dunes on Mars have been studied for decades and are known to comprise fine particles, most probably dust size but also potentially including larger grain sizes (Breed et al., 1979; Hayward et al., 2014). Dunes occur over a wide range of elevations, latitudes, and terrain types. Outside of the North Pole region, most dune fields reside in craters (Figure 5.1). For example, in the equatorial region, 428 out of 547 dune fields (72% by area) are in craters (Hayward et al., 2014). At least some of the observed dunes appear to be active today, such as the dunes seen within Endurance crater (Chojnacki et al., 2011) and Gale crater (Bridges et al., 2017). Many of these dunes show prominent orientations and Hayward et al. (2007, 2014) investigated the azimuth of dune fields with respect to the center of the crater. They found that dune field position within a crater and the relative area of the dune field as a function of crater size and morphology are most likely indicative of prevailing wind conditions at the time of dune formation. Hayward et al. (2007) found that dune azimuths generally correlate with regional wind patterns and noted that crater size appears to be a critical factor in the correlation between dune orientation and vectors determined from global climate models.

5.2.2 Flat Floors

The interiors of numerous Martian craters have been filled to a considerable depth to create flat floors (Figure 5.2). Some of the processes by which this might occur include eolian sedimentation, infilling by volcanism, or glacial and fluvial activity. Physical differences can exist between these mechanisms (Pan et al., 2021), with lava flows and clastic rocks as the potential infilling materials for craters with high thermal inertia. High-resolution thermal inertia measurements using THEMIS TIR as well as TES compositional data show that such craters contain some of the rockiest materials on the planet (Edwards et al., 2014). Because of the extensive distribution of fine material, out of a sample of 119 craters with high thermal inertia, Pan et al. (2021) concluded that explosive (pyroclastic) volcanic activity may have been widespread on early Mars.

Some crater floors that were presumably flat at the time of initial modification can be seen to have undergone a second episode of modification and subsidence. A particularly fine example of this can be seen in Figure 5.3, which illustrates a filled crater in Amazonis Planitia at the distal end of Marte Vallis. Here the interior of the crater displays numerous concentric rings that may be successive high stands of fluid (water or lava?). But the central third of the crater has a ridged pattern that lacks any prominent tectonic fabric. Perhaps significantly,

Figure 5.1 (Top) Soon after crater formation, eolian processes start to work on the pristine morphology of a crater. Here we see eolian dunes that have formed on the floor of Endurance crater in Meridiani Planum (2.01°S, 354.50°E), as imaged by the Opportunity Rover on May 21, 2004. The crater is 130 m in diameter. JPL image PIA06866. Bottom: Close-up view of the Endurance crater dunes, as imaged by the Opportunity Rover. These dunes are less than 1 m high. Illumination direction is from the left. JPL photojournal image PIA06754.

the transition between these floor units is marked by a low escarpment, which may be a few tens of meters high. This escarpment may have been created when subsurface material in the center of the crater was withdrawn to a greater depth.

5.2.3 Chaotic/Fractured Texture

Highly fractured floor deposits are sometimes observed in craters (Figure 5.4). Lunar floor-fractured craters typically do not occur in floor deposits and are attributed to stretching and fracturing of the floor by intrusions of magma (Wichman and Schultz, 1995; Jozwiak et al., 2012). Typical Martian floor fractures generally display different morphologies from the lunar examples. Martian floor fractures have variously been proposed to have an origin linked to volcanic activity (Greeley and Spudis, 1978; Schultz, 1978, 1979; Schultz and Orphal, 1978; Bamberg et al., 2014), tectonism (Smrekar et al., 2004; Hanna and Phillips, 2006; Bamberg et al., 2014), glacial processes (Bamberg et al., 2014), groundwater (Zegers et al., 2010; Bamberg et al., 2014), and surface fluvial activity (Sato et al., 2010).

5.2.4 Concentric Crater Fill

Impact craters in the mid-latitudes of Mars are commonly filled to varying degrees with a combination of ice, dust, and rocky debris, forming concentric surface features (Figure 5.5). Pearce et al. (2011) identified curvilinear lineations and ridges that are most likely glacial deposits within large (> 5 km diameter) craters in Utopia Planitia. They observed that the depth/diameter ratio of these craters consistently decreases over the latitude range 30°N to 60°N. Hypotheses accounting for the formation of this concentric crater fill (CCF) range from ice-free processes (e.g., eolian fill) to ice-assisted talus creep to debris-covered glaciers (Levy et al., 2010). It is likely that the CCF represents one component of significant Amazonian-aged mid-latitude glaciations on Mars that may have occurred as recently as 40 Ma ago (Weitz et al., 2018).

Other landforms within craters are also believed to have a glacial origin and may be genetically related to the CCFs (Figure 5.6). Several small craters with layered interior deposits exist in Utopia Planitia. This infill shows multiple near-horizontal layers suggestive of mantling (Figure 5.7). Using global climate models, Hartmann et al. (2014) proposed

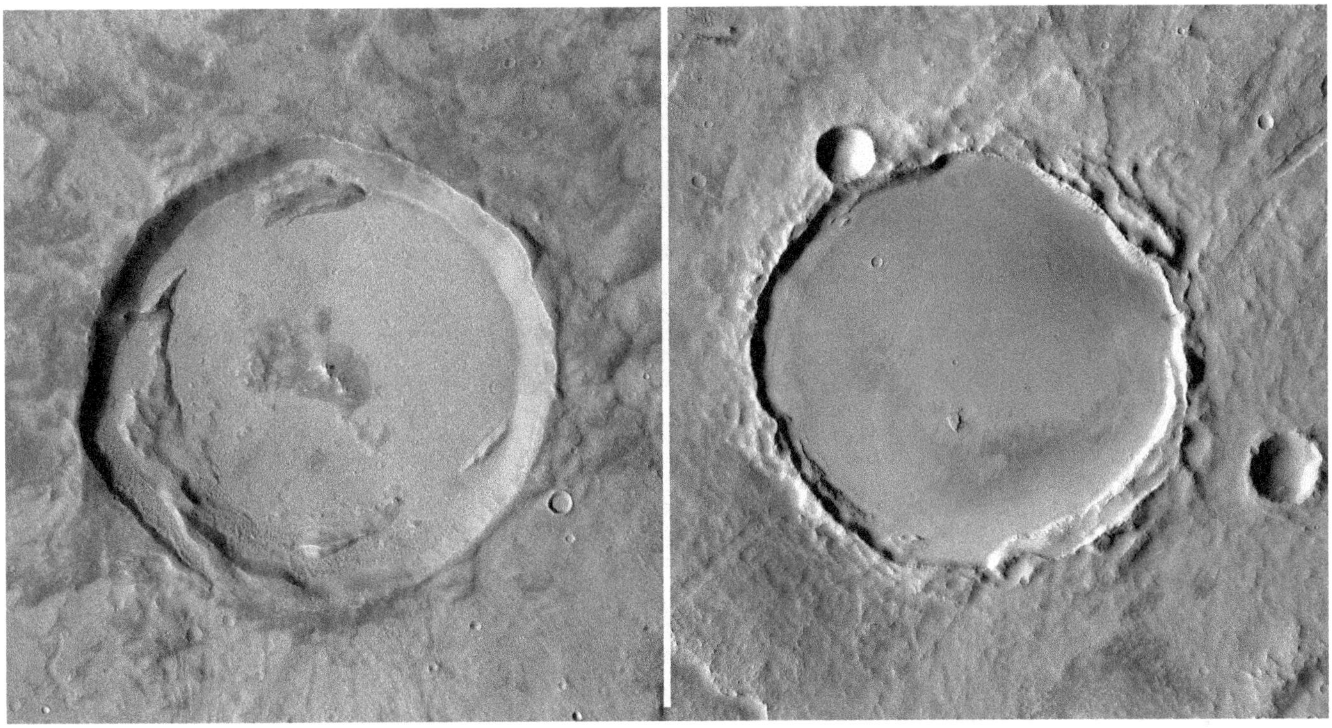

Figure 5.2 Examples of flat-floored craters. These differ from the example shown in Figure 4.2 because craters are larger and have been flooded by material to only leave remnants of the terraces and a central peak. In particular, the crater on the right displays lobate flow margins that are strongly suggestive of lava flows that have been linked to a fissure eruption east of the volcano Jovis Tholus (Wilson et al., 2009). At left, the crater is 13.2 km in diameter at 33.20°N, 15.76°E. CTX image P16_007413_2122. On the right, a crater 18.3 km in diameter located at 18.28°N, 245.85°E. CTX image J03_046146_1985.

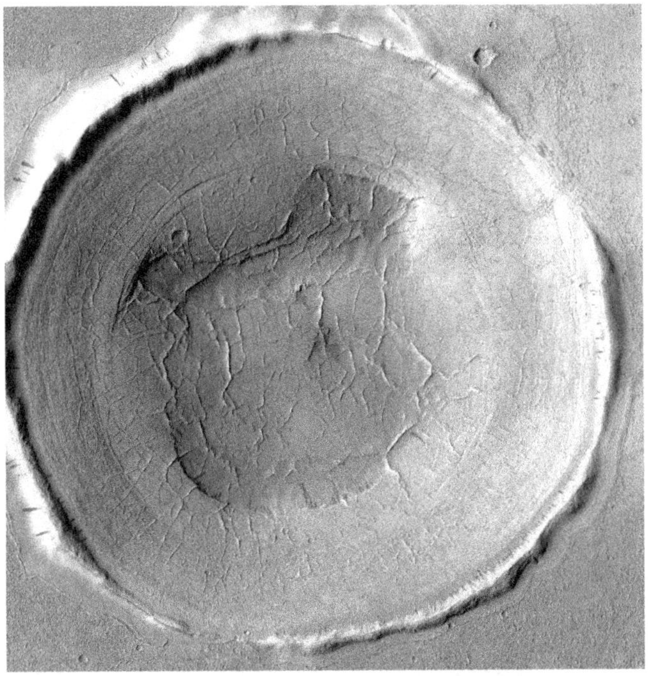

Figure 5.3 This 5.5 km diameter crater in Amazonis Planitia was evidently nearly filled to the rim by lava flows at the distal end of Marte Vallis (24.70°N, 198.74°E). After the cessation of this volcanism, the central portion of the floor subsided producing the "spider-like" pattern of ridges on the floor. HiRISE image ESP_017032_2050.

that Greg crater (38.5°S, 113.1°E) lies in one of two nonpolar areas with extremes of ice deposition during high-obliquity epochs. Further analysis conducted by Tsibulskaya et al. (2020) suggested that this crater is relatively old (>1–3 Ga) and has glacial features with decameter-deep surface layers, estimated to be ~5–15 Ma old.

5.2.5 Exposures of Megabreccia

HiRISE images have revealed some impressive exposures of impact megabreccia in Holden crater (Grant et al., 2008). Outcrops within the crater walls are rounded and poorly sorted. Chaotically arranged blocks up to 50 m across are unconformably overlain by sediments deposited during two subsequent Noachian-age phases of aqueous activity. In the case of the 79 km diameter Ritchey crater, Ding et al. (2015) were able to map small dikes and veins within the central uplift and identified ten different units with some containing three types of megabreccia. These observations were used to help constrain the relationship between breccia, pre-existing megabreccia, and other rock units. Comparable exposures of megabreccia exposed on the floor of Toro crater are illustrated in Figure 5.8.

5.2.6 Massive Sedimentary Deposits

Layered, massive, and thin mesa units within craters were discussed by Malin and Edgett (2000). These outcrops (Figure 5.9)

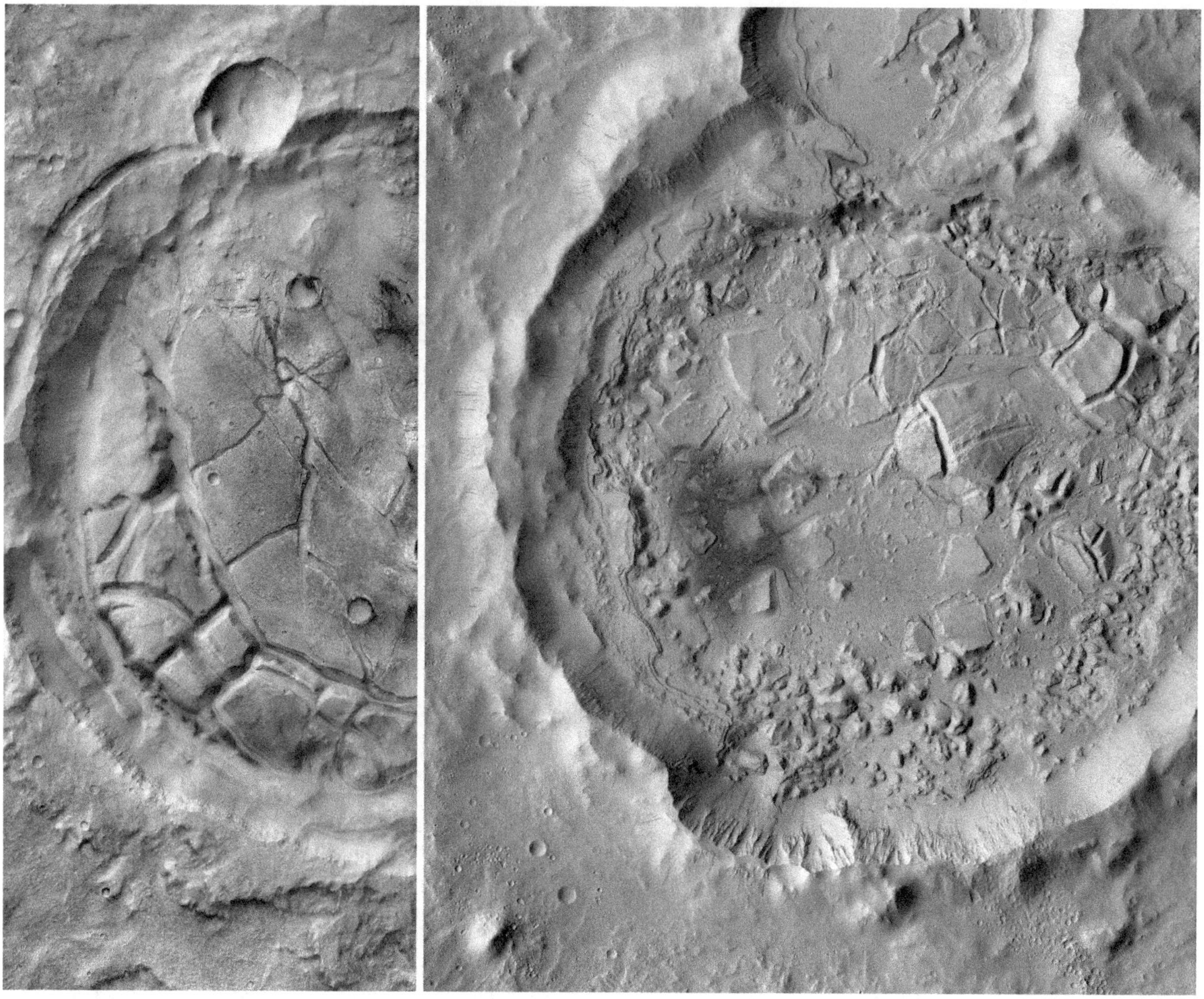

Figure 5.4 At left, classic floor fractures can be seen in a crater (51.5 km diameter) on the western rim of Miyamoto crater (2.32°S, 351.57°E). CTX image P17_007770_1794. At right, the floor fractures in this 26.8 km diameter crater (at 12.84°N, 174.85°E) appear to have been confined to the deformation and erosion of a later sediment fill. In this case, some of the fractured blocks have been embayed by smoother material. CTX image F04_037221_1929.

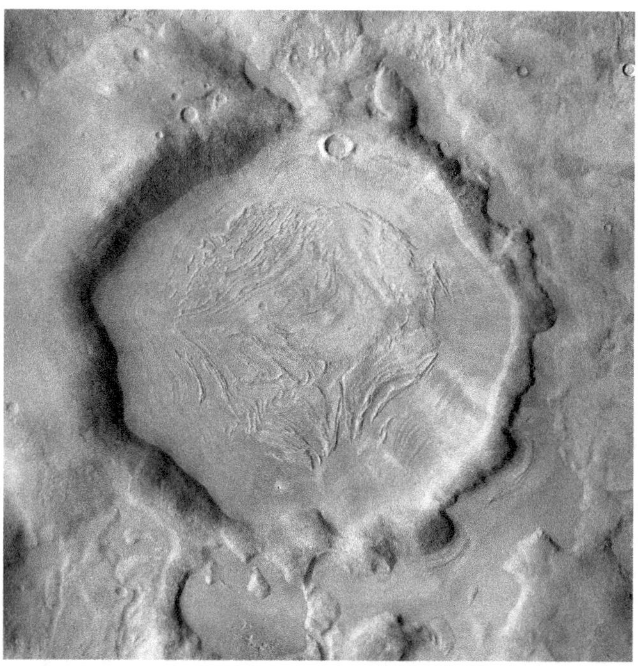

Figure 5.5 Concentric fill within an unnamed 20 km diameter crater located at 34.23°N, 75.59°E. CTX image P17_007569_2143.

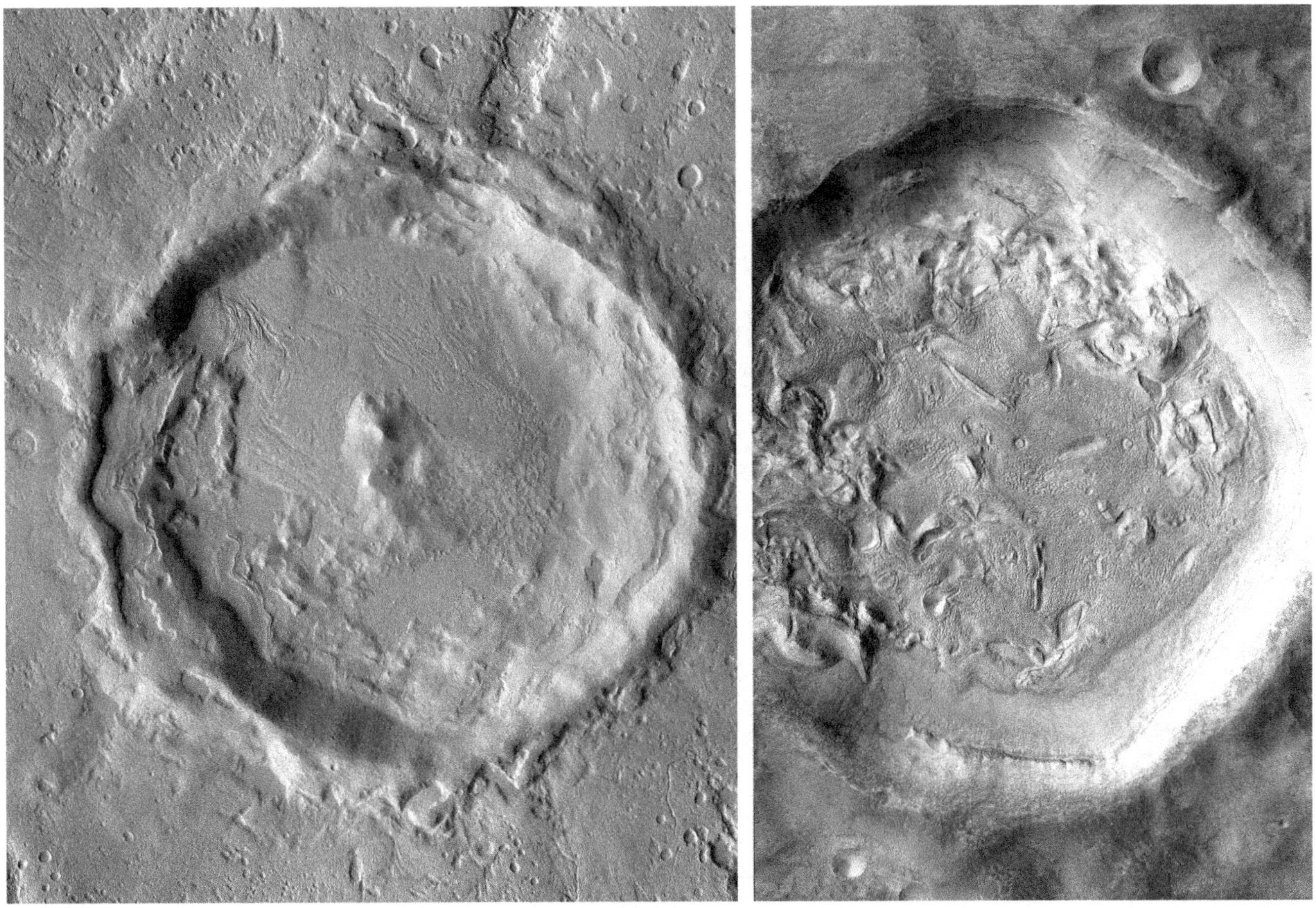

Figure 5.6 At left: Glacial features in the form of lobate deposits partially surround the northern part of the central peak of this 24.0 km diameter crater within Arabia Terra. 39.51°N, 49.21°E. CTX image G04_019635_2220. At right: Pits, hollows, and viscous flow can be seen within this 6.5 km diameter crater at 35.76°N, 2.10°E. HiRISE image ESP_060858_2160.

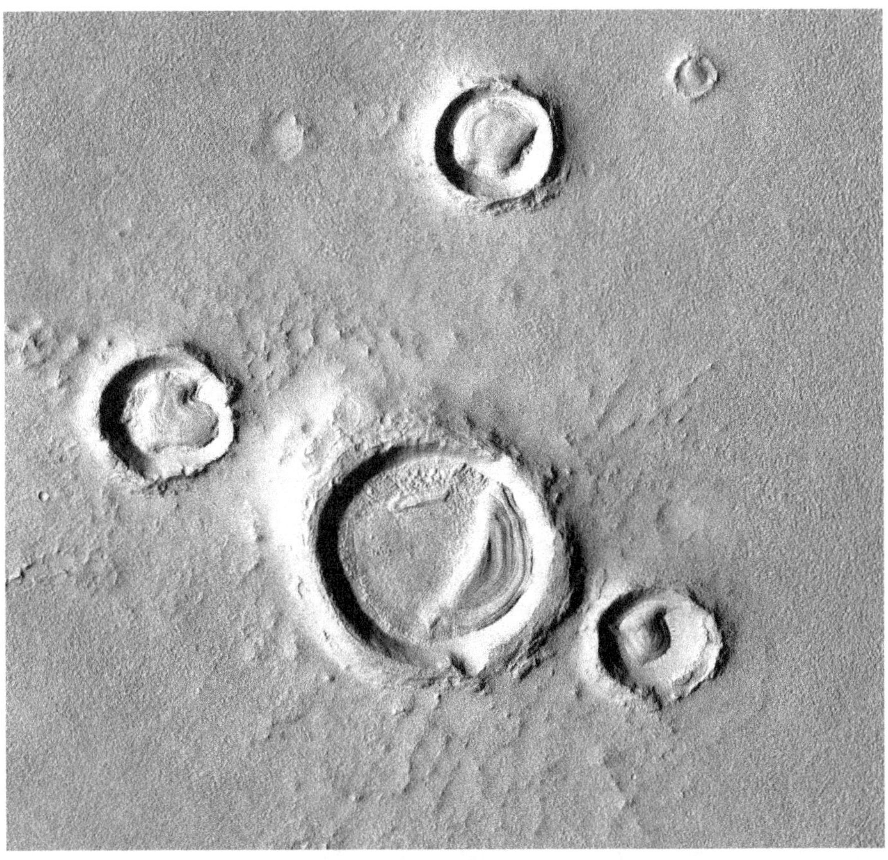

Figure 5.7 Many small craters with layered interior deposits exist in Utopia Planitia. This infill shows multiple near-horizontal layers suggestive of mantling. The largest crater in this scene is ~1 km in diameter. Image centered at 37.07°N, 82.83°E. HiRISE image PSP_008452_2175.

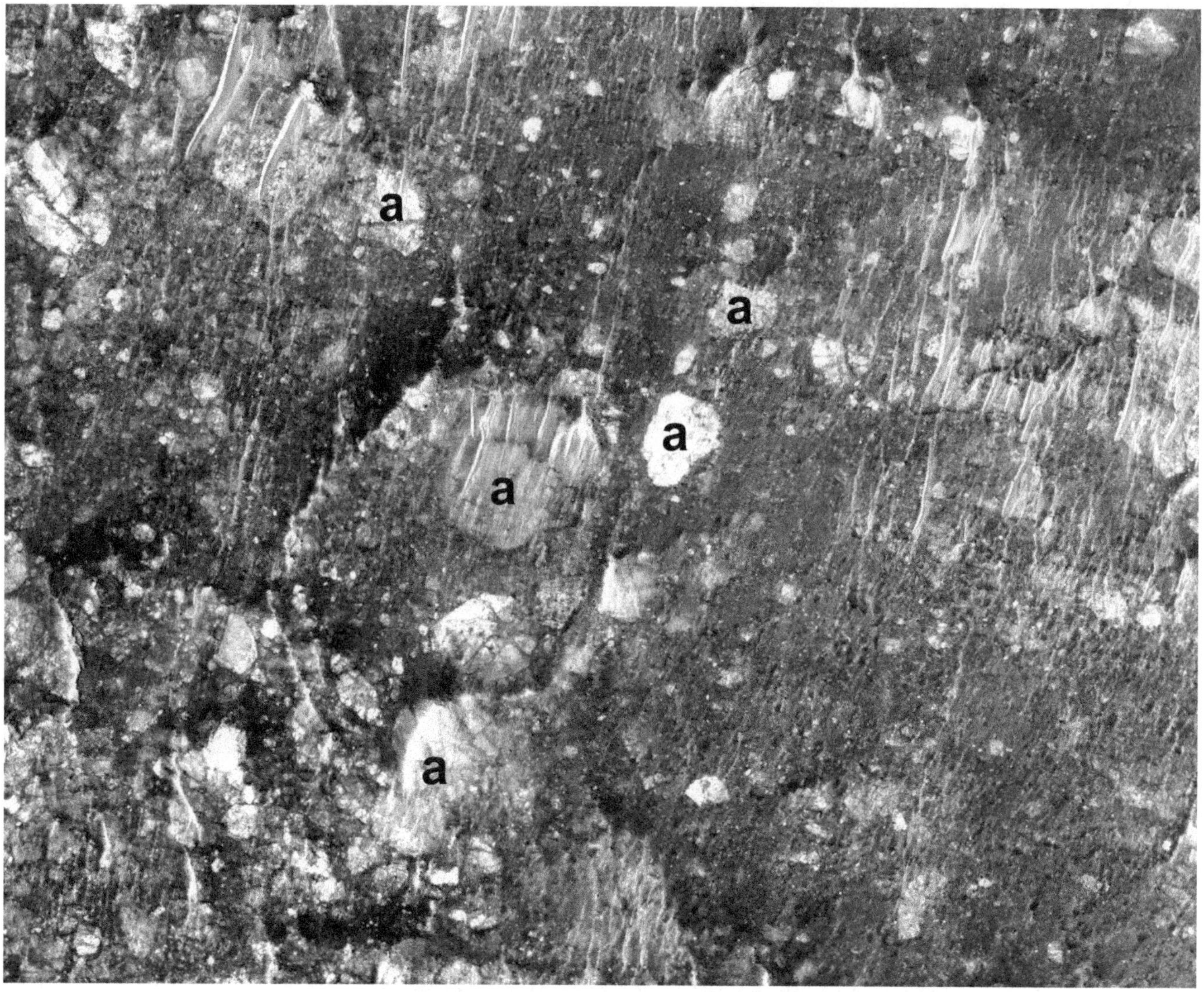

Figure 5.8 Megabreccia exposed on the floor of Toro crater (43 km diameter) in Syrtis Major, located at 17.04°N, 71.81°E. Some of the larger individual blocks are identified by "a." Notice the irregular sharp-edged shape of these blocks as well as their diverse colors, suggesting that they experienced extensive fragmentation and came from different parts of the target material. HiRISE image PSP_005842_1970_MRGB. Image width is ~500 m. Note: A black and white version of this figure will appear in some formats. For the color version, please refer to the plate section.

have attributes of sedimentary material because they exhibit evidence of being fine-grained and bedded. The 174 km diameter crater Terby (Figures 5.10 and 5.11), located on the northern rim of Hellas Basin, is a fine example of this type of crater fill, displaying an unusual interior morphology, including multiple flat-lying light-toned layered deposits as much as 2 km thick (Wilson et al., 2007; Ansan et al., 2011). Layers are stacked with the same strata pattern forming >1 km long and >100 m thick sedimentary wedges, in places truncated by erosional surfaces or unconformities. The geometry of these layers is consistent with sediments that settled mainly in a subaqueous environment. Eolian processes and a permanent cold environment most likely promoted viscous flow of some of the materials indicative of ice. A contemporaneous shallow (ice-covered?) lake covering the floor of Terby crater that transported material from the crater into the greater Hellas basin has also been hypothesized (Wilson et al., 2007).

Large central mounds, of which Mt. Sharp within Gale crater (where the *Curiosity rover* landed) is the best-known example, may also be an unusual example of sedimentary deposits within craters. Gale crater contains a 5.2 km-high central mound of layered material (Figure 5.12). Results from the rover *Curiosity* (Grotzinger et al., 2015) confirm that the lower portion of Mt. Sharp was formed via sedimentation within a shallow lake, but the upper reaches are too high to have a similar origin (Figure 5.13). One idea is that the upper portion of Mt. Sharp could be genetically related to the Medusae Fossae Formation (MFF) (Thomson et al., 2011) and imply that the MFF was once much more extensive than the coverage observed today. However, there is still uncertainty as to the origin of the MFF (Mandt et al., 2008; Mouginis-Mark and Zimbelman, 2020), and so the resolution of the origin of the layered material at the summit of Mt. Sharp awaits possible future discoveries by the *Curiosity* rover during its climb.

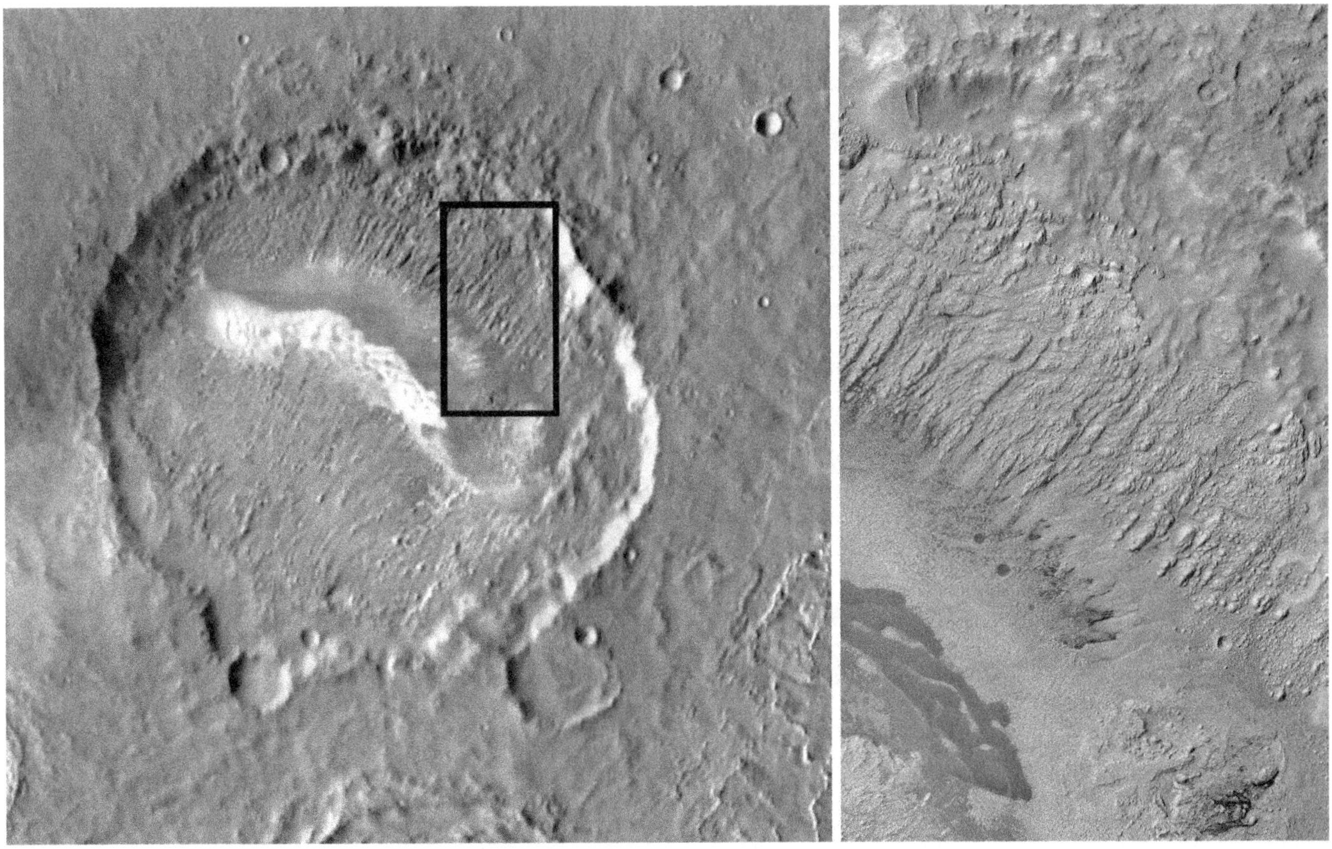

Figure 5.9 Unusual sedimentary fill within Danielson crater (64 km diameter) located at 8.00°N, 352.93°E. On the left is an overview of the crater (THEMIS daytime TIR mosaic), with box showing detail at the right. CTX image P18_008205_1880.

Figure 5.10 The complicated sedimentary layers lie within Terby crater (27.88°S, 74.04°E; 174.0 km diameter) are possibly similar to the eroded remnants of crater fill within the mounds of Gale and Nicholson craters (see Figures 5.12 and 5.14). THEMIS daytime TIR mosaic. The black band at the top right is due to a lack of image coverage. Box indicates the location of Figure 5.11.

Gale crater is not the only crater with a large central mound on Mars. Bennett and Bell (2016) conducted a global survey of central mounds within large (>25 km diameter) craters and found that more than 60% of the 50 mounds are in Arabia Terra. Thirty-two mounds rise above the host crater's lowest rim height, but only two mounds rise above the entire crater rim, namely Nicholson crater (Figure 5.14) and an unnamed crater in Arabia at 3.6°N, 24.8°E. Bennett and Bell (2016) concluded that the mound sediments were most likely to have once filled the host craters and that they were partially eroded by the wind, as most mounds are offset from the center of the host crater in the same direction as the present-day regional winds. Arabia Terra may therefore once have been covered by a more extensive regional sedimentary deposit and the mounds are just the remaining vestiges of this deposit.

5.2.7 Deltaic Fans

Eberswalde Crater has been the site of persistent aqueous activity on Mars, with Malin and Edgett (2003) and Irwin et al. (2015) documenting fan-shaped deposits with meandering channelized deposits (Figure 5.15). These fan deposits have been interpreted to be evidence for sustained liquid water early in Mars history with a paleolake formed during the Noachian period (Mangold et al., 2012). Subsequent analysis of these deposits by Irwin et al. (2015) considered several different modes of origin, one of which required an intermittent rainfall rate of ~1 cm/day and seasonal snowmelt consistent with a deposition

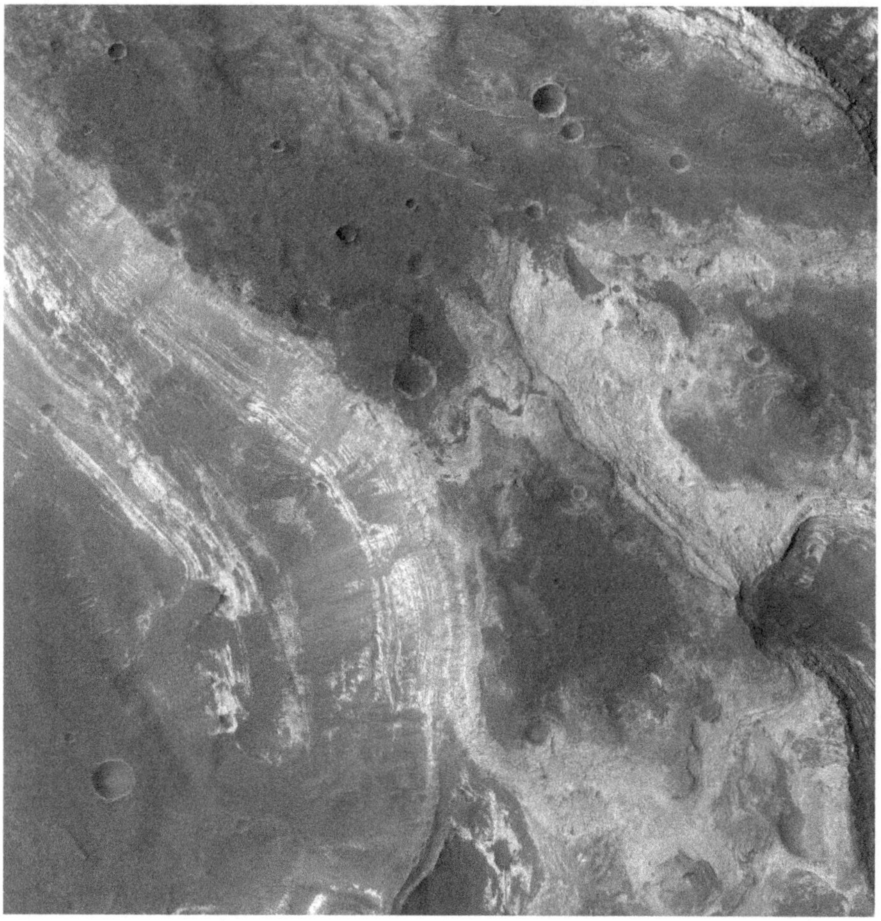

Figure 5.11 Details of the layering within Terby crater. See box in Figure 5.10 for location. Illumination is from the left. CTX image B09_013160_1528.

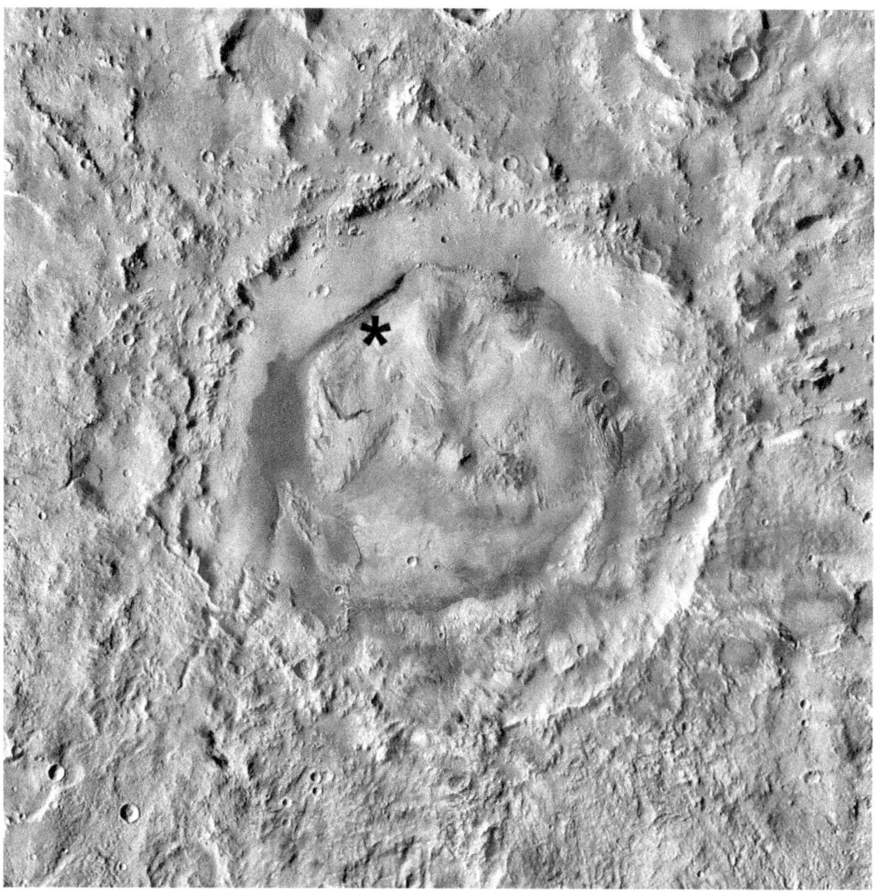

Figure 5.12 Gale crater (155.3 km diameter, 5.27°S, 137.73°E), site of the *Curiosity Rover* mission, also has a massive central mound. The approximate location where the rover acquired the image used in Figure 5.13 is indicated by the star. THEMIS daytime TIR mosaic.

Figure 5.13 Mt. Sharp within Gale crater is one of the largest central mounds on Mars. The rover *Curiosity* has spent about a decade climbing the lower levels of the mound but so far has only found lake sediments. The upper layers of Mt. Sharp are more enigmatic and may even be equivalent to the material found within the Medusae Fossae Formation. In this image the rover is looking eastward toward the summit of Mt. Sharp, which rises more than 4 km above the rover when imaged by the MastCam instrument on the rover on October 13, 2019, the rover's 2,555th day on Mars. JPL Photojournal image PIA23898.

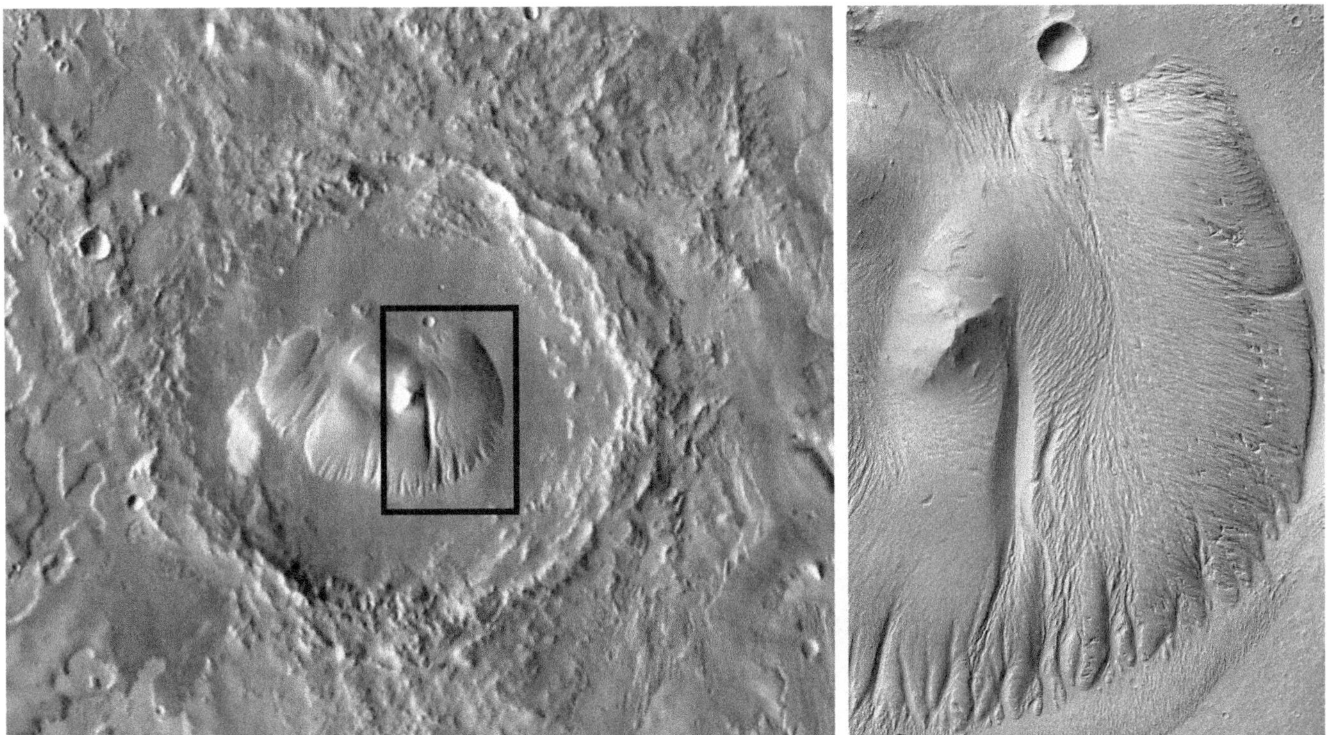

Figure 5.14 Central mound within Nicholson crater (102.5 km diameter, 0.28°N, 195.47°E). Left: THEMIS daytime TIR mosaic. Box indicates the location of image at right. Right: Extensive erosion of the central mound is evident here, with both wind-produced pits on the eastern lower slopes and, potentially, fluvial gullies on the southern flank. CTX image G22_027001_1801.

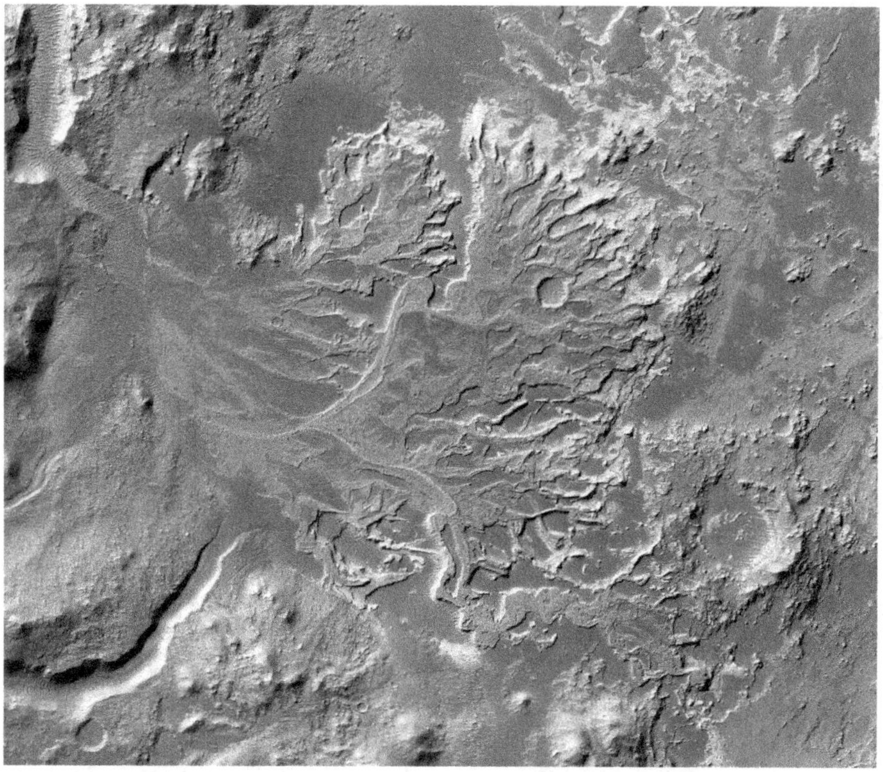

Figure 5.15 Braided delta deposit on the northwest side of the floor of Eberswalde crater (65.3 km diameter). Much of the delta now displays "inverted topography," where erosion of the surrounding material has left the channel floors standing higher than the surrounding terrain. The paleo-direction of flow was from left to right in this image. The image is centered at 23.83°S, 326.38°E. Image width is ~18 km. CTX image B21_017911_1559.

Figure 5.16 Jezero crater (49.0 km in diameter), the site of the *Perseverance* rover landing, possesses a fine example of a deltaic deposit on the western crater floor. The largest crater in this image is Belva crater at 18.49°N, 77.37°E. At left is CTX image P03_002387_1987. The white star shows the approximate location of the rover when the images on the right were taken. Image width is ~10.5 km. On the right are two views of this delta, as viewed from the ground by the rover. Top right is an image taken on June 13, 2022, showing the edge of the delta. Perseverance NavCam image N0260756ZCAM08487_1100LMJ. On the lower right, the rover has driven onto the front of the delta to reveal dozens of thin layers, presumably created by separate discharge events from the source of the delta. Perseverance NavCam image N0260850NCAM13474_01_195J, taken on June 20, 2022.

timescale totaling ~10^4–10^6 years. Comparable evidence of episodic alluvial activity comes from fans in Noachian-aged materials within Harris crater in the Terra Tyrrhena region (Williams et al., 2011). Jezero crater (Figure 5.16), the landing site of the 2020 *Perseverance Rover*, has a well-defined deltaic system on the western floor that has variously been estimated to have taken 10^6–10^7 years (Schon et al., 2012) or at least 90–550 years (Salese et al., 2020) to form. The Jezero crater delta fed a throughflow lake driven by significant surface runoff (Horvath and Andrews-Hanna, 2024). Interestingly, it has been inferred that the lake formed by this delta system was different from that driving the hydrology of Gale crater (Grotzinger et al., 2015), implying that basin hydrology could be different within different large craters on Mars (Horvath and Andrews-Hanna, 2024).

5.2.8 Hydrothermal Deposits

It is widely expected that hydrothermal systems might have been produced during the formation of complex craters (>~5–10 km in diameter) on Mars (Abramov and Kring, 2005; Osinski et al., 2013). Newsom et al. (1996) had earlier hypothesized the potential existence of crater lakes within Mars craters soon after their formation, and that these lakes may not have totally frozen for thousands of years. The SNC meteorites also show evidence for impact-generated hydrothermal alteration in the Nakhlite meteorite (Bridges and Schwenzer, 2012). Rathbun and Squyres (2002) and Abramov and Kring (2005) modeled several parameters that might influence the formation and preservation of a transient lake (such as the host rock permeability and the near-surface temperature). For craters ~30 km in diameter, such lakes might have existed for ~67,000 years, 290,000 years for a 100 km diameter crater, and potentially ~10 Myr for a Hellas-sized impact basin. While high-resolution multispectral data are still needed to confirm the widespread distribution of these hydrothermal deposits, the quest to find them is accentuated by the possibility that they could have served as transient habitats for the possible evolution of early life on Mars (Osinski et al., 2013). But out of a sample of 144 craters >7 km in diameter, Turner et al. (2016) found just three craters 20–62 km in diameter that show compelling spectral features suggesting the presence of hydrated silicate minerals.

Toro crater (41 km diameter, 17.05°N, 71.81°E) is one example of a crater that shows spectral and morphologic evidence of impact-induced hydrothermal activity. Spectroscopic observations by Marzo et al. (2010) were used to identify extensive hydrated silicate deposits, including chlorites, smectites, and opaline material, a suite of phases that frequently results from hydrothermal alteration in terrestrial craters and is also expected on Mars from geochemical modeling of hydrothermal environments. These deposits appear to be predominantly associated with the central uplift and with portions of the northern part of the crater floor. Detailed geologic mapping of these deposits by Marzo et al. (2010) reveals geomorphic features that are consistent with hydrothermal activity that followed the impact event, including vent-like and conical mound structures and a complex network of tectonic structures caused by fluid interactions such as fractures and joints.

5.3 WALL FEATURES

5.3.1 Gullies

Gullies are common on the inner walls of degrading craters (Figures 5.17 and 5.18). In one analysis, 188 craters out of 225 studied in the Phaethontis Quadrangle (MC-24) were found to display gullies (Berman et al., 2005). Harrison et al. (2015) conducted a global review of gullies and made inferences about their mode of formation. For at least some of the craters, Johnsson et al. (2014) proposed that the gullies were formed on the Sunward-facing interior walls of the crater by melting of very recent snow deposits after the termination of the last Martian glacial period, rather than being the remnants of volatiles released by impact-related heating. Malin et al. (2006) hypothesized that some of these gullies are still forming today, and subsequent imaging by higher resolution cameras (Dundas et al., 2019) has confirmed this. They observed several phenomena associated with erosion and transport of material

Figure 5.17 Extensive gullies have formed on the poleward-facing inner wall of Talu crater (40.34°S, 20.08°E, 10.0 km diameter). Here the gullies originate from a lower substrate, suggesting a volatile-rich layer below the topmost unit, which presumably is comprised of crater ejecta. Compare with Figure 5.18 to see a different place of origin for the gullies with respect to the crater rim. HiRISE image ESP_011817_1395.

Figure 5.18 Gullies can also be associated with crater wall alcoves in the mid-latitudes, as with this 6.6 km diameter crater located at 34.04°S, 134.51°E. Here the gullies extend almost to the rim crest (compare with Figure 5.17). Under appropriate climate conditions, such alcoves can then serve as cold traps that provide sources of meltwater that begin at the crest of the slope face (Dickson and Head, 2009). HiRISE image PSP_005985_1455.

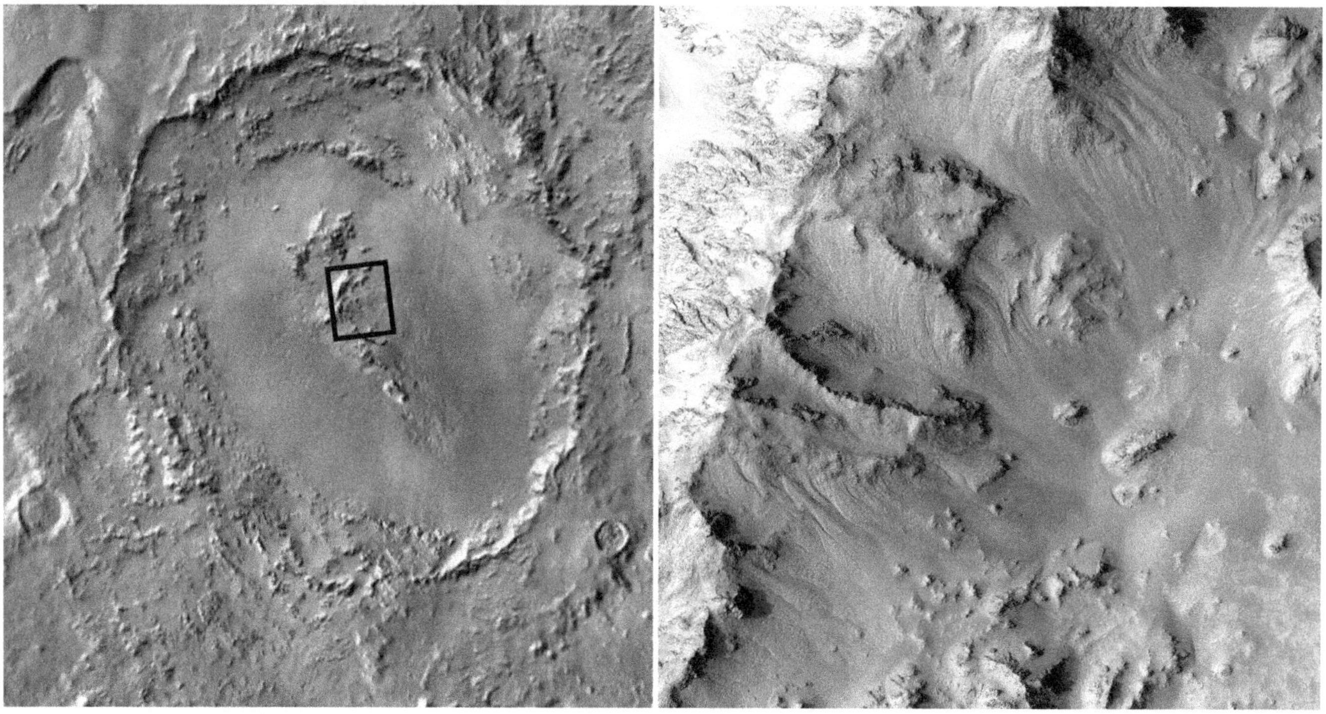

Figure 5.19 Gullies in the central peak of Hale crater (150 × 125 km; 35.62°S, 323.62°E). The image on the left is a THEMIS daytime TIR mosaic showing the whole crater (box denotes area at right). At right is a detailed view of some of the gullies on the central peak that indicates that volatiles must have been within the uplifted peak material at the time of crater formation or that there was a process of "volatile charging" the peak after its formation. HiRISE image ESP_012663_1440.

within the gullies; new terraces having formed freshly eroded channel segments as well as channel abandonment and the formation of lobate deposits.

Local differences in the occurrence and azimuthal distribution of gullies have been identified (Reiss et al., 2009). Hale crater (~125 × ~150 km diameter) possesses gullies (Figure 5.19) while the adjacent Bond crater (~115 km diameter) does not (Reiss et al., 2009). These local differences have been interpreted as an argument for the formation of gullies related to groundwater. It is also possible that local (or regional) gully distribution on Mars likely varies due to differences in crater topography and surface material properties as well as possible effects of the changes in obliquity.

5.3.2 Debris Flows

Debris flows can occur on the steep inner walls of craters. Reiss and Jaumann (2003) describe the morphology of erosion

Figure 5.20 Extensive debris aprons extend from the western side of this triplet crater located at 38.16°N, 9.25°E. Numerous degraded fluvial channel systems lie just to the west of this crater, and it is possible that one of these channels contributed to the sediment load within the crater interior. Image width is ~19 km. CTX image J02_045403_2180.

features on the dune field within Russell crater that resemble terrestrial viscous slurry flows (mudflows) and suggested that a flow of fine-grained material mixed with liquid water might have been responsible for their formation. Debris flows are common in the Arabia region. Figure 5.20 illustrates a triplet of craters with extensive flows from the western rim that extend across most of the crater floor and partially bury the lower portions of the central peak.

One potentially unique flow exists on the southern inner wall of Tooting crater (Figure 5.21). Immediately following a large impact event on Mars, it is likely that large volumes of water-laden sediment may have flowed both into the crater cavity and down the exterior slopes. In either case, this modification most likely took place within years (days or decades?) of crater formation but may still be preserved on fresh craters. The identification of volatile-rich sediment flows at Tooting crater is important because it indicates that water existed within the target rock at the time of impact and that at least some of the ejecta that now lies beyond the rim crest contained water. The source region for this flow lies between two discrete terrace blocks and is characterized by a network of channels that flow toward the crater floor (Mouginis-Mark and Boyce, 2012). The flow is ~1,800 m long and its width varies between 250 and 500 m. The distal 600 m of the flow is composed of ~8 discrete overlapping lobes each ~10–15 m in width. Boulders can be seen on the surface of this flow, implying that it was emplaced at sufficiently high speed to transport large material. It seems possible that this is a sedimentary flow that formed immediately after crater formation when water remobilized sediments after volatile-rich strata were elevated to the crater rim as part of the crater cavity formation.

5.4 POLAR CRATERS

Craters formed near the poles offer an interesting contrast to those formed at lower latitudes (Figures 5.22 and 5.23). Not only might the target material be different (i.e., water or carbon dioxide ice) but also the erosional history is likely to be very different due to the periodic passage of the ice cap. Garvin et al. (2000) found that the geometry of even relatively fresh craters poleward of 57°N had cavity infill as high as 80% of the reconstructed, pre-erosion levels. They also found that craters associated with ice demonstrated unique cavity geometries relative to nonpolar craters. Garvin et al. (2000) suggested that burial of these craters either by the episodic advance of the polar cap margin or by continuous ice deposition could account for these characteristics. Ice-rich mounds have been identified in 18 craters north of 65°N (Conway et al., 2012). These mounds are found in deeper craters and may have formed due to specific microclimate factors such as atmospheric deposition of ice.

5.5 DEGRADATION OF LARGE CRATERS AND BASINS

Craters larger than ~150 km in diameter and basins as large as a few thousand kilometers present a different set of landforms on their floors compared to the smaller craters illustrated earlier in this chapter. Some of these largest craters are only recognizable from the analysis of MOLA topographic data (Frey et al., 1999; Frey, 2008). Indeed, the interiors of these largest landforms are all different from each other, reflecting their size, latitude, and time of formation. As a function of diameter, we now highlight some of these unique landforms.

5.5.1 Schiaparelli (2.71°S, 16.77°E) and Huygens (13.88°S, 55.58°E) Basins

These two small basins are each ~460 km in diameter and display wrinkle ridges on their floors (Figure 5.24). Within Schiaparelli basin, these wrinkle ridges are morphologically like those found on the lunar maria and suggest that Schiaparelli was once flooded by lava flows. Within Huygens basin, most wrinkle ridges are found within the central part of the floor, suggesting that subsequent volcanism only affected the lowest parts of the floor. For both basins, the ridges are believed to be associated with multiple distinct phases of compressional stress induced by mantle processes during the Noachian/Hesperian, followed by multiple regional volcano-tectonic events as part of the thermal evolution during the Hesperian (Bernhardt et al., 2016; Ruj and Kawai, 2021).

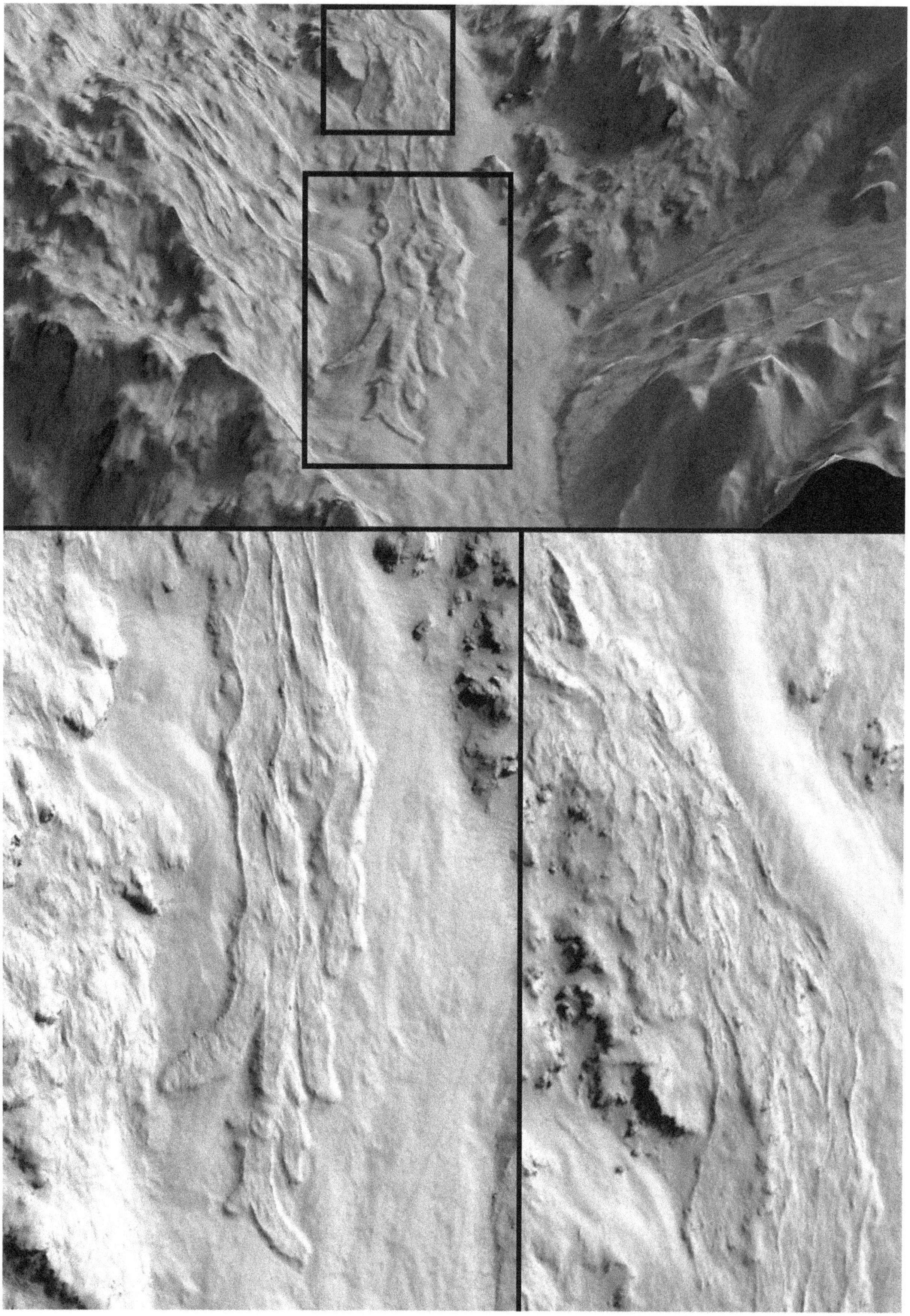

Figure 5.21 (Top): Oblique, computer-generated view of part of the southwest inner wall of Tooting crater, showing the location of a lobate flow identified by Morris et al. (2010). Boxes outline coverage in the lower pair of images. Image width is ~600 m. (Lower left): Nadir view of the details of the multi-lobed flow at its distal end. (Lower right). Further up-slope, the flow has numerous large boulders on the surface. Morris et al. (2010) measured some of these to be larger than 4 m in diameter. In all images, the direction of the flow was toward the bottom of the image, which are parts of HiRISE image PSP 003569 2035.

Figure 5.22 The polar crater Jojutla (18.6 km diameter) at 81.63°N, 189.80°E. Notice how the ejecta blanket has prevented the development of dunes (or maybe sastrugi?) from developing close to the crater rim. A prominent crater mound with multiple layers exposed on the right-hand side can also be seen. This mound may have formed through atmospheric deposition and subsequent partial erosion of icy sediments. CTX image P18_008131_2616.

5.5.2 Argyre Basin (49.7°S, 316.0°E)

Located in the southern highlands southeast of Tharsis, Argyre (~1,700 km in diameter) is one of the larger basins on Mars and formed in the Early Noachian. Faults and valleys that are both radial and concentric to the basin were created by the impact event (Dohm et al., 2007). It has been proposed that glacial and fluvial/lacustrine processes in conjunction with eolian modification were probably most important in the evolution of the basin interior (Hiesinger and Head, 2002). These processes included a significant input of sediments and a partial infilling of the basin with water during the Hesperian.

Figure 5.23 Extensive erosion of the NW rim and deposition of glacial deposits are visible at Korolev crater (72.81°N, 164.55°E, 84.2 km diameter). Using THEMIS thermal data, Armstrong et al. (205) proposed that these interior deposits are likely to be thick (several meters) high thermal inertia material, implying either a water ice or ice-rich regolith. Image width is ~15 km. CTX image P21_009042_2528.

5.5.3 Isidis Basin (13.25°N, 87.56°E)

Isidis Basin is located on the boundary between the cratered highlands and the northern plains. A ring of massifs, an outer scarp to the south, and small knobs (<10 km) to the north delineate this degraded multiring basin, which is ~1,500 km in diameter. Mustard et al. (2007) interpreted exposures in the Nili Fossae region located to the northwest as impact melt associated with the basin. The floor of Isidis exhibits a landform assemblage, nicknamed "thumbprint terrain" (Figure 5.25). Comprising arcuate ridges, aligned hills, and a network of sinuous ridges, linear depression, and mounds, this terrain has been interpreted as a glacial terrain inherited from a massive ice sheet that formed over the basin during the Hesperian (Guidat et al., 2015). Alternative ideas for the origin of the thousands of pitted hills within the thumbprint terrain may be mud volcanoes (Skinner and Tanaka, 2007) or volcanic vents (De Toffoli et al., 2021).

5.5.4 Hellas Basin (41.37°S, 67.87°E)

Hellas has an elliptic floor with a diameter of ~1,700 × 1,300 km (Tanaka and Leonard, 1995). Because of this size and Noachian

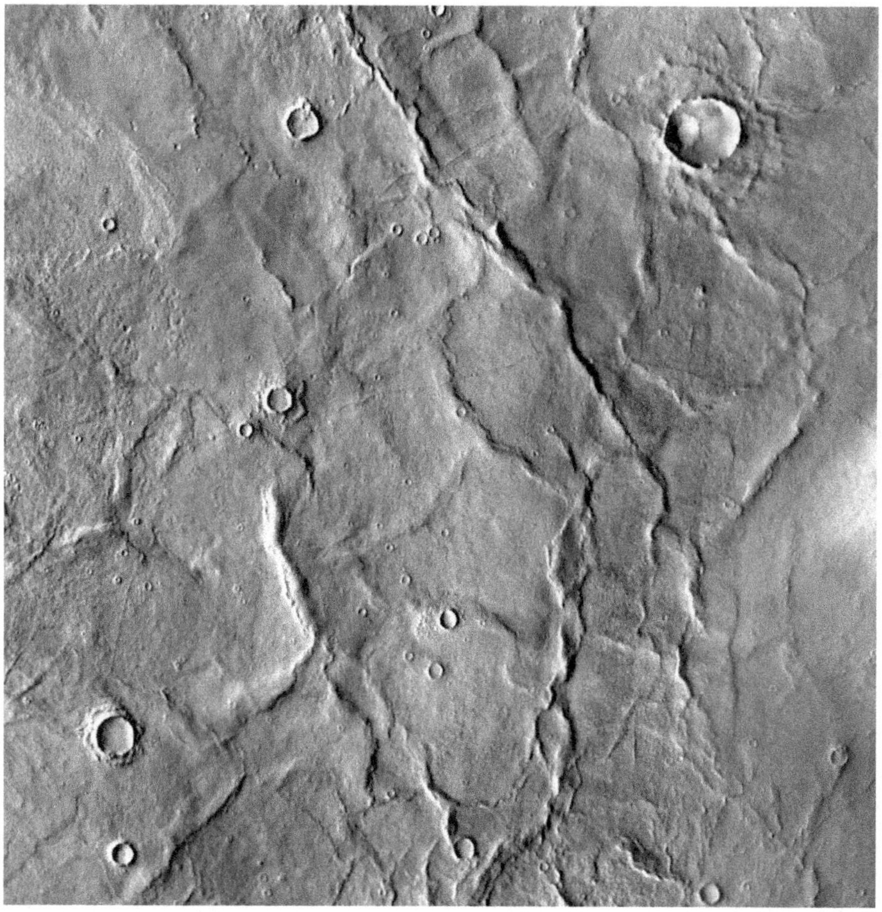

Figure 5.24 Schiaparelli Basin is one of the few large craters on Mars with a prominent set of wrinkle ridges on the floor (Mouginis-Mark et al., 1981), suggestive of volcanic infilling of the crater after its formation and subsequent tectonic activity. THEMIS daytime TIR mosaic, centered at 2.50°S, 16.62°E.

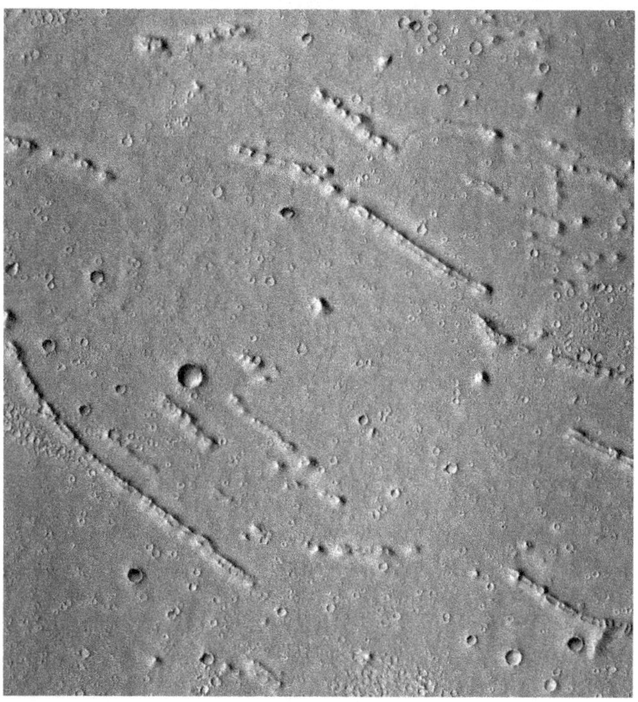

Figure 5.25 Characteristic thumbprint terrain, formed as a series of arcuate ridges and many pitted cones, in SW Isidis Basin. Width of image is ~22 km. CTX image P03_002400_1887, centered at 8.46°N, 83.79°E.

age, it is believed to have been a major sink for volcanic, glaciofluvial, and eolian materials (Diot et al., 2014; Bernhardt et al., 2016). The northern basin rim shows ample signs of extended periods of low-energy fluvial, and possibly lacustrine, activity as well as periglacial processes (Figure 5.26). But the majority of basin fill consists of volcanic materials that were emplaced as two distinct sequences (~3.8 and ~3.7 Ga) that include the formation of the wrinkle ridge plains units.

5.5.5 Utopia Basin (46.7°N, 117.5°E)

Utopia Basin may either be a single 1–3 km deep depression ~3,200 km in diameter or the combination of a few smaller impact features that merge into a single depression (McGill, 1989; Thomson and Head, 2001). Numerous circumferential features (tens of kilometers wide and hundreds of kilometers long) within the basin are most likely due to tectonic deformation. These features have subsequently been modified by sedimentation associated with the formation of the Vastitas Borealis Formation. Many quasi-circular depressions have also been identified using MOLA data (Buczkowsk et al., 2005) and are most likely to be buried impact craters. Utopia Basin has also been modified by mud volcanism (Skinner and Tanaka, 2007) and differs significantly in depth and degradation state from the deeper and fresher-appearing Hellas Basin despite their similar ages, most likely due to the extensive sedimentary and volcanic infilling history of the northern lowlands.

Figure 5.26 Diot et al. (2014) identified concentric linear bands on the floor of Hellas Basin, which no doubt reflect the complex fluvial and glacial history of basin modification. CTX image P17_007768_1371. Illumination is from the left. Image width is ~20 km and is centered at 42.35°S, 52.44°E.

5.6 SUMMARY

Eroded and modified impact craters on Mars can show a wide diversity of morphologies that sometimes make it difficult to infer the original geometry. Crater infill and erosion of the rim crest are commonly associated with craters of greater age. Depending upon the local geographic setting, infilling by eolian materials and/or younger lava flows may also occur. Particularly at the mid- to high latitudes, craters display strong evidence of modification by glacial processes on the Sunward facing inner walls. Several landforms remain enigmatic, as demonstrated by the extreme height and friable nature of the upper reaches of Mt. Sharp within Gale Crater.

The interior features of Martian craters appear to be common to most other craters of a similar size across the planet (polar craters excepted). Not so the exterior features on the ejecta blankets of fresh and modified craters. Indeed, the apparent diversity of fluidization of the ejecta led to decades of morphologic analysis of these features. In Chapter 6, we review the three dominant ejecta morphologies. Aspects of some of the modified ejecta blankets are then explored in Chapter 7.

6

Mars Crater Exterior Features – Major Types of Ejecta Blanket

6.1 INTRODUCTION

Although detailed studies of *Mariner* 9 images by Head and Roth (1976) had hinted at the unusual morphology of some ejecta blankets, it was only after seeing the high-resolution images acquired by the *Viking Orbiters* that it became clear that Martian craters are surrounded by ejecta deposits unlike those found elsewhere in the Solar System (Carr et al., 1977). Typically, on the planets, ejecta were thrown from the crater along ballistic trajectories, and upon impacting the surface, the movement of ejecta across the surface is dominated by friction between particles (Chapter 1). This produces hummocky deposits that rapidly thin outward to ~1 crater radius from the rim where they become discontinuous and eventually break up into clusters, strings, and chains of secondary craters. In contrast, the *Viking Orbiter* images showed evidence that upon striking the surface, Martian ejecta were emplaced as ground-hugging flows potentially similar to that typical of terrestrial geophysical flows such as long-runout landslides, debris flows, or pyroclastic density currents (Carr et al., 1977; Mouginis-Mark, 1979a, 1981; Strom et al., 1992; Barlow et al., 2000; Barnouin-Jha et al., 2005; Baratoux et al., 2005). Most researchers now think that this fluidization is caused by the inclusion of volatiles (e.g., water or ice) in the ejecta that act to reduce friction between particles. In addition, ejecta show subtle morphologic differences that may be correlated with size and location (e.g., latitude, elevation, and the geology; Mouginis-Mark, 1979a). As discussed in Chapter 3, this is consistent with the inference that volatiles on Mars are not distributed uniformly in space and time across the planet.

The morphologic characteristics of ejecta deposits are the basis for a standardized nomenclature system used to investigate Martian craters (Mouginis-Mark, 1979a; Barlow et al., 2000; Robbins and Hynek, 2012). The simplest of the systems developed for this purpose, and the one adopted in this book, was developed by the Mars Crater Consortium (Barlow et al., 2000). The types of craters whose ejecta show evidence of fluid flow are termed "layered ejecta craters" because their ejecta deposits occur as one or more thin sheets of continuous ejecta that extend radially outward around the crater (Barlow et al., 2000). Besides layered ejecta deposits, there are additional types of ejecta deposits that show little or no evidence of fluidization. Some of these ejecta deposits are like those on other planets (see Chapter 2), but some are unique to Mars, including radial, pancake, and pedestal ejecta deposits. These ejecta types may in part be the result of effects of the local Martian environment (e.g., repeated climate changes and removal/addition of glacial material) that modifies primary morphologic characteristics after the craters form (see Chapter 3). In addition, a few craters appear to have ejecta deposits that are hybrids between the different types and do not fit neatly into the systems used to aid in discussing and investigating Martian craters (Robbins and Hynek, 2012).

In this chapter, we first review the principal types of Martian ejecta morphology. We then delve more deeply into the attributes of these ejecta deposits. In Chapter 7, we will consider the other, less common, types of ejecta.

6.2 SINGLE, DOUBLE, AND MULTI-LAYERED EJECTA

The ejecta deposits of most Martian impact craters ranging from ~3 km to ~100 km in diameter show evidence of fluidization of their ejecta because the material traveled across the surface. Although most fluidized ejecta are superficially similar, there are differences that hint at multiple physical processes acting during emplacement. To a first order, the exterior deposits are classified according to the number of concentric "layers" or sheets of ejecta that surround the crater. These include "single-layered ejecta" (SLE), "double-layered ejecta" (DLE), and "multi-layered ejecta" (MLE) craters with a lobate perimeter possessing a distal ridge or rampart (Figure 6.1) (Barlow et al., 2000). Morphologically, the individual ejecta layers of SLE and MLE craters appear quite similar, while ejecta from DLE craters are different. Building on the initial classification of these three crater types, Barlow (2015) recognized that there are two distinct morphologic types of DLE craters, one termed "Type-1 DLE" craters that are the type originally defined as DLE craters by Mouginis-Mark (1979a) and Boyce and Mouginis-Mark (2006), and a second type termed "Type-2 DLE" craters that are morphologically similar to MLE craters but lack the topographic attributes of Type-1 DLEs (Boyce and Mouginis-Mark, 2021).

As discussed in Chapter 3 the "fluidized" appearance of the layered ejecta has been attributed to the incorporation of subsurface volatiles in the ejecta curtain (Carr et al., 1977; and many other subsequent studies), turbulent interaction of the ejecta curtain with the Martian atmosphere (Schultz and Gault, 1979), a combination of these two processes (Barlow, 2005), or mass movement/granular flow (Barnouin-Jha et al., 2005). The maximum distance of the ejecta blanket from the crater rim may be quantified through the ejecta mobility (EM) ratio (Mouginis-Mark, 1978; Barlow, 2006), which is the average extent of the ejecta blanket from the crater rim divided by the crater radius.

We will see in Section 6.4 that EM values for SLE craters in the northern latitudes range between 0.5 and 8.2 with an average value of 1.5. No statistically significant variation can be recognized for craters on different geological units, although a slight trend toward higher EM values is seen with increasing latitude (Barlow, 2006). For MLE craters, the EM value is calculated using this outermost layer and typically ranges from 1.2 to 5.4, but the average values increase from 2.5 within the 0–30°N latitude zone to >3.5 at latitudes >40°N. In her analysis of Mars craters, Barlow (2006) found that 506 of the 3,279 craters that she studied had MLE ejecta morphologies, and that these craters ranged in diameter from 5.6 to 90.7 km, but are most common at diameters >10 km.

Figure 6.1a Example of a single-layered ejecta (SLE) crater 11.7 km in diameter located at 13.29°N, 82.94°E. This crater shows no signs of multiple ejecta layers, placing it in a different category compared to MLE craters. CTX frame P22_009797_1934.

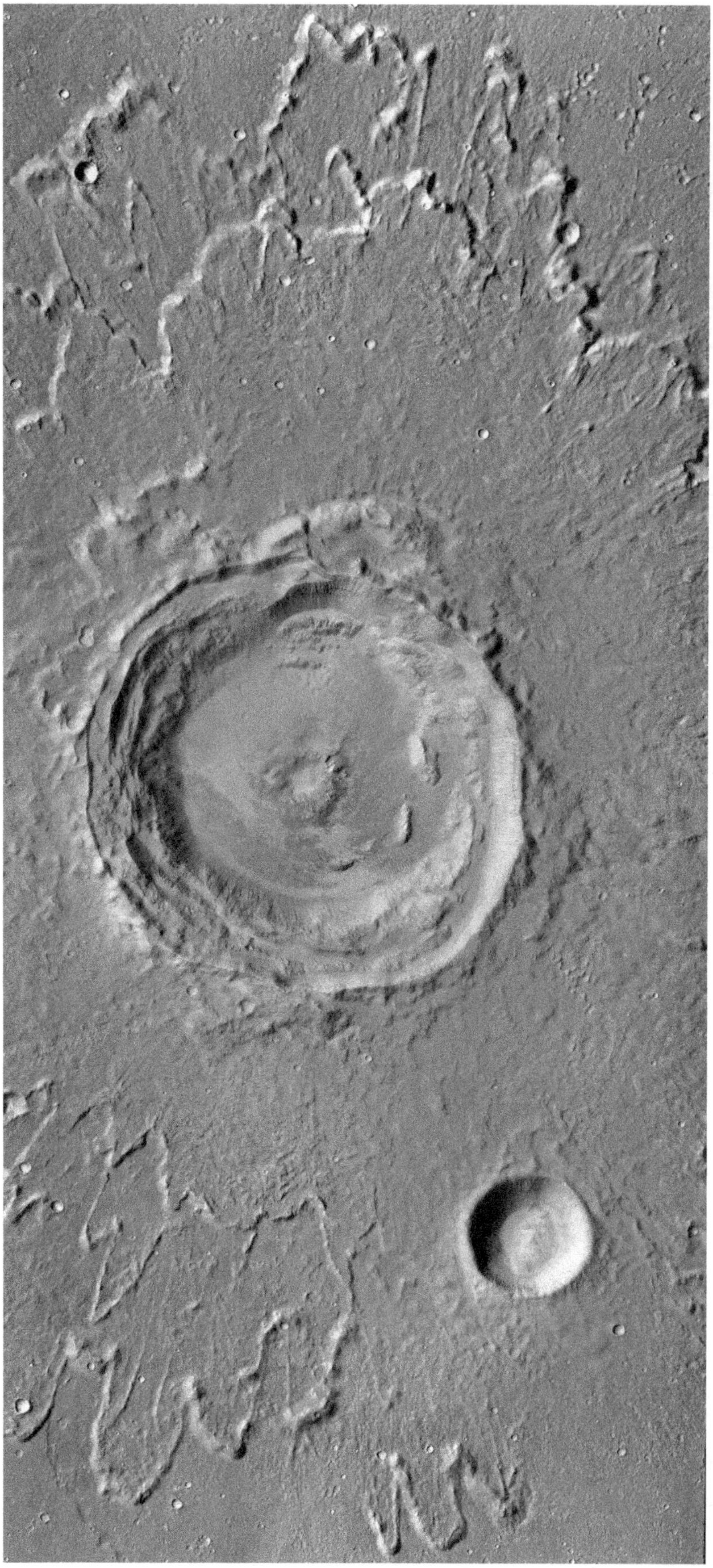

Figure 6.1b A 16.8 km diameter multilayered ejecta (MLE) crater, located at 19.92°N, 163.56°E. Several layers of ejecta exist here, each with a distal rampart, which distinguishes this crater type from an SLE crater. CTX image J02_045503_2001.

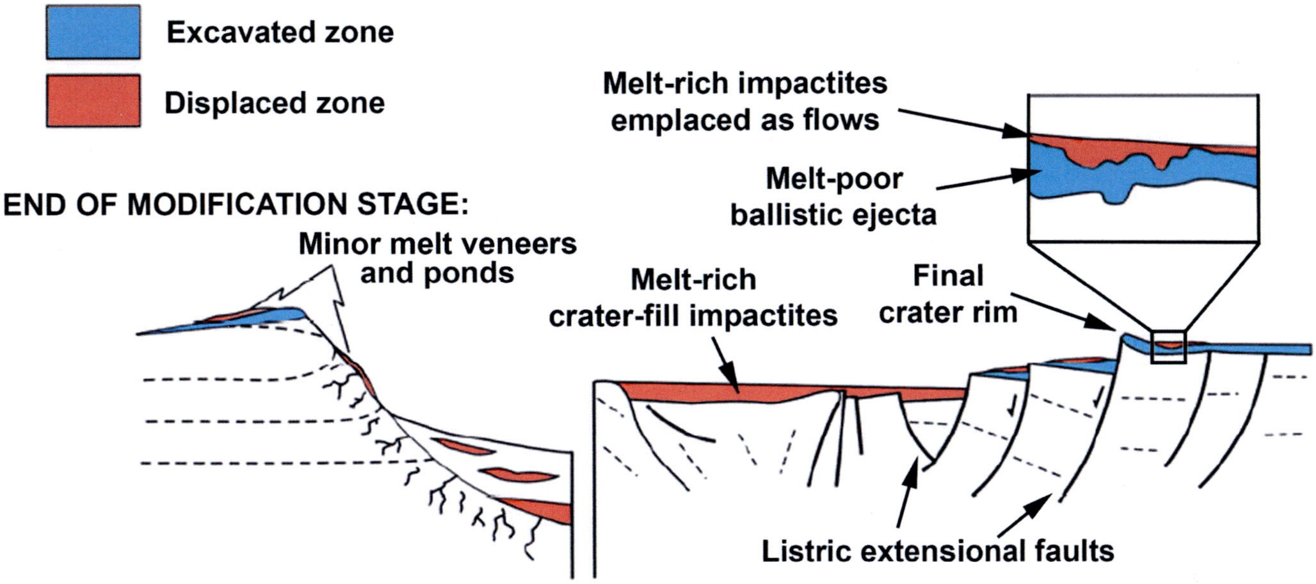

Figure 1.2 Representations of the final cross-section of simple craters (left) and complex craters (right), illustrating the relative positions of displaced and excavated materials. Reprinted from Osinski et al. (2011). "Impact ejecta emplacement on terrestrial planets," *Earth Planetary Science Letters*, vol. 310, pp. 167–181, Figure 7. With permission from Elsevier.

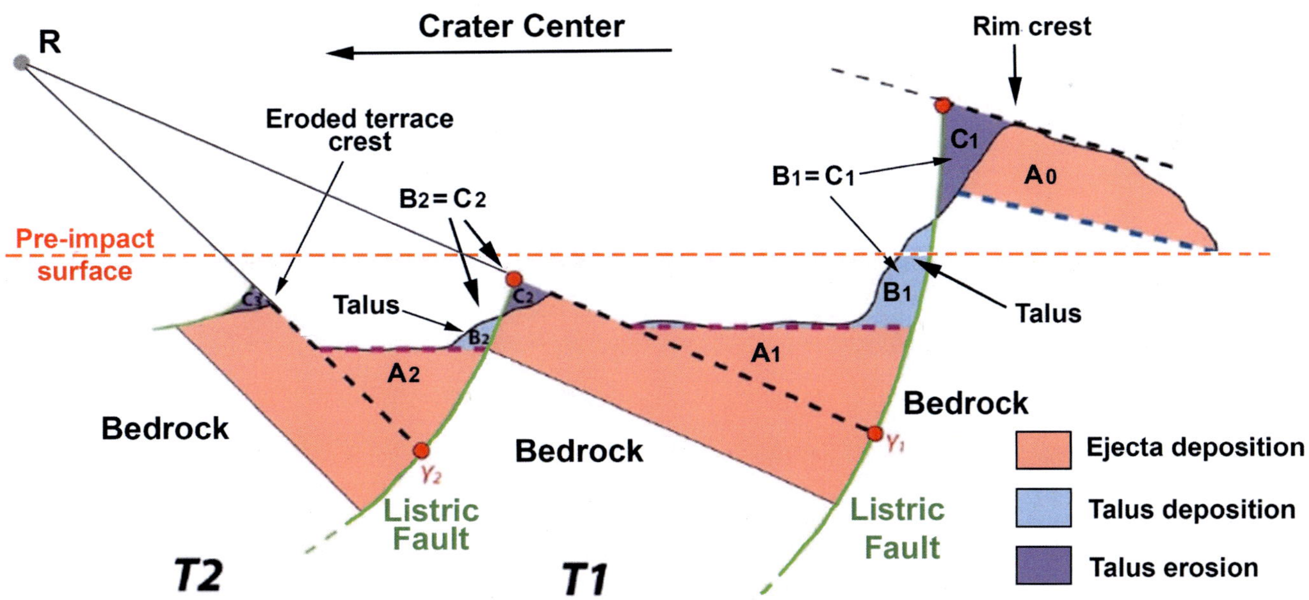

Figure 1.6 Illustration of how terrace blocks (T1 and T2) may form along listric faults and then may be eroded in a complex crater. Talus eroded from exposed blocks were rotated along these faults (C1 and C2), and then redeposited at the base of the blocks (B1 and B2). This diagram illustrates that one might find ejecta deposits (impact breccia and impact melt) at multiple places within the interiors of complex craters. Reprinted from Sturm et al. (2016) "Ejecta thickness and structural rim uplift measurements of Martian impact craters: Implications for the rim formation of complex impact craters," *Journal of Geophysical Research: Planets*, vol. 121, pp. 1026–1053, Figure 9. With permission of John Wiley and Sons, Inc.

Figure 2.2 Field photographs of Meteor crater, Arizona. See Figure 2.1 for image locations. Top left is a view of the crater's northern rim crest seen from the approach road. Top right shows the interior of crater. The crater is ~175 m deep. The bottom left image illustrates the inner southern wall. The bottom right illustrates one of the largest ejecta blocks on the SE rim. All photos by the authors.

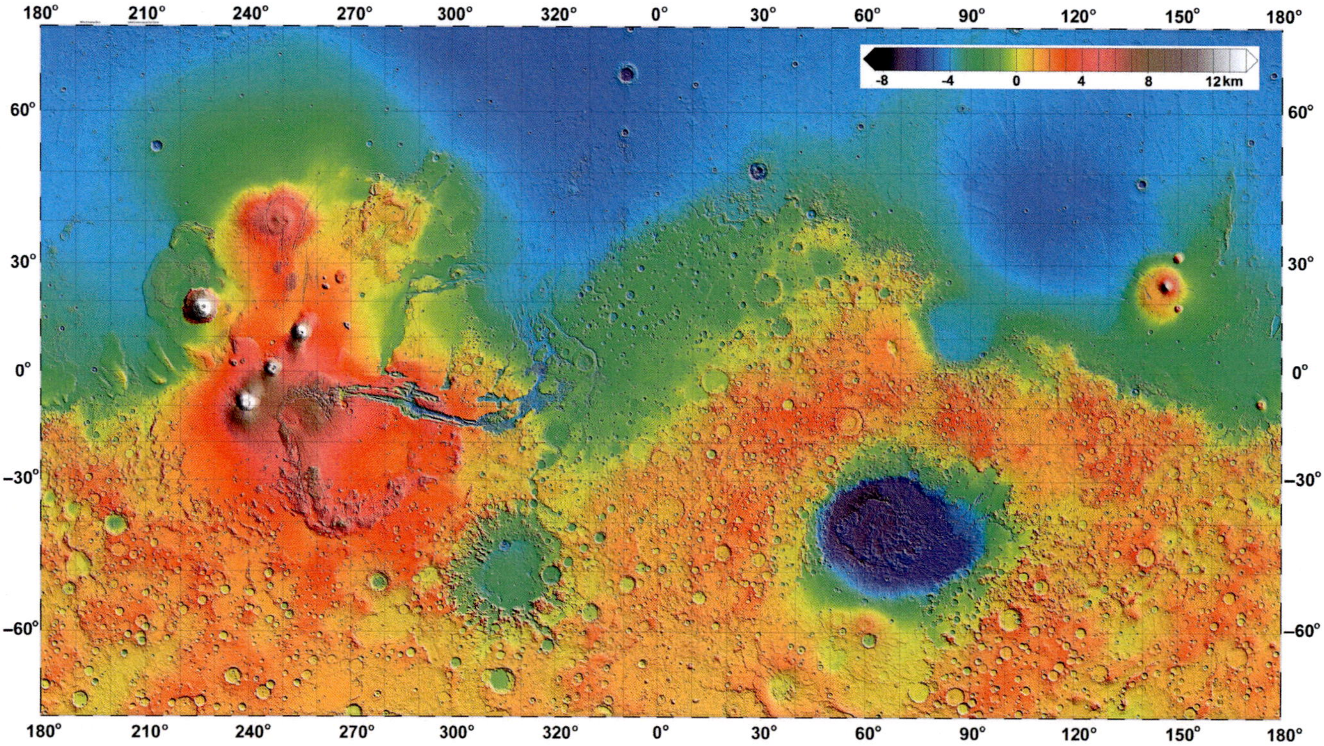

Figure 3.2 MOLA topography of Mars. The color bar at the top right shows the range of elevations from minus 8 to plus 12 km relative to the datum (Smith et al., 2001). Prominent here is the hemispheric difference between the low northern hemisphere and the high southern hemisphere. Hellas Basin is the deep depression at lower right. The Tharsis volcanoes form the high point (in white) to the left of center. The summits of the large volcanoes rise to elevations of ~20 km above the datum.

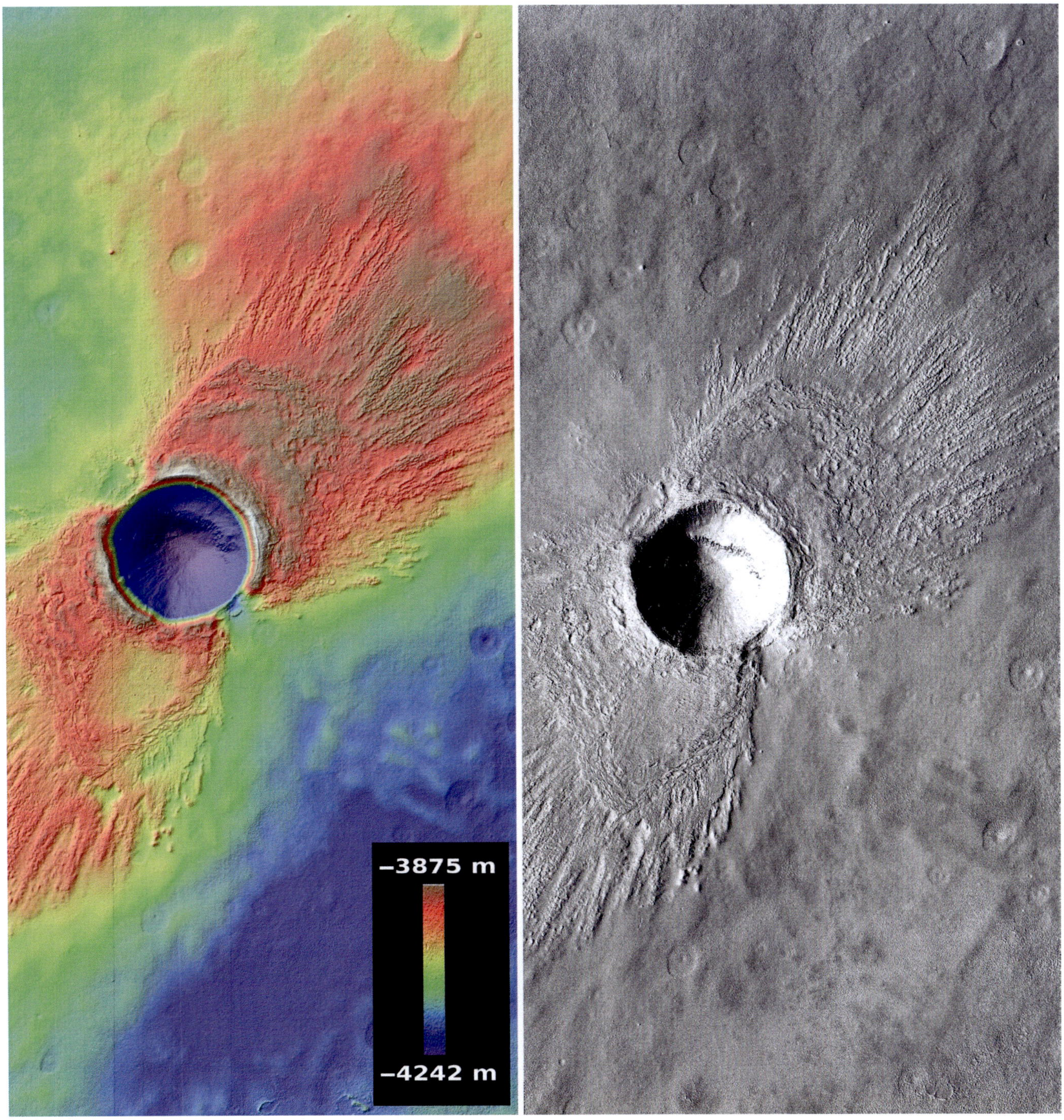

Figure 3.3 Topographic data for a spectacular small (~900 m diameter) crater within the northern plains (38.15°N, 191.58°E), generated from stereo HiRISE images. The elevation range is from −4,242 to −3,875 m relative to the datum and clearly shows the asymmetric distribution of the ejecta. Left: HiRISE image DTEEC_025735_2185_025801_2185. Right: A monoscopic view of the same crater, showing the prominent partially eroded continuous ejecta blanket. HiRISE image ESP_025735_2185.

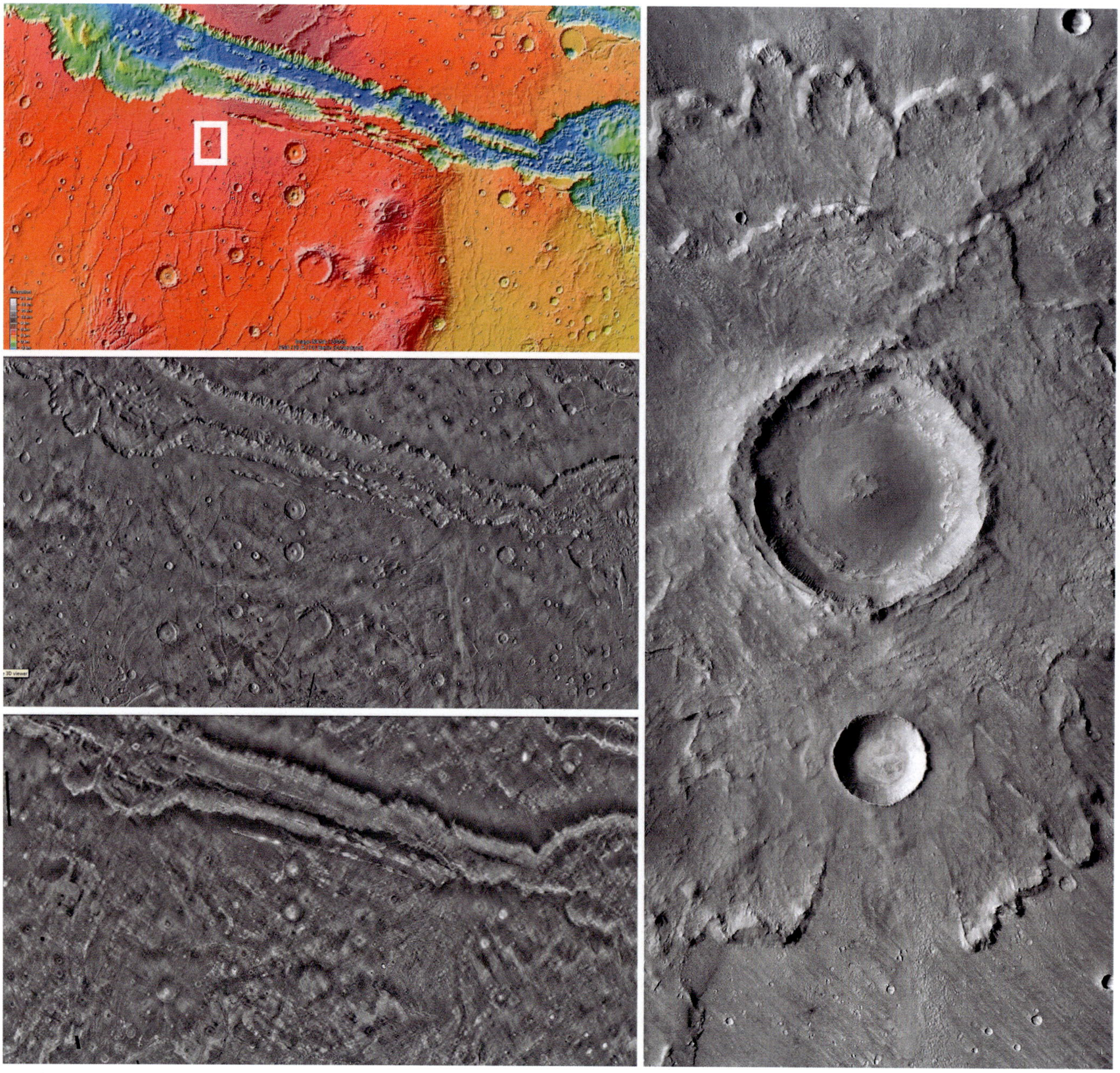

Figure 3.4 Comparison of four global data sets used extensively in this book. The top left image is a colorized rendition of the MOLA topography (color scale is the same as Figure 3.2). White box outlines coverage at right. Middle left is a THEMIS daytime IR image, and bottom left is THEMIS nighttime TIR image. Each image displays part of Thaumasia Planum and the eastern part of Valles Marineris (12.80°S–24.53°S, 282.48°E–309.30°E). At right is an example of a CTX image (frame G05_020312_1642). CTX provides near-global coverage at ~6 m/pixel resolution. The larger crater is 15.6 km in diameter and is located at 15.62°S, 292.41°E.

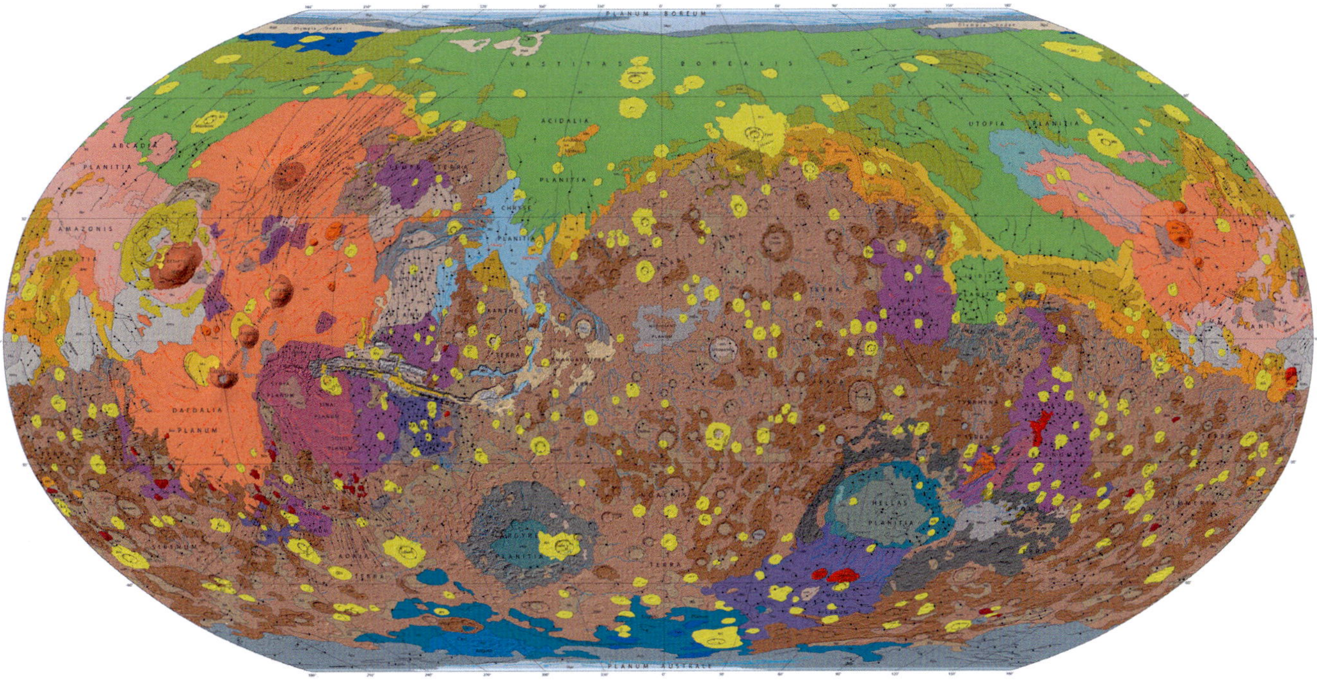

Figure 3.5 Tanaka et al. (2014) geologic map of Mars created at a scale of 1:20M using the Robinson pseudocylindrical projection. Of particular importance for the study of impact craters is the subdivision between the old Noachian-aged highland to the south (brown units) and the relatively young northern plains (green unit). Young volcanics in the Tharsis and Elysium regions are shown in orange and red. The younger craters are shown as yellow dots. Compare this map to the topographic data displayed in Figure 3.2.

Figure 3.12 A recently formed 20 m diameter crater on Mars. Repeat coverage with satellite images confirms that this crater formed during the past decade. Crater located at 44.20°N, 164.20°E. The white ejecta is water ice, which must be within a few meters of the surface at the present time. HiRISE image ESP_016954_2245.

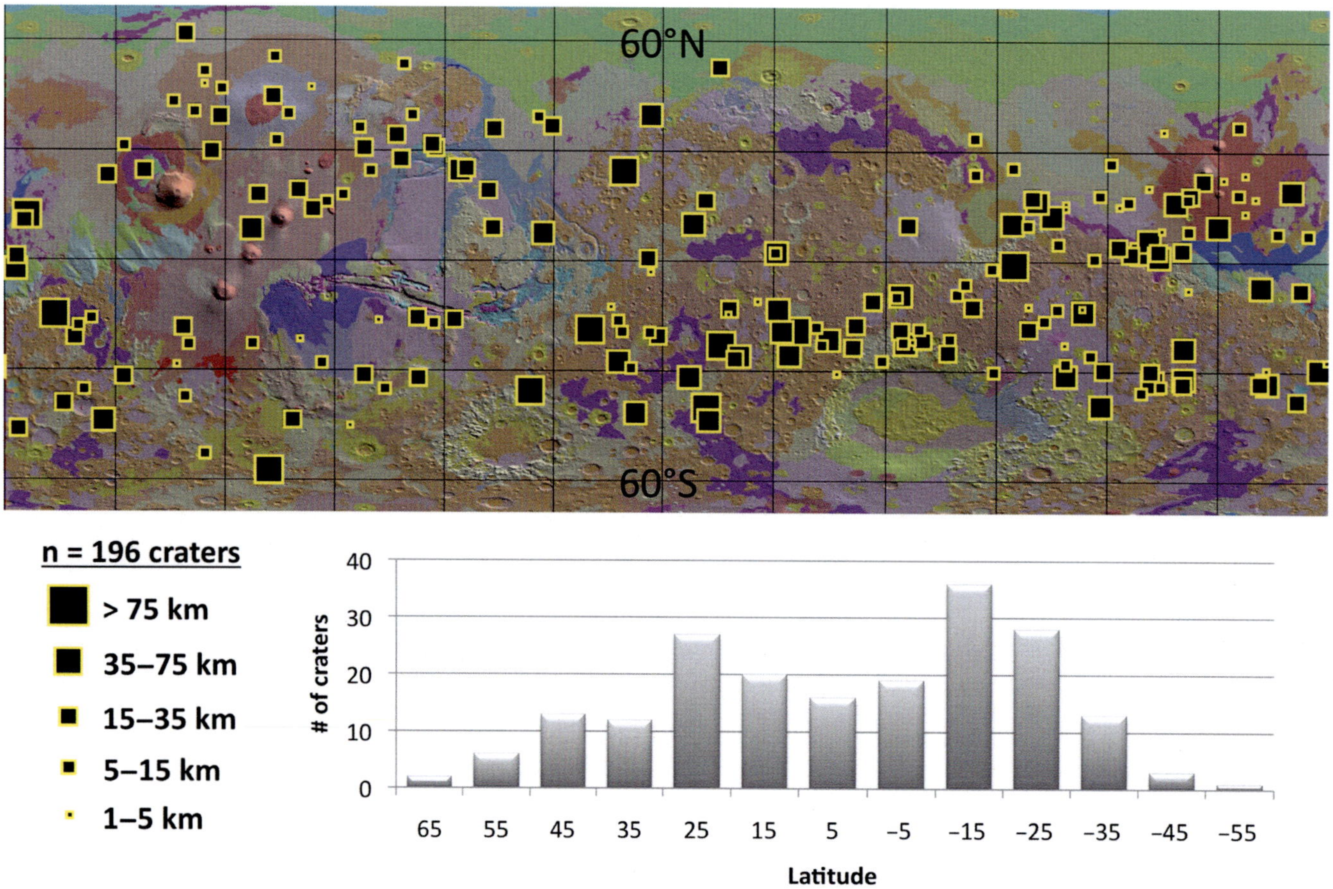

Figure 4.15 Global distribution of all pitted material-bearing craters identified by Tornabene et al. (2012), "Widespread crater-related pitted materials on Mars: Further evidence for the role of target volatiles during the impact process," *Icarus,* vol. 220, pp. 348–368, Figure 2. With permission from Elsevier. Craters range from ~1 to 150 km in diameter and show no obvious correlation with latitude or with target material. The interpretation is, therefore, that these pitted craters are the randomly distributed sample of the most recent large craters to form on Mars. If water is required to exist within the target material in order for the pits to form, then this global distribution implies that water was globally distributed in the relatively recent past.

Figure 5.8 Megabreccia exposed on the floor of Toro crater (43 km diameter) in Syrtis Major, located at 17.04°N, 71.81°E. Some of the larger individual blocks are identified by "a." Notice the irregular sharp-edged shape of these blocks as well as their diverse colors, suggesting that they experienced extensive fragmentation and came from different parts of the target material. HiRISE image PSP_005842_1970_MRGB. Image width is ~500 m.

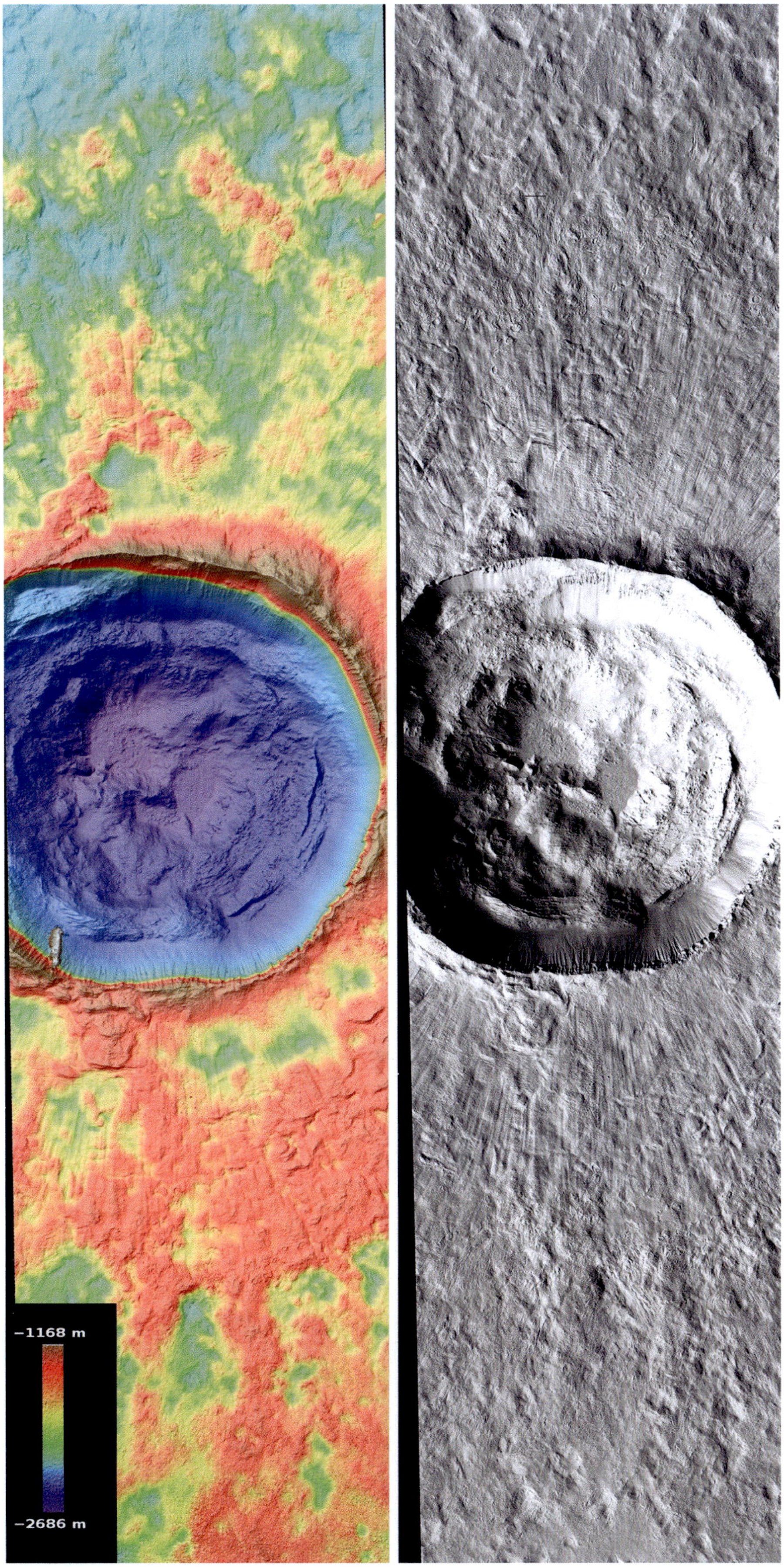

Figure 6.2 Topography of a fresh unnamed Type-1 DLE crater (6.5 km diameter 43.25°N, 225.78°E) in Northern Plains reveals the existence of a "moat" or topographic low close to the crater rim and a relatively high rampart on the edge of the inner ejecta layer. At left is a topographic image derived from stereo HiRISE images. The elevation range is from −2,686 m to −1,168 m relative to the datum. HiRISE image DTEEC_028714_2235_020156_2235. At right is one of the image pairs showing details of the crater. HiRISE image ESP_028714_2235.

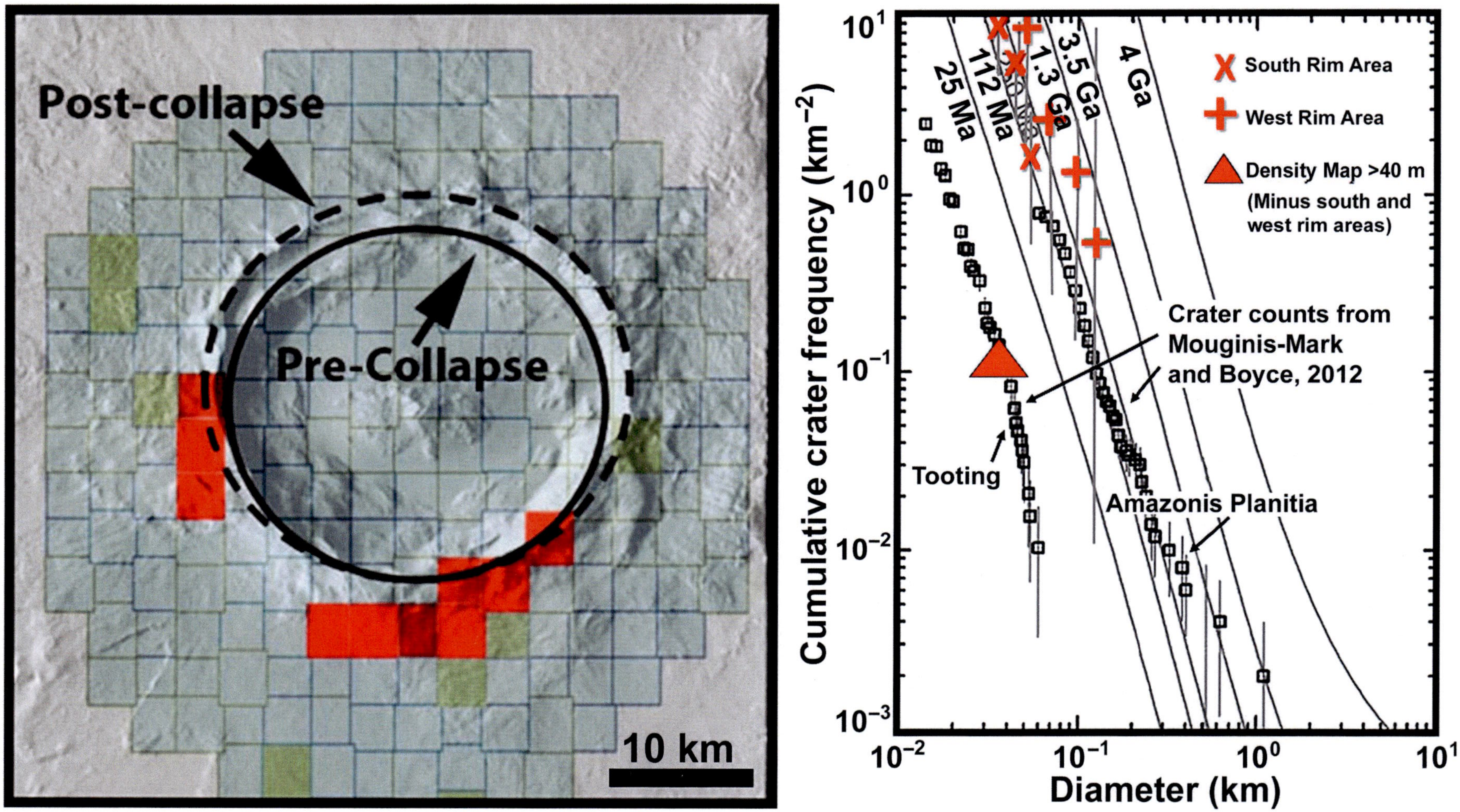

Figure 7.10 (a): Number density of craters >40 m diameter on the rim and outer wall of Tooting crater out to ~0.5 R from the crater rim. Each cell is ~3 km × 3 km, with the color of each cell a function of the number of craters in that cell. Light blue cells contain 2 ± 2 craters/cell, light green from 6 ± 1 crater/cell, burgundy from 9 ± 1 craters/cell, and red >11 craters/cell. The solid black circle is the approximate location of the rim before wall collapse and the formation of terraces. The dashed circle is the approximate location of the current rim. (b): Cumulative size frequency distribution of the craters on Tooting ejecta, the Amazonian plains surrounding Tooting, and the densely cratered areas on the south (red X) and west (red +) rim of Tooting. The cumulative number of craters >40 m diameter (normalized to 1 km^2) in the light blue and green cells is consistent with the density of craters on Tooting ejecta. The cumulative number of craters in the burgundy and red cells is consistent with the density of the fallback craters in this figure.

Figure 6.1c Type-1 double-layered ejecta (DLE) craters possess an inner and an outer continuous ejecta layer. This is Steinheim crater (11.1 km, 54.62°N, 190.63°E), which is one of the rare DLE craters with abundant secondary craters. See also Section 6.7. Mosaic of CTX images P17_007736_2349 and G21_026302_2344.

Of course, one parameter that most likely will be important, and yet is impossible to determine from image data, is the age of the crater with respect to the stage of the obliquity cycle. As we saw in Section 3.5.2, significant variations in the obliquity of Mars over geologic time could have profound effects on the distribution of near-surface (or surface) ice, which would in turn influence the mobility and EM value of the ejecta flows. Barlow (2004) attempted to determine if subsurface volatile concentrations had changed measurably over time, given the variations in its obliquity cycle (Jakosky and Carr, 1985;

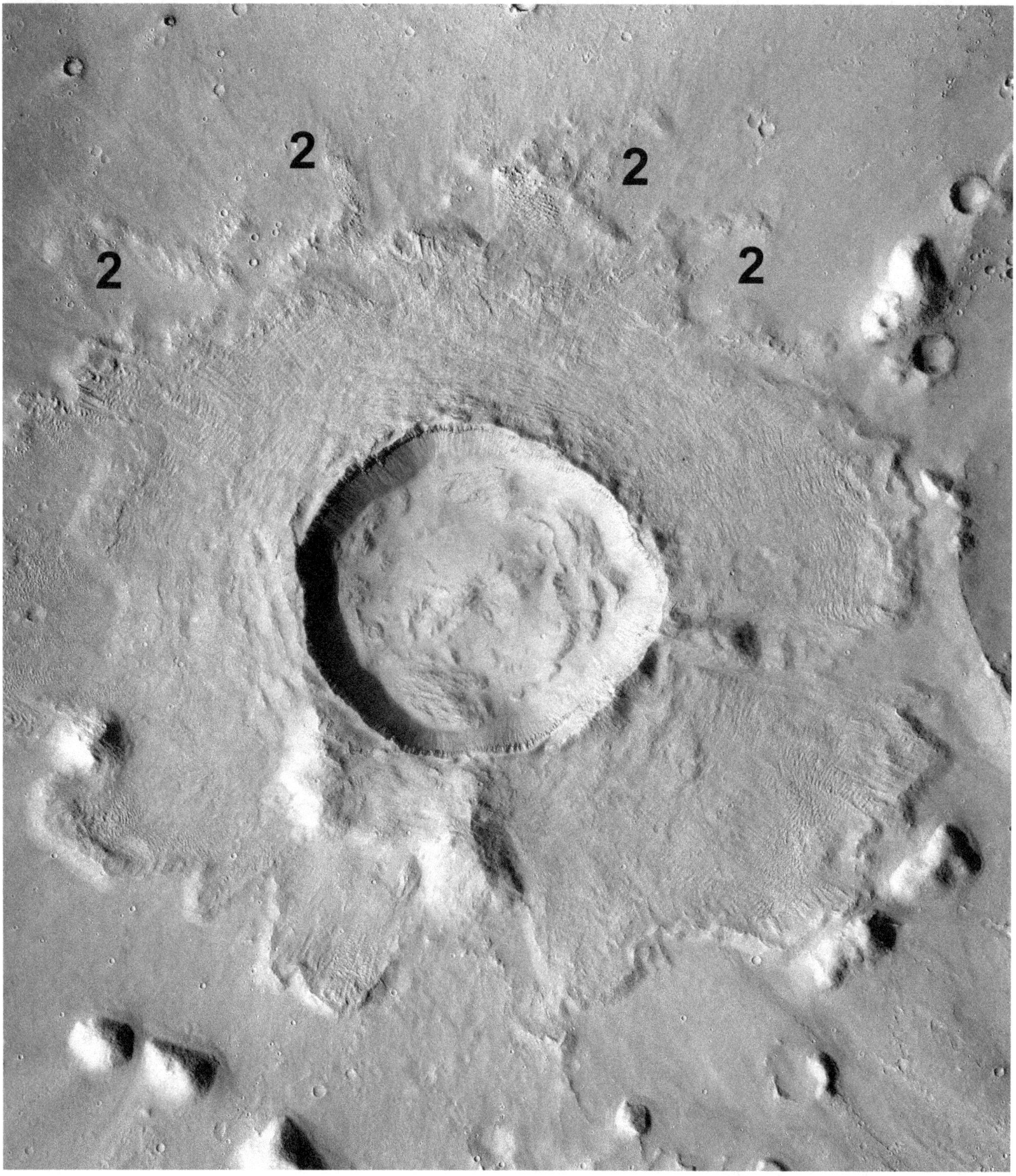

Figure 6.1d Nutak crater, an 11.2 km diameter Type-2 DLE crater, is located at 17.43°N, 329.73°E. While much of the ejecta is like that of SLE craters, the second (outer) layer is evident north of the crater and is indicated by "2." CTX image G07_020772_1967.

Jakosky et al., 1995). Given the difficulty in obtaining a precise crater age, Barlow (2015) found no strong evidence for such a temporal variation in the EM values. At least some craters of all types appear to have formed in the Amazonian (Tanaka et al., 2014; Kirchoff and Grimm, 2017). This implies that the volatiles responsible for the mobility of the ejecta must have persisted until the present day near the surface even at low latitudes (see also Section 9.5).

The majority of DLE craters are classified as Type-1 DLE craters. Steinheim crater (Figure 6.1c) and many other DLE craters possess radial grooves on the inner layer that are not deflected by topography. DLE craters were quickly recognized as being different in several ways from the SLE and MLE classes (Mouginis-Mark, 1979a, 1981). Whereas SLE and MLE craters display relatively thin, lobate ejecta layers, the DLE craters are characterized by a thick, almost circular inner layer and a thinner, more sinuous outer layer (Boyce and Mouginis-Mark, 2006). This inner layer typically has a low point or "moat" close to the crater rim with the ejecta thickness then increasing radially away from the crater until a wide rampart terminates this layer. The moat is most clearly seen in digital elevation data for Type-1 DLE craters (Figure 6.2) and, in some cases, may be tens to hundreds of meters lower than the more distal ejecta layer (Boyce and Mouginis-Mark, 2006). There is also a set of radial grooves on the inner ejecta layer very close to the outer crater rim, while the flow pattern on the outer layer is more meandering. In contrast, Type-2 DLE craters (Figure 6.1d) have two ejecta layers but are like MLE craters with the ejecta thinning rapidly from the rim. Grooves on the ejecta blanket of Type-2 DLE craters are deflected by topography and stop abruptly at the layer edge (see Section 6.5). The perimeter of Type-2 DLE craters is quite similar to the distal rampart seen around MLE craters. Sturm et al. (2013) suggested that the Ries crater in Germany may be a good terrestrial analog to Type-1 DLE craters, with a massive and continuous rampart structure at 1.45–2.12 crater radii from the crater center and a depression closer to the crater rim.

The relative number of layers around a crater is partly a function of its size (Mouginis-Mark, 1979a; Barlow, 2006), but the primary cause is still controversial. Several models for the formation of the layers have been proposed including: (1) increased impact energy resulting in surge deposits that form the layers (Wohletz and Sheridan, 1983); (2) the role of ice in the formation of SLE and DLE craters versus liquid for MLE craters (Barlow and Bradley, 1990); (3) the creation of atmospheric vortices by the formation of the larger MLE craters (Schultz, 1992; Barnouin-Jha and Schultz, 1996; Barnouin-Jha et al., 1999a, b); and (4) projectile penetration through the cryosphere for the formation of the MLE morphology (Weiss and Head, 2014). To date, there is little consensus on the mode of formation of the different layers.

Figure 6.3 displays the global distribution of the ejecta types within the northern hemisphere of Mars. Note that while SLE and MLE craters can be found from the Equator to near the North Pole, DLE craters occur predominately between ~30° and 65°N. All three crater types have been found in the southern hemisphere, but no complete inventory for this hemisphere has been completed to date. However, DLE craters also have a limited range of latitudes in the South, being restricted to ~40°–60°S (Barlow and Perez, 2003; Barlow, 2015). This may imply that certain attributes of the target, such as the existence of extensive layers of fine sediments, are less common here

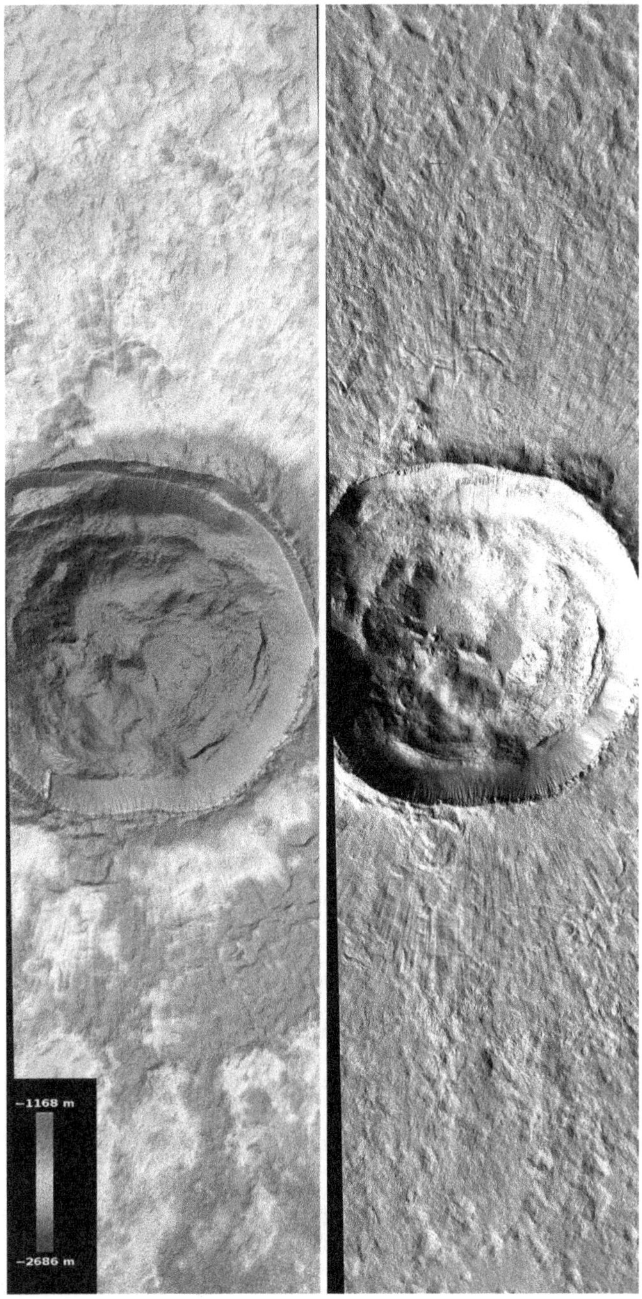

Figure 6.2 Topography of a fresh unnamed Type-1 DLE crater (6.5 km diameter 43.25°N, 225.78°E) in Northern Plains reveals the existence of a "moat" or topographic low close to the crater rim and a relatively high rampart on the edge of the inner ejecta layer. At left is a topographic image derived from stereo HiRISE images. The elevation range is from −2,686 m to −1,168 m relative to the datum. HiRISE image DTEEC_028714_2235_020156_2235. At right is one of the image pairs showing details of the crater. HiRISE image ESP_028714_2235. Note: A black and white version of this figure will appear in some formats. For the color version, please refer to the plate section.

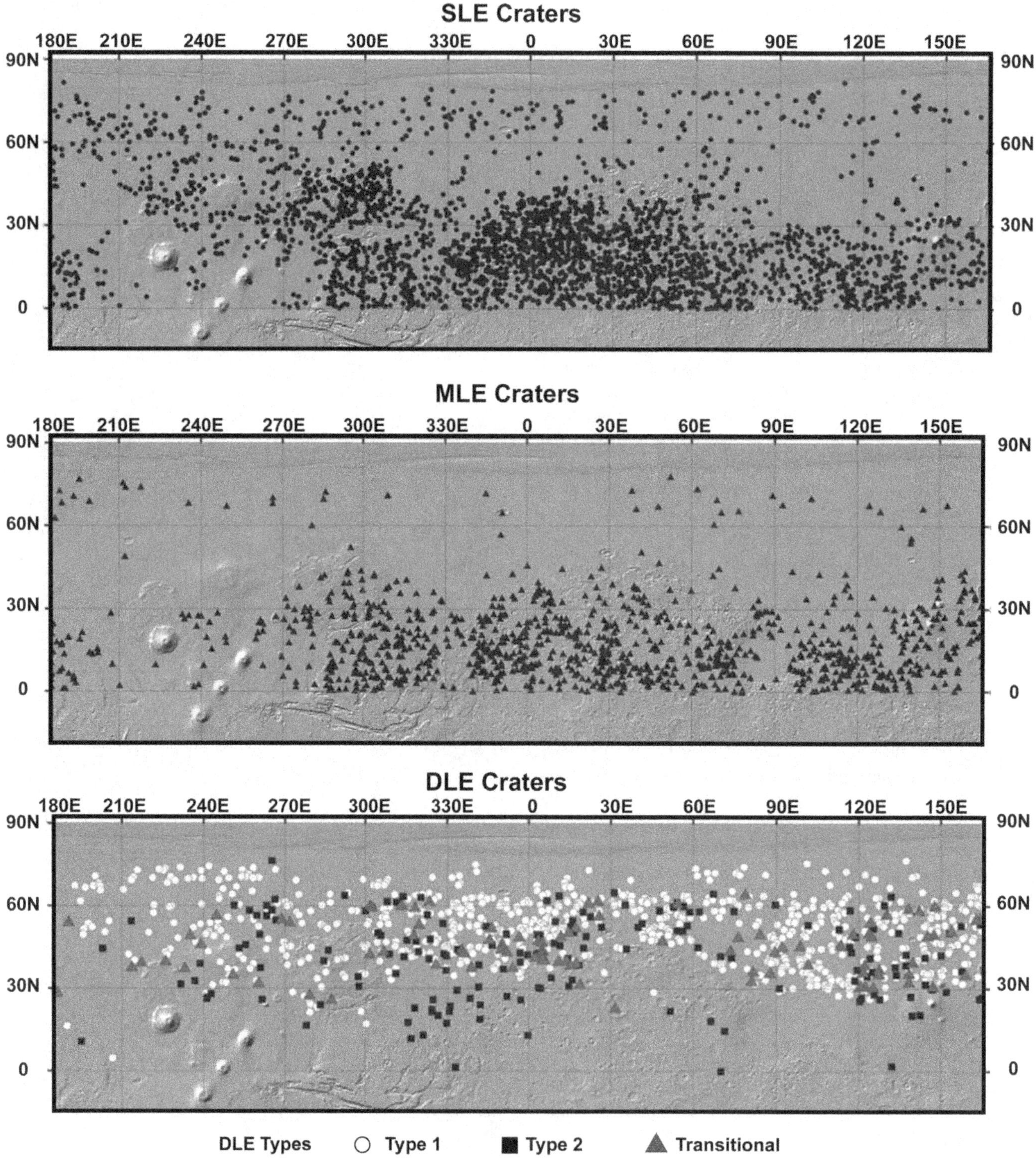

Figure 6.3 Northern hemisphere distribution of single-layered (SLE), multi-layered (MLE), and Type 1 (open circles) and Type 2 (black squares) double-layered (DLE) craters in the northern hemisphere. Notice the relative lack of MLE craters between 40° and 60°N and the almost complete absence of Type-1 DLE craters south of 30°N. Reproduced from Barlow (2015), "Characteristics of impact craters in the northern hemisphere of Mars," in Osinski and Kring (eds), *Large Meteorite Impacts and Planetary Evolution V*, Geological Society of America Special Publication 518, pp. 31–63, Figures 7 and 8. Used with permission of the Geological Society of America.

compared to the northern plains with the widespread occurrence of the Vastitas Borealis Formation (VBF).

Type-1 DLE craters often display other unique characteristics not seen at SLE or MLE craters, including the lack of distal ramparts (see Section 6.6) at the edge of the outer ejecta layer and straight radial striations on both the inner and outer ejecta layers. These striations do not mimic the direction of primary ejecta flow at low relief obstacles and do not fan or widen at the ejecta lobes (Boyce and Mouginis-Mark, 2006). While the greatest concentrations of Type-1 DLE craters are seen in topographically low regions of Mars, which are thought to contain large amounts of fine-grained materials, Type-1 DLE craters are also found at higher elevations with higher thermal inertia values that imply more coherent target rocks (Barlow and Perez, 2003). For example, a few examples can be found on other terrain types, such as the young lava flows from the volcano Alba Mons (Figure 6.4). Furthermore, Type-1 DLE craters are most often found in the mid-northern latitudes in both hemispheres (Barlow and Perez, 2003; Barlow, 2006). This suggests that elevation and/or fine-grained surface material are not the sole contributors to the morphology. The radial grooves

Figure 6.4 While most Type-1 DLE craters are found in the northern lowlands of Mars where sedimentary deposits such as the Vastitas Borealis Formation are found, rare examples of these craters can also be found at higher elevations and on relatively fresh lava flows. This 10.2 km diameter unnamed crater on the southern flanks of Alba Mons (37.11°N, 248.84°E) is an example. This crater formed at an elevation of ~5.2 km above the datum. CTX image B08_012639_2177.

that characterize the craters may have resulted from this landslide process. Viola et al. (2017) concluded that Type-1 DLE craters contain a significant quantity of excess ice at the time of formation and that this ice may have persisted even until the present time.

Weiss and Head (2013, 2018) proposed a model for Type-1 DLE formation (Figure 6.5) that requires that surface ice, perhaps decameters thick, was present at the time of the impacts that formed the DLE craters irrespective of target rock and/or elevation. This mechanism would have significant implications for Type-1 DLE craters formed in areas not previously believed to have surface ice, such as high on the volcano Alba Mons (Figure 6.4), or at low latitudes on Elysium Planitia. Conversely, a supersonic base surge mechanism was proposed by Boyce and Mouginis-Mark (2006) to have generated the straight groove on Type-1 DLE crater ejecta. While not precluding the idea of ice-assisted sliding, the base surge model requires a fuel-coolant interaction between the ejecta and volatiles within the crater cavity. This implies the presence of more volatiles than required for the formation of pitted material reported by Tornabene et al. (2012) and Boyce et al. (2012), and yet Type-1 DLE craters do not show areas of pitted material. The identification of pitted material within Type-1 DLE craters with grooves would be a valuable observation that contradicts our current understanding of Type-1 DLE crater formation (Boyce and Mouginis-Mark, 2006; Wulf and Kenkman, 2015; Weiss and Head, 2018). To date, no such material has been identified.

Four alternative ideas for the formation of Type-1 DLE craters have been proposed and further research is needed to definitively identify the process:

1. *A layered target model* (Mouginis-Mark, 1981). The meteorite excavates through an upper volatile-poor layer into an underlying volatile-rich layer. The inner layer forms from material ejected from the upper volatile-poor layer. The deeper volatile-rich region results in material ejected at higher angles, which collapses outside the crater rim to produce a ground flow that emplaces the outer layer and carves the observed radial texture into the inner layer.
2. *Subsurface and atmospheric volatile model* (Boyce and Mouginis-Mark, 2006). The inner layer is emplaced by a combination of ballistic and ground flow processes, like the formation of SLE craters. Strong winds are generated through the interaction of the ejecta curtain with the atmosphere, causing a high-velocity outflow that emplaces the outer layer. This outflow is erosive near the crater rim where the velocity is supersonic and erodes the moat within the inner layer. The flow becomes depositional at greater radial distances once the velocity drops to subsonic speed.
3. *Layered volatiles model* (Black and Stewart, 2008). Excavation into near-surface ice-rich materials results in a lower ejection angle and higher velocity of material, producing the outer ejecta layer. Excavation into a deeper ice layer results in higher ejection angles and lower velocities, producing the inner layer.
4. *Glacial substrate model* (Weiss and Head, 2013, 2014). The impactor penetrates through a decameter-thick ice-rich surface deposit to produce the crater and emplaces the

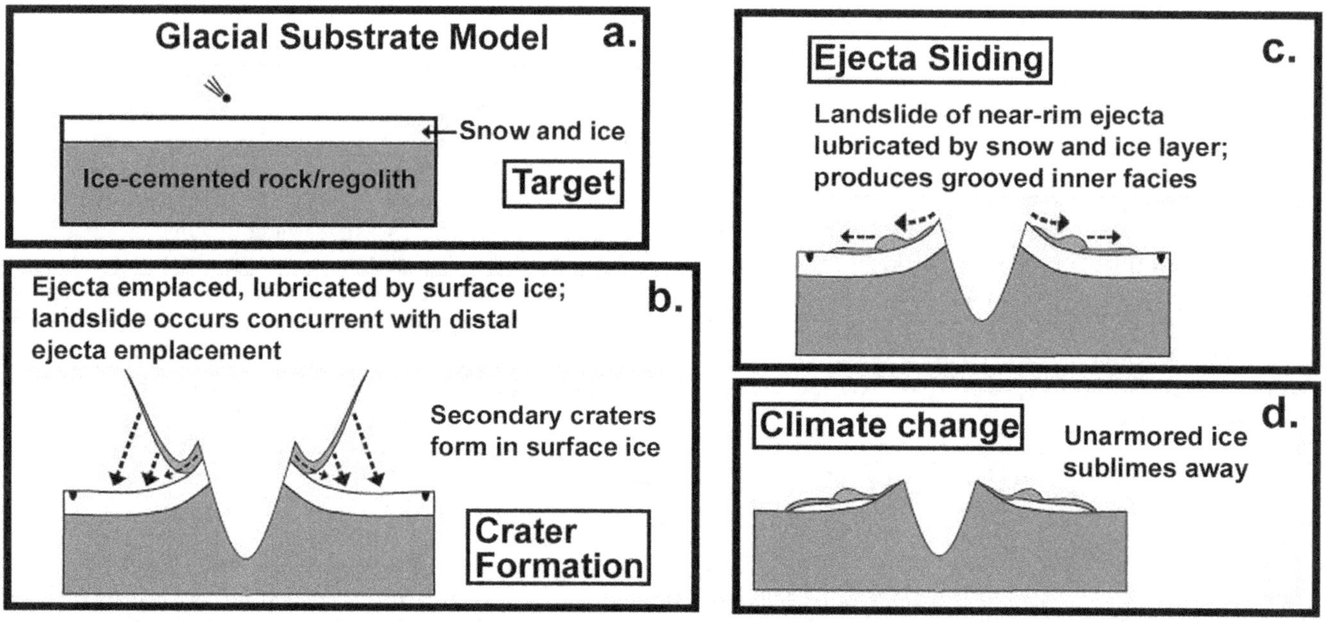

Figure 6.5 Possible four-stage model of ejecta emplacement wherein they propose that the ejecta layers of Type-1 DLE craters formed via landsliding of the ice-lubricated rim materials. The key step in the formation of the grooves on the inner layer is shown in part "C." Reprinted from Weiss and Head (2018), "Testing landslide and atmospheric effects models for the formation of double-layered ejecta craters on Mars," *Meteoritics and Planetary Science*, vol. 53, pp. 741–777, Figure 5. With permission from John Wiley and Sons.

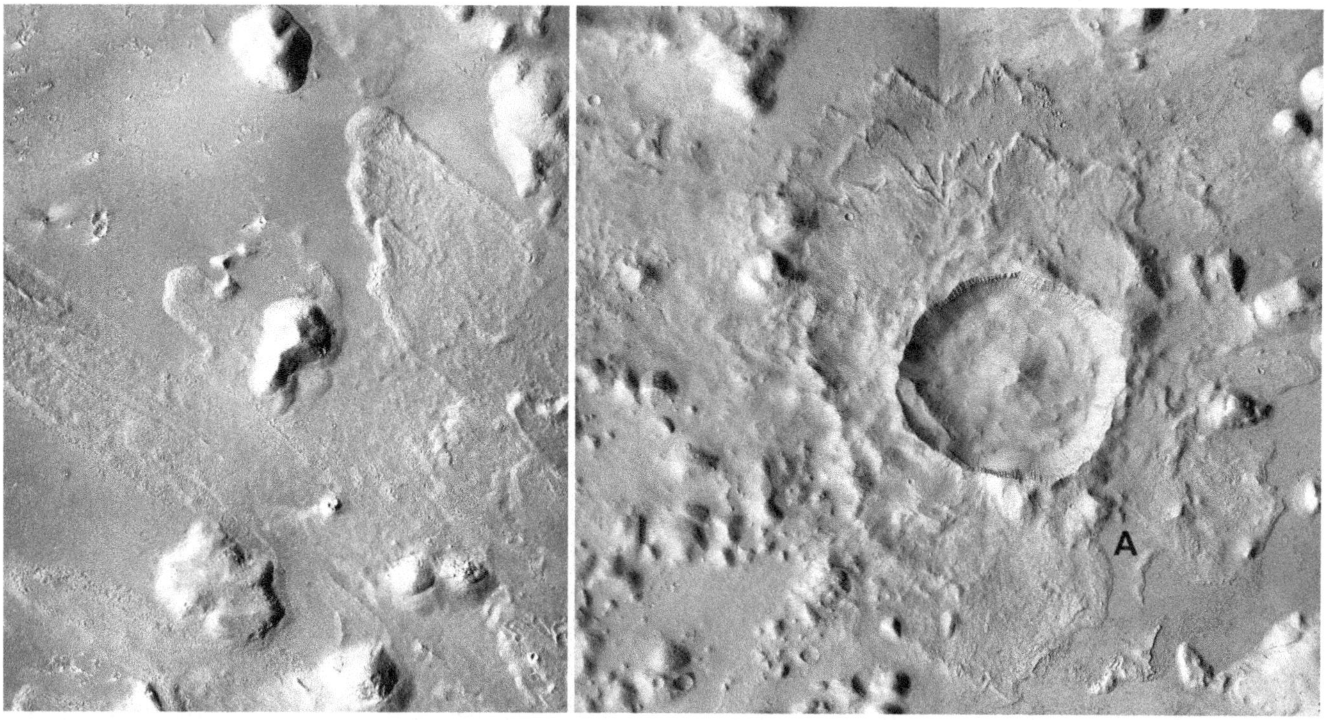

Figure 6.6 The role of pre-existing topography in controlling the path of the ejecta flow at MLE craters is shown here. Left: Outer ejecta layer from Kotka crater (19.5 km, 19.28°N, 169.86°E). The direction of ejecta flow was from lower right to top left in this image. Note the massif in the middle of the image, where a lobe of ejecta appears to have overtopped the obstacle and continued as a flow on the down-range side. CTX image B02_010519_2002. Right: An MLE crater (11.8 km diameter) located at 23.21°N, 173.43°E. Note the diversion of an ejecta layer around the pre-existing topographic obstacle at "A" to the SE of the crater rim. Mosaic of CTX images F04_037287_2039 and P14_006669_2050.

outer ejecta layer in a manner comparable to the formation of SLE deposits. Collapse of the rim due to the high percentage of ice in the near-rim materials in the later stages of ejecta emplacement subsequently produces a landslide that moves across the surficial ice deposits to form the inner ejecta layer and the grooves on the surface.

6.3 RADIAL EXTENT OF EJECTA

The radial distance reached by the ejecta has long been interpreted to be an indicator of the amount of volatiles contained within the ejecta (Mouginis-Mark, 1978; Woronow, 1981; Barlow and Pollack, 2002). As an example, the crater Kotka displays some remarkable attributes of the mobility of ejecta layers from MLE craters (Figure 6.6). To the northwest of the crater, the outer ejecta layer was emplaced within a field of isolated massifs that form part of the eroded landscape. Some parts of the ejecta had an easy passage when flowing around these massifs so that they extended a considerable distance (~3.3–4.2 crater radii) from the rim of Kotka. In other places the ejecta encountered a massif and deposited a layer of ejecta on the up-range part of the topographic high. There is even one example of a low-relief massif where an ejecta flow appears to have overtopped the obstacle and created an ejecta lobe down-range of the massif (Figure 6.6).

One way to quantify the radial extent of the ejecta is to calculate the EM. EM has been defined as the maximum range of the ejecta normalized to the crater diameter (Mouginis-Mark, 1978; Barlow, 2006). EM appears to vary with latitude, ranging from an average of 1.43 near the Equator to over 2.0 between 40° and 50°N. EM shows little variation between different terrain units, although elevation does appear to have an influence (Mouginis-Mark, 1979a). The sinuosity of the ramparts (i.e., the variability of the range of ejecta as a function of azimuth around the crater) has been quantified by a parameter called "Lobateness" (Γ), which is defined in terms of ejecta area (A) and perimeter (P) (Barlow, 1994), $\Gamma = P/(4\Pi A)^{1/2}$. Table 6.1 lists the representative values of EM for SLE, MLE, and Type-1 DLE craters. Few investigations have exploited the sinuosity of the ejecta blankets as a parameter that might provide clues to the physical properties of the ejecta. This is somewhat surprising as a comparison of values for similar crater types on various terrains might reveal useful information on the role of small-scale (meter to decameter) local topography in controlling the spatial distribution of ejecta. Sinuosity might also be related to the angle of impact of the projectile that produced the crater. A comparison between sinuosity and the distribution of secondary craters might also be informative.

6.4 INFLUENCE OF PRE-EXISTING TOPOGRAPHY ON EJECTA EMPLACEMENT

Unique insights can be gained on the emplacement process of Martian ejecta from inspection of how the ejecta layers interacted with the pre-existing topography and the target. Ejecta from all three main crater types (SLE, MLE, and Type-1 DLE) can be seen to be influenced by large topographic highs. In one instance, a Type-1 DLE crater formed within a larger impact structure causing the ejecta layers to encounter the older crater rim (Figure 6.7a). Here the outer layer was unable to rise to the pre-impact elevation, so there is no outer ejecta layer at this azimuth. But topographic obstacles do not appear to have influenced the formation of the radial grooves on the inner layer (Figure 6.7b). While the grooves did not form on top of a pre-existing mound, the grooves formed immediately as the elevation dropped to the same value as the main part of the ejecta layer. In some instances, where this surface flow encounters pre-existing topography, pressure ridges can be found up-slope from the obstacle (Figure 6.8).

Table 6.1 *Ejecta mobility (EM) and lobateness (Γ) for the three main types of ejecta morphology.*

	Av. EM	Max EM	Min EM	Av. lobate	Max lobate	Min lobate
SLE	1.53	6.60	0.20	1.10	1.00	3.57
DLE						
Inner	1.49	3.30	0.40	1.04	1.38	1.00
Outer	3.24	10.60	1.20	1.14	2.27	1.01
MLE	2.17	4.70	0.30	1.18	1.74	1.02

EM = Ejecta mobility
Lobate = lobateness
SLE = single-layered ejecta
DLE = double-layered ejecta
MLE = multi-layered ejecta
Source: Data from Barlow (2005), "A review of Martian impact crater ejecta structures and their implications for target properties," in T. Kenkmann, F. Horz, A. Deutsch, eds. *Large Meteorite Impact III*, pp. 433–442, Table 1, Geological Society of America Special Publication 384. Used with permission of the Geological Society of America.

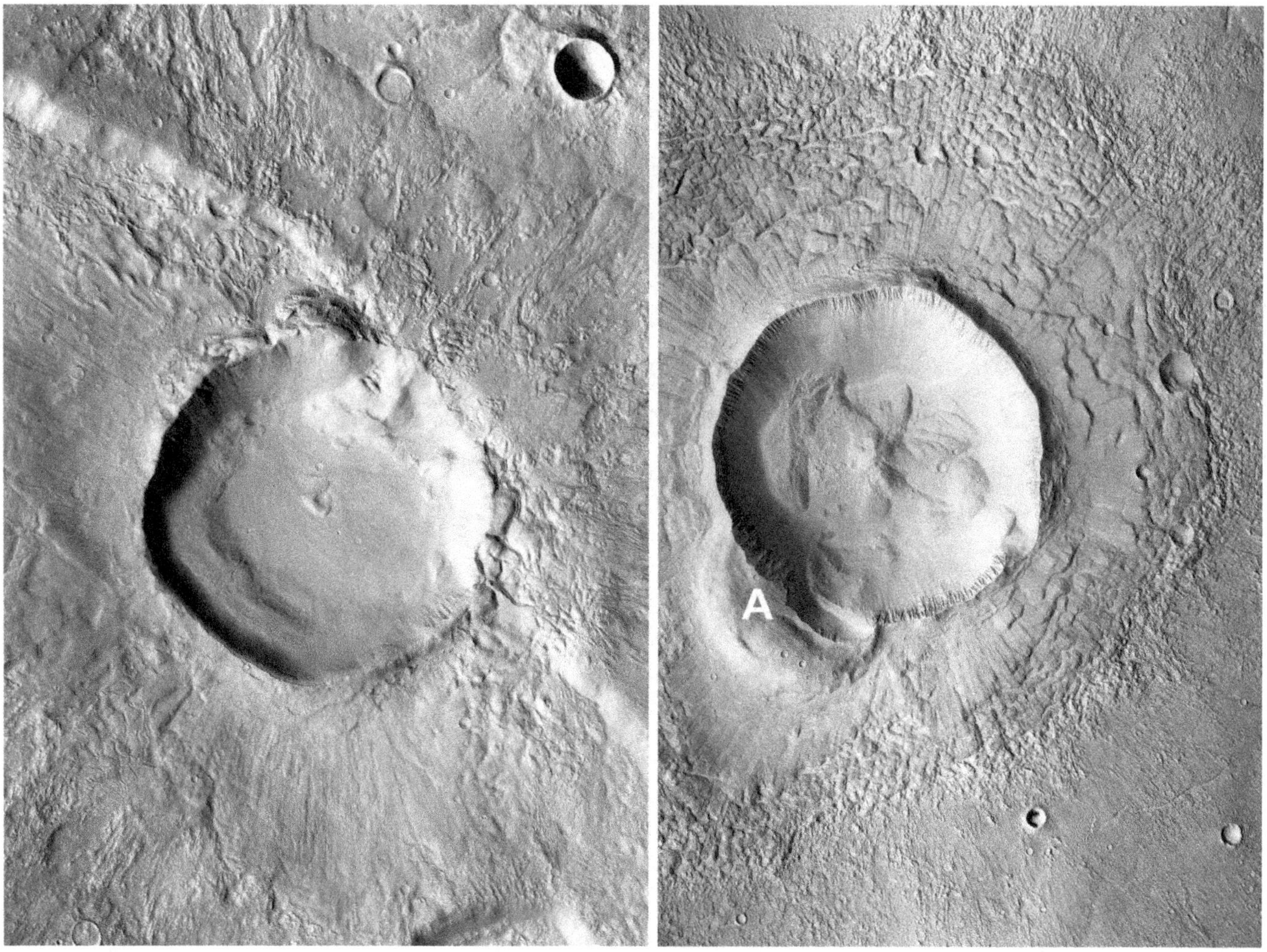

Figure 6.7 The role of pre-existing topography in controlling ejecta flow at Type-1 DLE craters. Left: Unnamed crater, 7.2 km diameter, on the rim of Timoshenko crater, 41.23°N, 295.24°E, illustrating disrupted ejecta at the rim of older pre-existing crater. CTX B18_016791_2212. Right: Unnamed crater, 8.9 km diameter, at 28.03°N, 116.67°E. Pre-existing high topography ("A") lacks the radial grooves that characterize the rest of the ejecta blanket at this radial distance from the rim. CTX image P15_006816_2100.

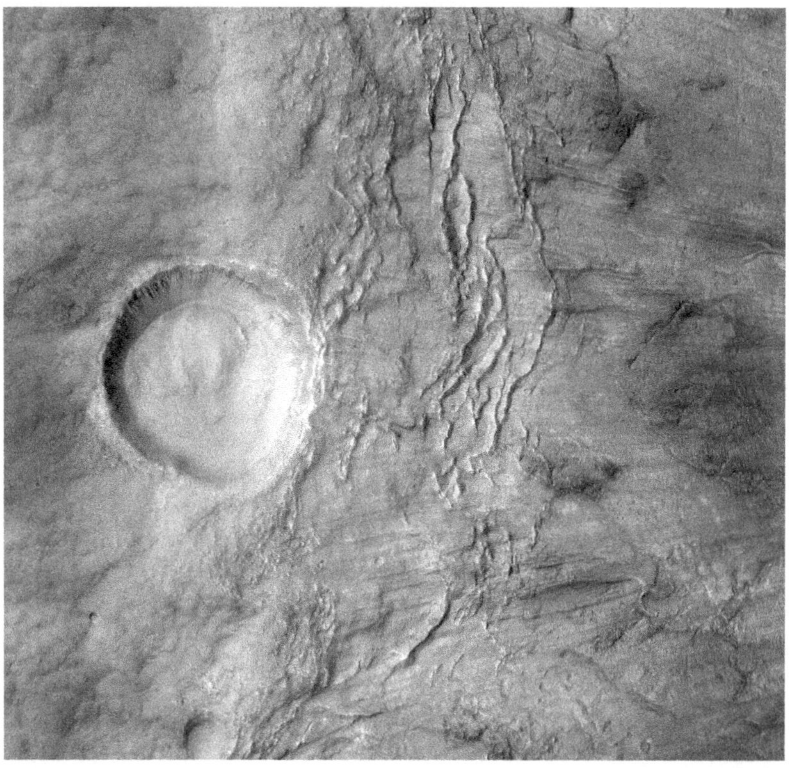

Figure 6.8 Pressure ridges on the inner ejecta layer of the Arandas crater. Image centered at 42.45°N, 344.26°E. The direction of ejecta flow was from right to left in this image. Notice that the radial grooves appear unaffected by the presence of the pressure ridges. The crater just left of the image center is ~5.2 km in diameter. CTX image B18_016776_2226.

6.5 EJECTA THICKNESS

As discussed in Chapter 4 (Section 4.4.2), Sturm et al. (2016) conducted a detailed analysis of the amount of structural uplift of the crater rim during crater formation. This enabled them to determine the thickness of the ejecta on the rim. Few previous studies have investigated the radial decay of thickness, most likely because of the highly variable topography upon which the ejecta were emplaced (Garvin and Frawley, 1998). Black and Stewart (2008) recognized that there are several hundred craters, predominately within Utopia Planitia, with anomalously high ejecta volumes compared to crater cavity volumes, varying by factors of 2.5–5.8. These craters are typically Type-1 DLE craters, although one of the examples used by Black and Stewart (2008) is an MLE crater. The reported excess volumes correspond to ejecta thicknesses of ~20–100 m averaged over the continuous ejecta blanket.

To better understand the differences in ejecta thickness for the different crater types, it is valuable to compare the typical radial profiles for the ejecta blankets of SLE, Type-1 DLE, and MLE craters (Figure 6.9). Boyce and Mouginis-Mark (2006) had recognized that the moat seen near the rim

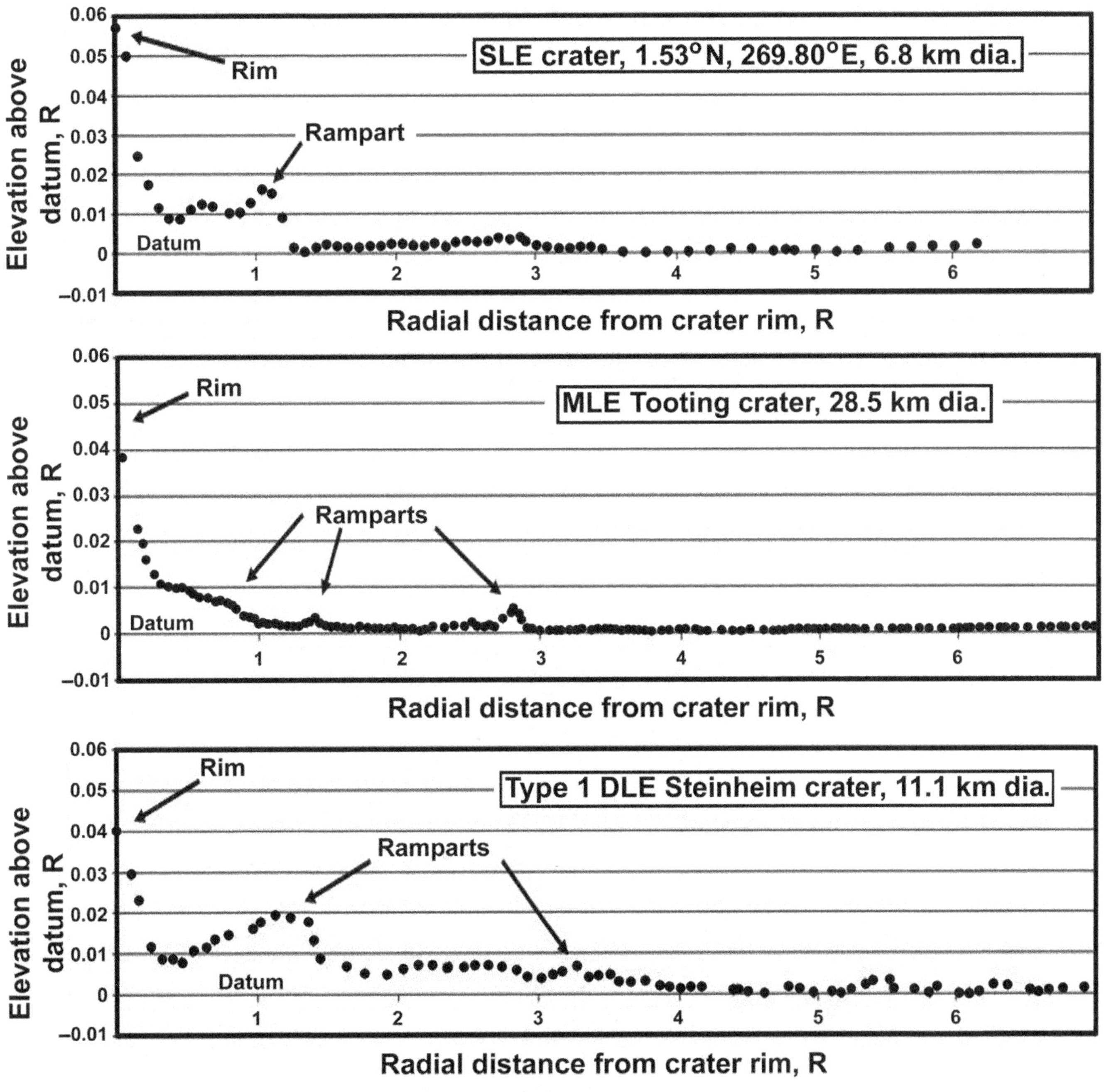

Figure 6.9 Contrasting radial profiles, normalized to crater rim diameter, for the three main crater types on Mars. The ejecta rapidly thins with increasing radial distance for both SLE and MLE craters, while thick ramparts on Type 1 DLE craters leave a "moat" or low point surrounding the outer crater wall.

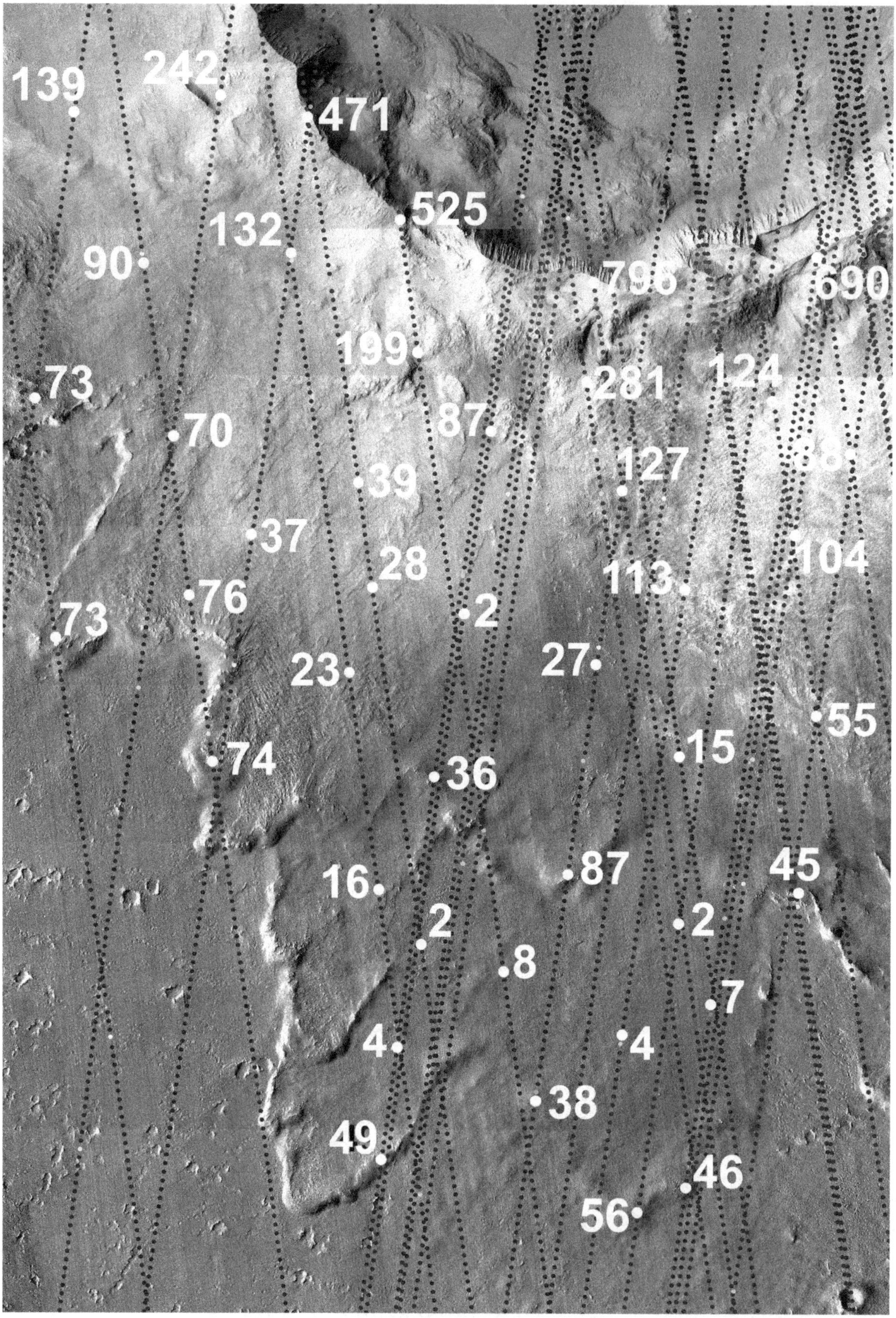

Figure 6.10 Representative thickness measurements (in meters) of the southern ejecta blanket of the 28.0 km diameter crater Tooting. Here it is assumed that the pre-existing topography was flat with an elevation of minus 3,872 m relative to the Mars datum (see Mouginis-Mark and Garbeil, 2007). Dotted black lines show the location of individual MOLA elevation measurements. These data show that most of the ejecta layer is only a few meters thick, while the distal ramparts are of the order of 50–70 m high. Image centered at 22.69°N, 207.64°E. The base image is a THEMIS VIS mosaic from the JMARS website.

of Type-1 DLE craters may be ~100–200 m deep compared to the ejecta layers further from the rim crest. While a distal rampart does exist at the edge of this inner layer, the rampart is much wider and has shallower slopes compared to the rampart of SLE and MLE craters. For SLE and MLE craters, the dominant feature is the topographic rise associated with ramparts at the leading edge of the ejecta lobes. For MLE craters, smaller ejecta layers within the perimeter of the entire deposit also display ramparts that are local topographic highs. These smaller ejecta layers do not extend around the entire perimeter of the crater.

Tooting crater (23.4°N, 207.5°E) is one of the very rare examples of a fresh crater that has both formed on an almost flat surface and is of sufficient size (28 km diameter) to have hundreds of MOLA laser shots falling upon the ejecta blanket. A total of 24,201 MOLA shots from 90 individual orbits lie on the ejecta blanket beyond the rim crest of this crater out to a radial distance of 66.5 km. Using these topographic data, Mouginis-Mark and Garbeil (2007) conducted point-by-point studies of the ejecta thickness that revealed some surprising features of the ejecta blanket. South of the crater rim (Figure 6.10) the rim is ~500 m above the surrounding terrain. But the ejecta layers thin rapidly with increasing radial distance (<10 km) with total thickness rapidly approaching ~30 m. Further from the crater, the ejecta can be remarkably thin. There are places where the ejecta layers are only 3–5 m thick at radial distances of ~11 km from the rim crest, or ~30% of the maximum radial distance of the layer in that azimuth.

6.6 DISTAL RAMPARTS

Ramparts (i.e., narrow ridges) are common at the distal edge of the layers of Martian fluidized ejecta, as well as along the margins of some longer layers (Carr et al., 1977; Barlow et al., 2000) (Figure 6.11). Indeed, early in the *Viking Orbiter-Era* analysis of Martian craters, they were often called "Rampart Craters" (Mouginis-Mark, 1978; Mutch and Woronow, 1980). Although originally hypothesized to be features formed by wind erosion (Arvidson et al., 1976), soon after the start of the *Viking Orbiter* missions, the preferred mode of formation of the ramparts of SLE and MLE craters (Carr et al., 1977; Wohletz and Sheridan, 1983) was that they are deceleration ridges, akin to the distal rampart observed on certain terrestrial landslides such as the Blackhawk slide in California, USA (Shreve, 1968). Ramparts formed at either the distal margin of the ejecta blanket or as lateral ramparts at the side of some ejecta lobes (Baloga et al., 2005; Mouginis-Mark and Baloga, 2006; Boyce and Mouginis-Mark, 2025). A further characteristic of ramparts is that some of the freshest examples possess numerous pits along the crest of the ridge on SLE and MLE craters (Figure 6.12). These pits suggest that some material was entrained within the rampart during its formation and that this material (ice?) has since been removed.

A key characteristic of the ramparts is that they are topographic highs, having greater elevation not only over the terrain surrounding the parent crater but also that they are elevated with respect to the ejecta immediately toward the crater rim. Using nighttime THEMIS TIR data, Baratoux et al. (2005) recognized that for craters in Syrtis Major, the distal ramparts had

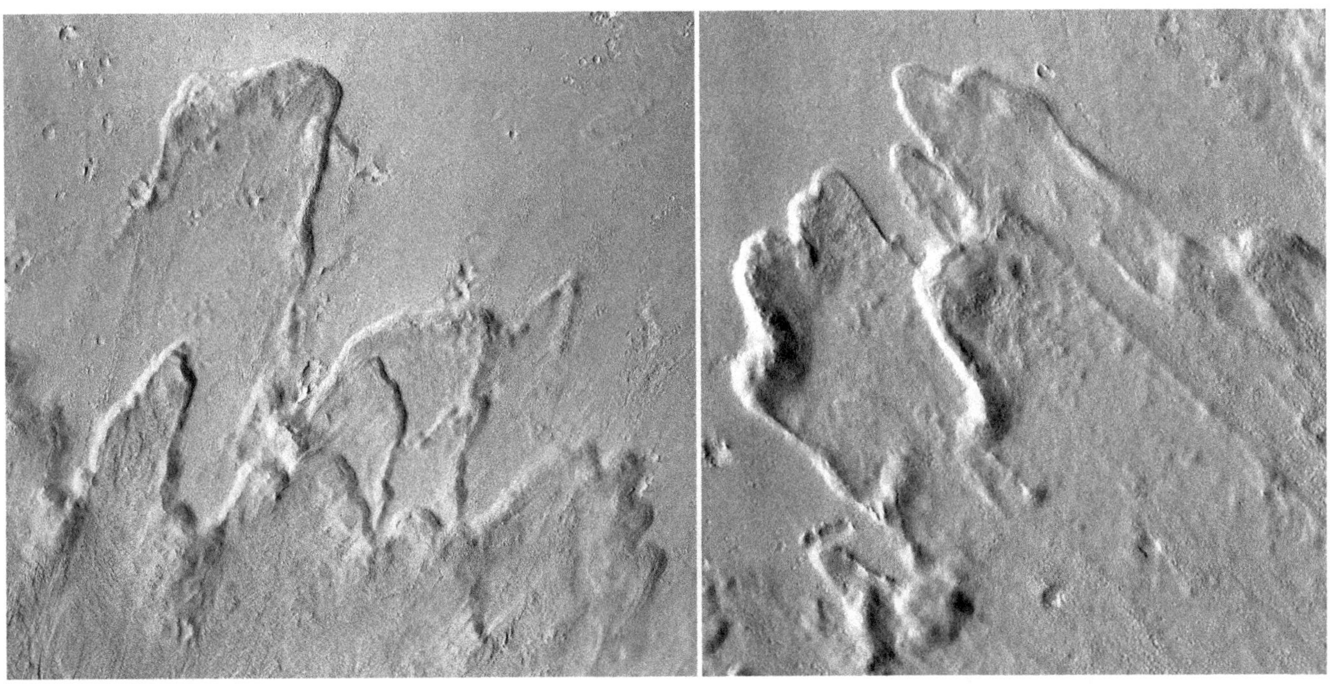

Figure 6.11 Distal and lateral ramparts are common on the ejecta blanket of Tooting crater, illustrating physical characteristics of the ejecta layers. (1) Multiple ejecta lobes showing that the rampart formed as surges of material rather than as a single flow event. The direction of the flow was from the lower left. Image width is 22 km centered at 24.12°N, 208.09°E. CTX image B17_016425_2037. (2) Several of the ejecta lobes have ramparts along the sides of the lobes in addition to the distal ramparts. The direction of the flow was toward the top left. Image width is 13 km centered at 24.04°N, 207.06°E. THEMIS image V11152006.

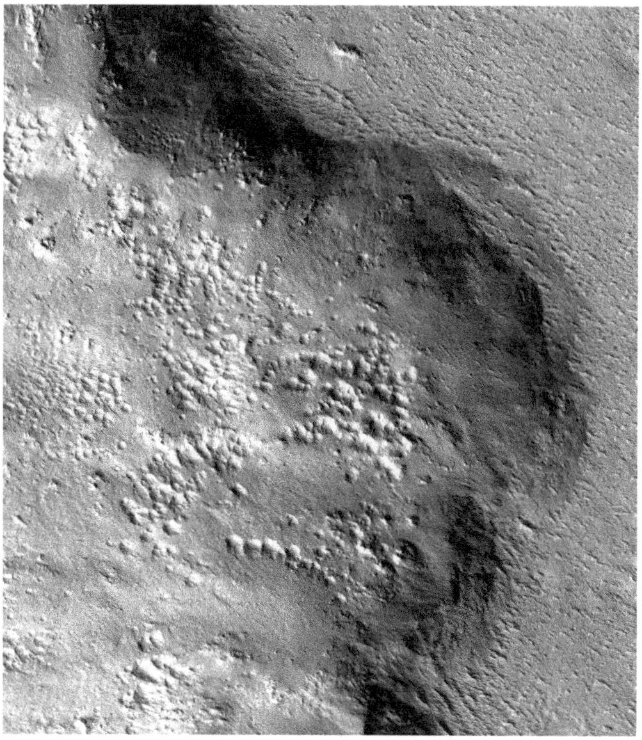

Figure 6.12 Pits at the crest of a distal rampart at the 28 km diameter Tooting crater, which lies to the left of this image. Illumination direction is from the left in this image. Width of the image is ~1.6 km. Part of HiRISE image PSP_003714_2030, centered at 23.10°N, 208.61°E.

a systematically higher nighttime temperature than the rest of the ejecta blanket. This they interpreted to be due to the accumulation of larger particles at the ejecta flow front, possibly due to a kinetic sieving process akin to that observed in terrestrial pyroclastic flows.

Few studies have investigated the geometry of the distal ramparts. Indeed, the spatial variability of rampart height and width, even at a single crater, is poorly known for most craters. One investigation that attempted to address this uncertainty was conducted by Mouginis-Mark and Baloga (2006). Using photoclinometry applied to THEMIS VIS images, and tied to individual MOLA ground tracks, they found that ramparts around the crater Toconao varied in height from ~35 to 175 m, and in width from ~300 m to 2.8 km (Figure 6.13). Mouginis-Mark and Baloga (2006) also investigated the potential correlation between rampart height and run-out distance of the ejecta layer from the rim of the parent crater. The lack of correlation between the height of the rampart and the radial distance from the parent crater implies that the ejecta blanket must be very thin upslope from the rampart, and that the rampart was constructed from most of the material contained within the ejecta flow. This can be confirmed with the use of a digital elevation model (DEM) for an ejecta blanket; for example, measurements made on a SLE crater in Utopia Planitia (Figure 6.14) show that the rampart is often ~3× the height of the ejecta thickness moving toward the primary crater.

This interpretation is supported by measurements of the proximal and distal heights of the terrain on either side of the rampart (Figure 6.14). Most of the ejecta blankets mapped from *Viking Orbiter* images (Wohletz and Sheridan 1983;

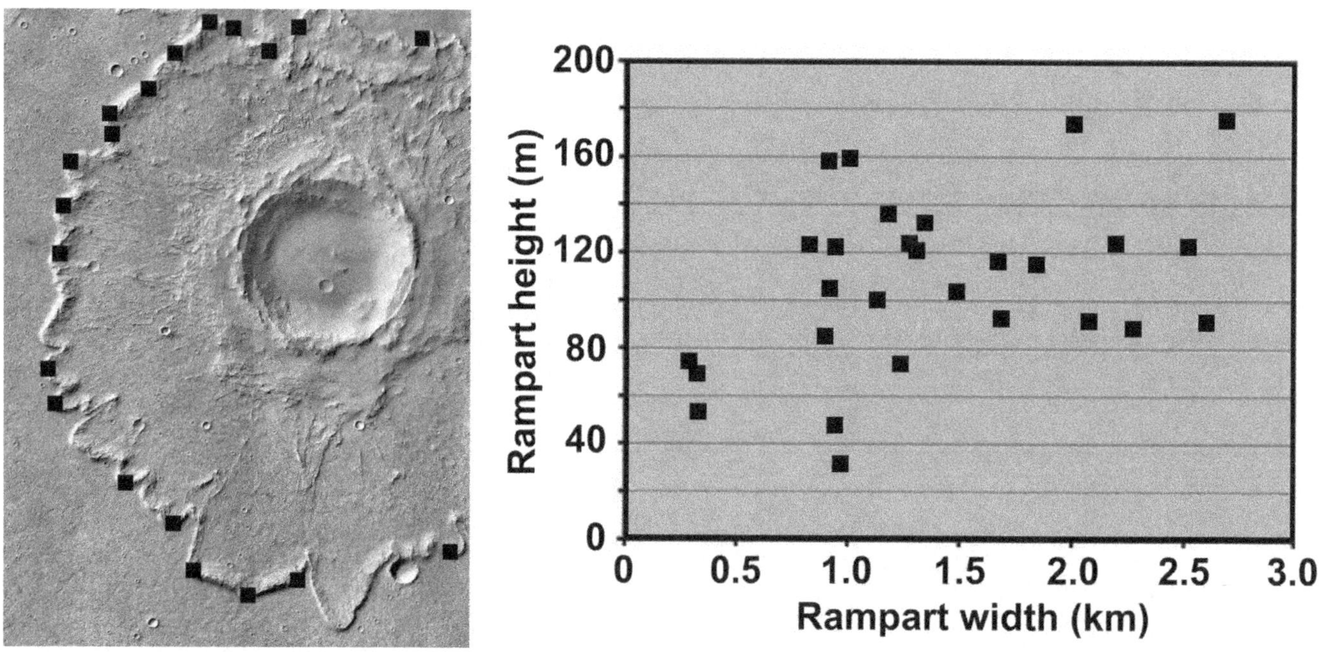

Figure 6.13 The 17.7 km diameter crater Toconao (20.81°S, 285.27°E) has a prominent distal rampart to the west and south of the crater. Mouginis-Mark and Baloga (2006) used photoclinometry keyed to individual MOLA laser shots to determine the variation in rampart height and width at this crater. The black squares in the left-hand image show where the data presented at right were collected.

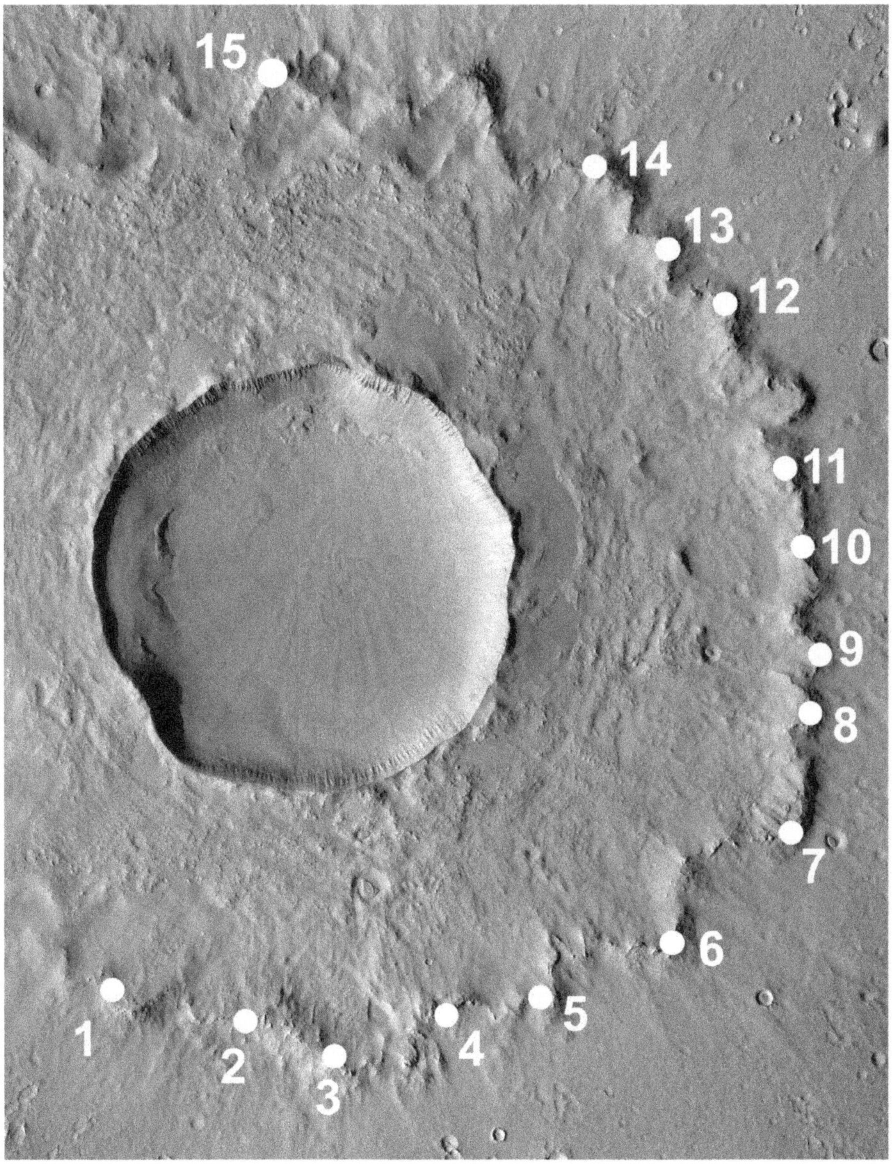

Pt. #	Rampart Ht.	Elev. diff
1	107 m	27 m
2	97 m	29 m
3	110 m	47 m
4	116 m	38 m
5	86 m	48 m
6	95 m	12 m
7	102 m	30 m
8	108 m	5 m
9	94 m	3 m
10	91 m	-17 m
11	119 m	-3 m
12	113 m	12 m
13	117 m	42 m
14	114 m	18 m
15	107 m	0 m

Figure 6.14 Rampart heights (in meters) and the difference between the proximal ejecta elevation just beyond the rampart (in meters) for a crater 11.7 km in diameter located at 36.05°N, 80.43°E. The table at right gives 15 pairs of measurements for this slightly degraded crater, which is also shown in Figure 3.6. Data from a stereo-derived DEM from CTX images B04_011287_2163 and B05_011564_2163.

Barlow and Bradley 1990; Barlow and Perez, 2003) appear to be mainly thin veneers (a few meters to tens of meters) on the Martian landscape. Only the easily recognizable distal rampart allows the extent of the ejecta blanket to be identified. Few structures within the ramparts, such as layering in the deposit or additional evidence for size-sorting of the entrained particles in the decimeter- to meter-scale, have so far been detected in HiRISE images.

6.7 SECONDARY CRATERS

Fluidized ejecta are not the only materials thrown from craters. As seen on the Moon (Oberbeck and Morrison, 1973; Allen, 1979; Dundas and McEwen, 2007), impact events produced fields of secondary craters formed from ejecta blocks reimpacting the surface (Figures 6.15 and 6.16). Studies of the ~2.9 km diameter Zumba crater (Chuang et al., 2016) and the 10.4 km diameter Zunil crater (Preblich et al., 2007) show that these craters produced thousands of secondary craters >15 m in diameter. Secondary craters from Zunil can be identified on THEMIS nighttime TIR images to occur up to 1,700 km from the parent crater and are typically associated with rays visible in nighttime THEMIS TIR images. Secondary craters up to ~1 km in diameter have been identified around larger primary craters on Mars over 100 km in diameter (Robbins and Hynek, 2011).

The type of target material has clearly played a role in the formation of secondary crater fields around impact craters on Mars. Secondary craters appear to have formed mainly around

Figure 6.15 Unnamed 6.5 km diameter crater with many secondaries. Crater located at 43.27°N, 225.77°E. CTX image B18_016543_2235.

Figure 6.16 An abundance of secondary craters was produced by the formation of this unnamed 27 km diameter crater, which formed within the Olympus Mons aureole materials (24.44°N, 217.79°E). Note the absence of fluidized ejecta, even from the primary crater. CTX image B18_016517_2025.

SLE and MLE craters, but not around most Type-1 DLE craters. This situation is possibly due to target properties associated with Type-1 DLE craters precluding the formation of large ejecta blocks that subsequently reimpacted the surface. A review of many MOC and THEMIS VIS images by McEwen et al. (2005) led them to conclude that many small (~20–100 m) craters distributed across the planet are secondary craters as they are often clustered in space and time (or age) and are shallow and flat-floored. Additional studies of Martian secondary craters have been conducted. Preblich et al. (2007) mapped more than 7×10^7 secondary craters > 15 m in diameter from Zunil crater. Quantin et al. (2016) studied the secondary crater field around the young primary crater Gratteri (7 km) and observed that the crater density in rays tends to peak around 120–130 km from the primary (~20–30 D) and reaches ~30–70 times the inter-ray crater density.

Most Type-1 DLE craters lack secondary craters, suggesting that the large ejecta blocks that normally produce such secondary craters may have been absent during ejecta emplacement. Possibly this was because the blocks were fragmented due to the presence of large amounts of water in the target materials or because the target material comprised fine-grained sediment before the crater was formed. Steinheim crater (54.60°N, 190.62°E) is different (Figure 6.17). It appears to be a very fresh Type-1 DLE crater and shows attributes that (potentially) have been eroded from older examples. Although pitted terrain is absent on the floor of Steinheim crater (interpreted as a sign of a recently formed crater, see Chapter 3), Steinheim crater is

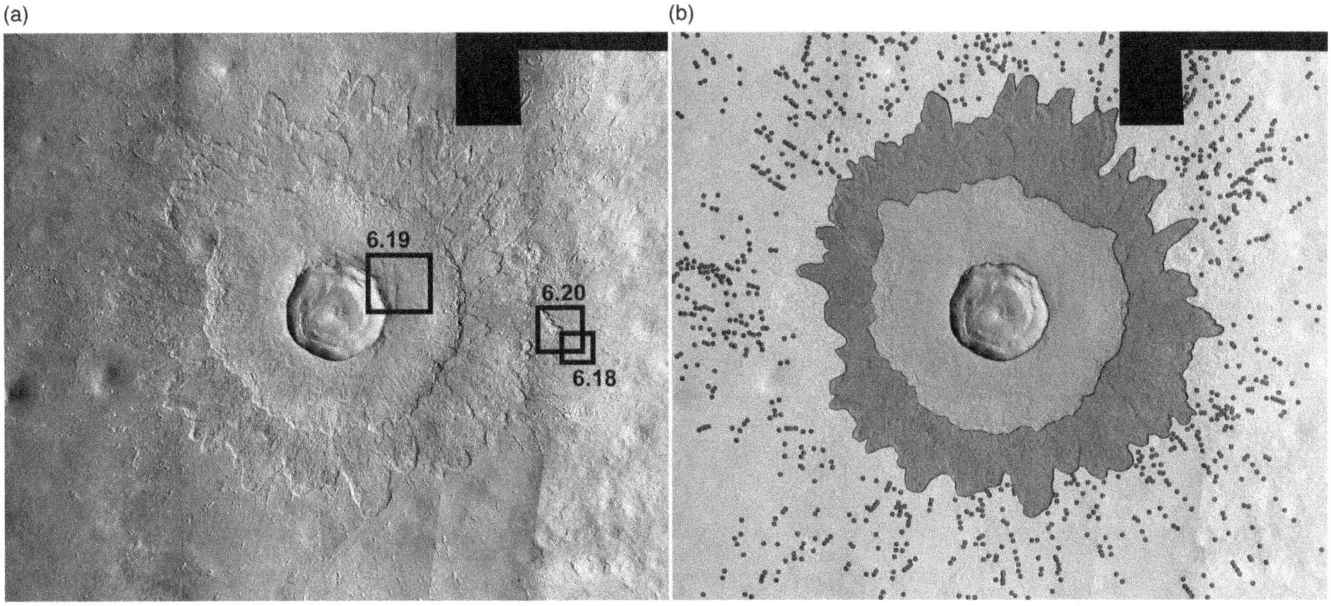

Figure 6.17 Distribution of secondary craters around Steinheim crater (11.1 km diameter), located at 54.60°N, 190.62°E. (a) Mosaic of CTX images G01_018799_2324, G21_026302_2344, and P18_007947_2348. (b) Interpretative sketch of the crater, showing the inner (light gray) and outer (dark gray) ejecta layers and the prominent secondary craters (black dots). Not all the smaller secondary craters are mapped. Dot size is not representative of the size of each secondary crater. Mapped from the image in (a). Boxes show the locations of Figures 6.18–6.20.

Figure 6.18 Details of some secondary craters ~20 km east of the rim crest of Steinheim crater, which lies to the left of this scene. See Figure 6.17 for location. Part of HiRISE image ESP_027014_2350.

a rare example of a Type-1 DLE crater with many secondary craters. Here we explore in more detail the morphology of the Steinheim ejecta layers to try to determine if these secondary craters (Figure 6.18) can be attributed to any unusual attributes of the target material, or if such secondaries were once more common around other Type-1 DLE craters but that they have been eroded away after their formation.

For most Type-1 DLE craters, is it possible that the target material was too friable to produce ejecta blocks that then create secondary craters when they reimpact the surface? Wulf and Kenkmann (2015) identified strings of boulders on the surface of the inner ejecta layer of Steinheim crater. They noticed that these boulders did not leave any indentation on the surface, and so most likely did not impact the surface at high speed. With HiRISE images, it is possible to search for boulders at the distal end of the ejecta layers since the radial flow of the ejecta may have concentrated boulders at this locality. Inspection of the eastern edge of the outer ejecta layer reveals many blocks a few meters in diameter (Figure 6.19). Even though blocks of this size would be unlikely to form secondary craters, their occurrence does illustrate that the target material had some strength at the time of crater formation. The existence of blocks at the leading edge of the ejecta layer (Figure 6.20) also suggests that meter-sized blocks within the ejecta layer become concentrated at the distal rampart.

A question raised by this large number of secondary craters is the potential uniqueness of the Steinheim crater. Is it possible that at other Type-1 DLE craters virtually all the secondary craters have been removed by processes such as the partial removal of material within the northern plains, such as the VBF? As suggested by Weiss and Head (2018), it seems quite likely that the secondary crater field has been preserved at Steinheim crater because the VBF has not been eroded away, while at other craters this erosion has continued post-crater formation. Of course, it raises the question of why other Type-1 DLE craters found beyond the VBF do not display secondary craters. The crater illustrated in Figure 6.4, for example, formed on the southern flanks of the volcano Alba Mons, where one can confidently predict that the target material comprised competent lava flows, and yet even here there are no prominent secondary

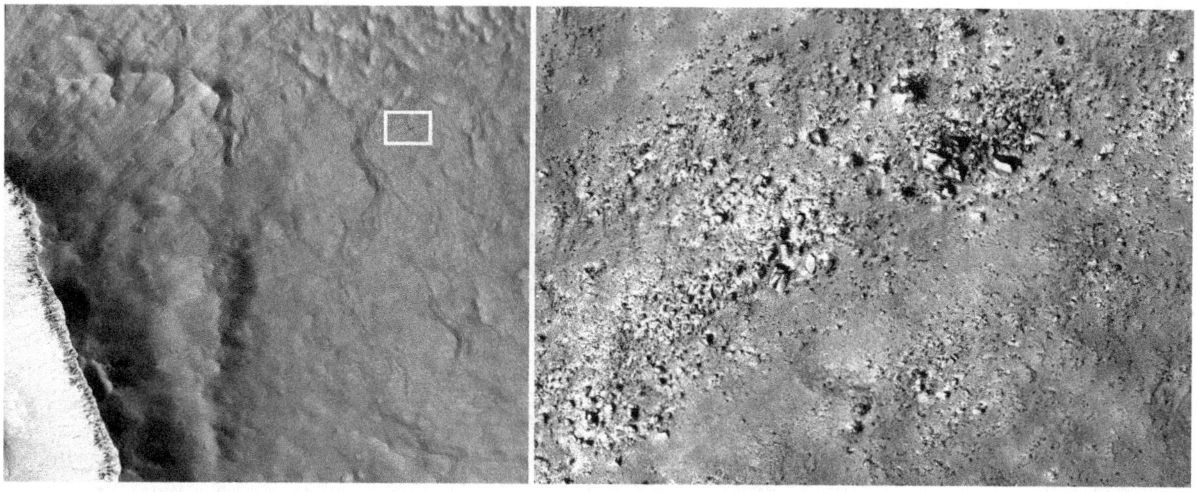

Figure 6.19 Blocks on the northeast segment of the inner ejecta layer for Steinheim crater. See Figure 6.17 for location. The box in the left image marks the location of the higher resolution view at right. The left image shows the crater's rim crest at lower left. Both images are parts of HiRISE image PSP_007736_2345.

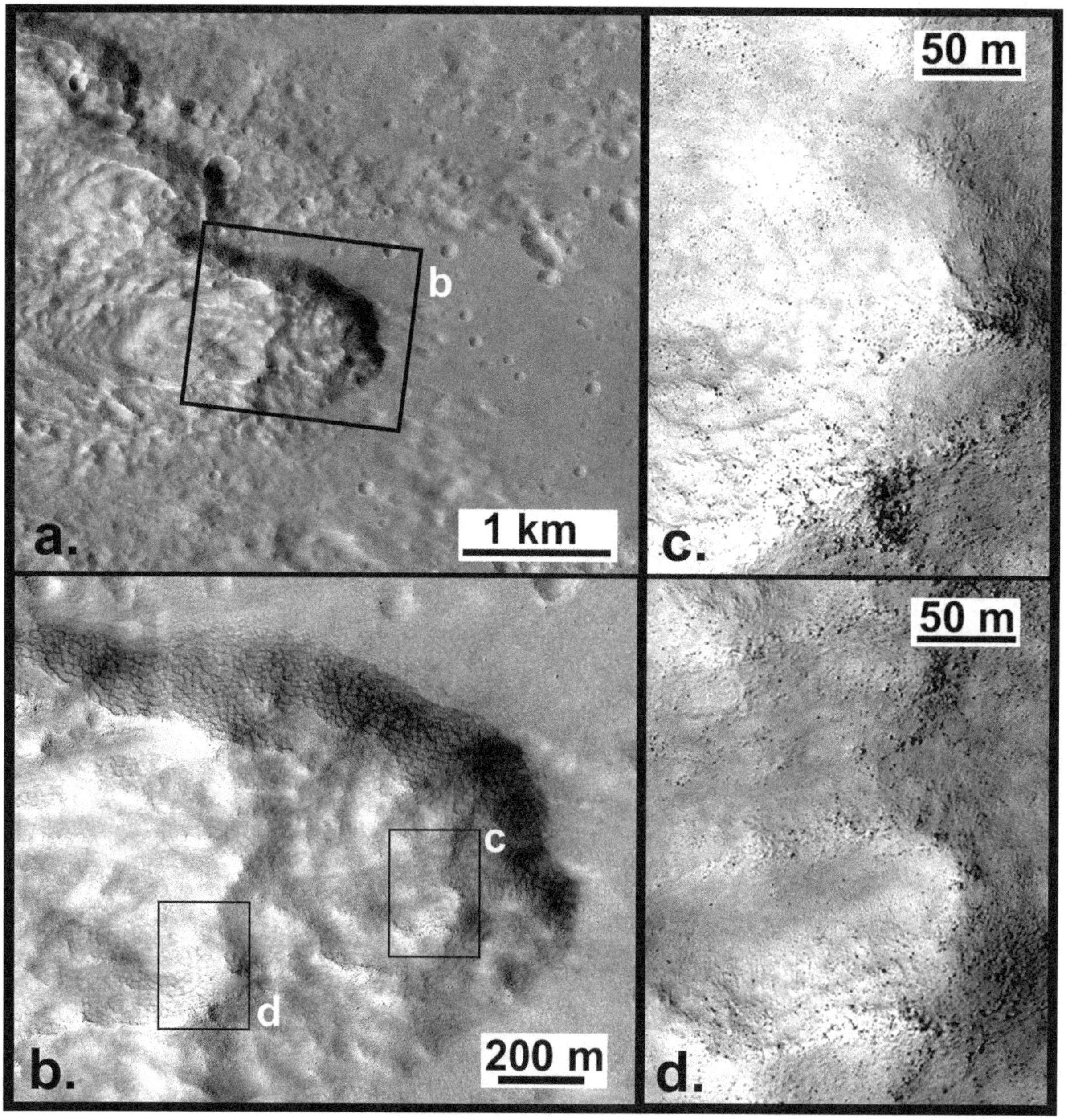

Figure 6.20 Details of the ejecta layers to the east of Steinheim crater, showing progressive detail of the distal ramparts. In (c) and (d) it is evident that boulders a few meters in diameter are concentrated at the rampart. See Figure 6.17 for location. All images are segments of the HiRISE image PSP_008303_2345.

craters. Thus, Steinheim crater remains an enigmatic outliner to the morphology typically associated with Type-1 DLE craters.

6.8 SUMMARY

In this chapter we have investigated the three main types of ejecta blankets seen around fresh and partially eroded impact craters on Mars. Striking differences exist between MLE and Type-1 DLE craters even if they have the same diameter (and, by inference, the same depth of excavation at the time of formation). Moving radially away from the crater rim, the topography and, hence, the inferred thickness of the ejecta layers are quite different. Grooves characterize fresh Type-1 DLE craters, while laminar flow with the production of distal ramparts was most common for SLE and MLE craters. These three ejecta types are not the only ones observed on Mars. In Chapter 7, we will explore other attributes of ejecta blankets.

7

Mars Crater Exterior Features – Other Ejecta Attributes

7.1 INTRODUCTION

In Chapter 6, we examined the diversity of the three principal types of fresh ejecta blankets on Mars, namely single-layered ejecta (SLE), Type-1 double-layered ejecta (DLE), and multi-layered ejecta (MLE). While these constitute the most common types, there is a much wider range of morphologies, particularly when smaller diameter craters or craters formed in the Northern Plains are considered. Oblique impacts or unusual target characteristics have also produced diverse ejecta types, as first identified by Mouginis-Mark (1979a) and elaborated upon by Barlow et al. (2000). In this chapter, we therefore explore more of this ejecta diversity.

7.2 SMALL, FRESH CRATERS

Craters smaller than ~1 km in diameter often have a similar appearance to their lunar counterparts (see Chapter 2). Craters smaller than a few hundred meters in diameter can have a blocky ejecta blanket and a simple interior with the inner walls covered in scree (Figure 7.1), probably because the impact did not excavate deep enough to reach a volatile layer. However, high-latitude craters of this size have been observed to have excavated water ice (Dundas et al., 2014). Slightly larger craters measuring a few kilometers across display a more distinct ejecta blanket with rays (Figure 7.2). At a diameter of ~3 km, the ejecta can resemble that of lunar craters with dunes of

Figure 7.1 Small craters on Mars (<~500 m diameter) often lack fluidized ejecta, most likely because the cavity did not penetrate the target to a sufficient depth to include enough volatiles in the ejecta. Instead, as in the case of this crater, a blocky ejecta blanket is seen that is more akin to a lunar crater (Figure 2.16) or Meteor Crater (Figure 2.1). HiRISE image ESP_038455_1330, centered at 46.78°S, 338.90°E.

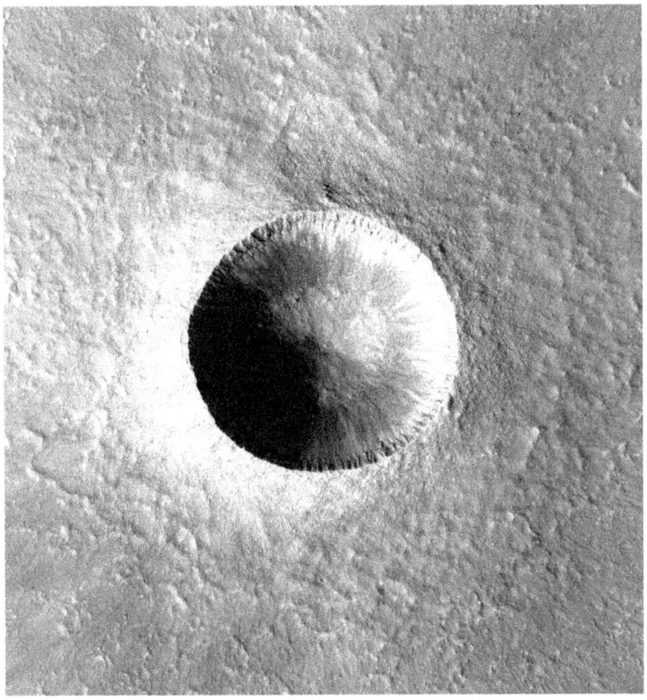

Figure 7.2 Fresh small unnamed crater with a hummocky ejecta blanket. The crater is ~1.3 km in diameter and is located at 10.68°N, 62.24°E. HiRISE image ESP_012791_1910.

Figure 7.3 Zumba crater, at 3.3 km in diameter, has an ejecta blanket close to the rim comprising dunes of material. Crater located at 28.67°S, 226.92°E. CTX image P03_002118_1512.

material close to the crater rim (Figure 7.3). For these larger craters, the ejecta blanket can display greater signs of fluidized ejecta, even if there is no coherent layering to the deposit.

7.3 RADIAL EJECTA CRATERS

The ejecta of some Martian craters display radial striations but lack the characteristics of either Type-1 DLE or MLE craters (see Chapter 6). First identified by Mouginis-Mark (1979a), this type of crater (Figure 7.4) suggests that while ejecta were thrown from the parent crater along ballistic trajectories, upon striking the surface some did not behave as ground-hugging flows during emplacement. They show no distal ramparts

Figure 7.4 Poona crater has a pronounced radial pattern to its ejecta blanket. This 19 km diameter crater (located at 23.77°N, 307.67°E) formed on the southern rim of Kasei Valles. Note that the canyon had already formed before the crater, as ejecta from the crater extend onto the floor at the top of the image. Thus, the substrate close to Kasei Valles may have lost most of the near-surface volatiles prior to crater formation. CTX image J09_048293_2040.

comparable to the ones associated with SLE and MLE craters. The ejecta of radial craters form a continuous hummocky blanket of material that thins rapidly away from the crater rim to ~1–2 crater radii. No radial grooves can be identified (see Section 6.5), but there is a pronounced radial texture to the ejecta blanket. Further away from the crater rim these deposits become discontinuous and eventually die out to be replaced by secondary craters and rays of material. Mouginis-Mark (1979a) proposed that these radial ejecta craters formed in the absence of volatiles within the target material. In such circumstances, the geologic setting must be unusual, such as the impact taking place close to a steep slope from which near-surface volatiles might easily escape over time.

7.4 LARLE CRATERS

An unusual type of crater, coined a low-aspect-ratio layered ejecta (LARLE) crater by Barlow et al. (2014), is characterized by a normal layered ejecta pattern (i.e., a SLE or MLE crater) surrounded by an extensive outer deposit that terminates in a sinuous, almost "flame-like" material (Figure 7.5). This outer ejecta for LARLE craters may extend more than 10 crater radii from the rim crest, that is, much further than that observed on SLE, Type-1 DLE, or MLE craters. Perhaps significantly, most LARLE craters larger than 1 km in diameter (91% of a total population of 140) are located poleward of 40°S or 35°N and all appear to occur within thick mantles of fine-grained deposits that are likely to be ice-rich (Figure 7.6). In addition, the majority of LARLE craters (83%) are smaller than 5 km in diameter. Barlow et al. (2014) and Boyce et al. (2015) proposed that LARLE craters were formed by impacts into ice-rich fine-grained deposits that produced a dusty base surge cloud (similar to that produced by explosion craters) that deposited dust and ice particles to produce the extended ejecta layers. After crater formation, salts emplaced by the upward migration of water through the LARLE deposit produced a surficial duricrust

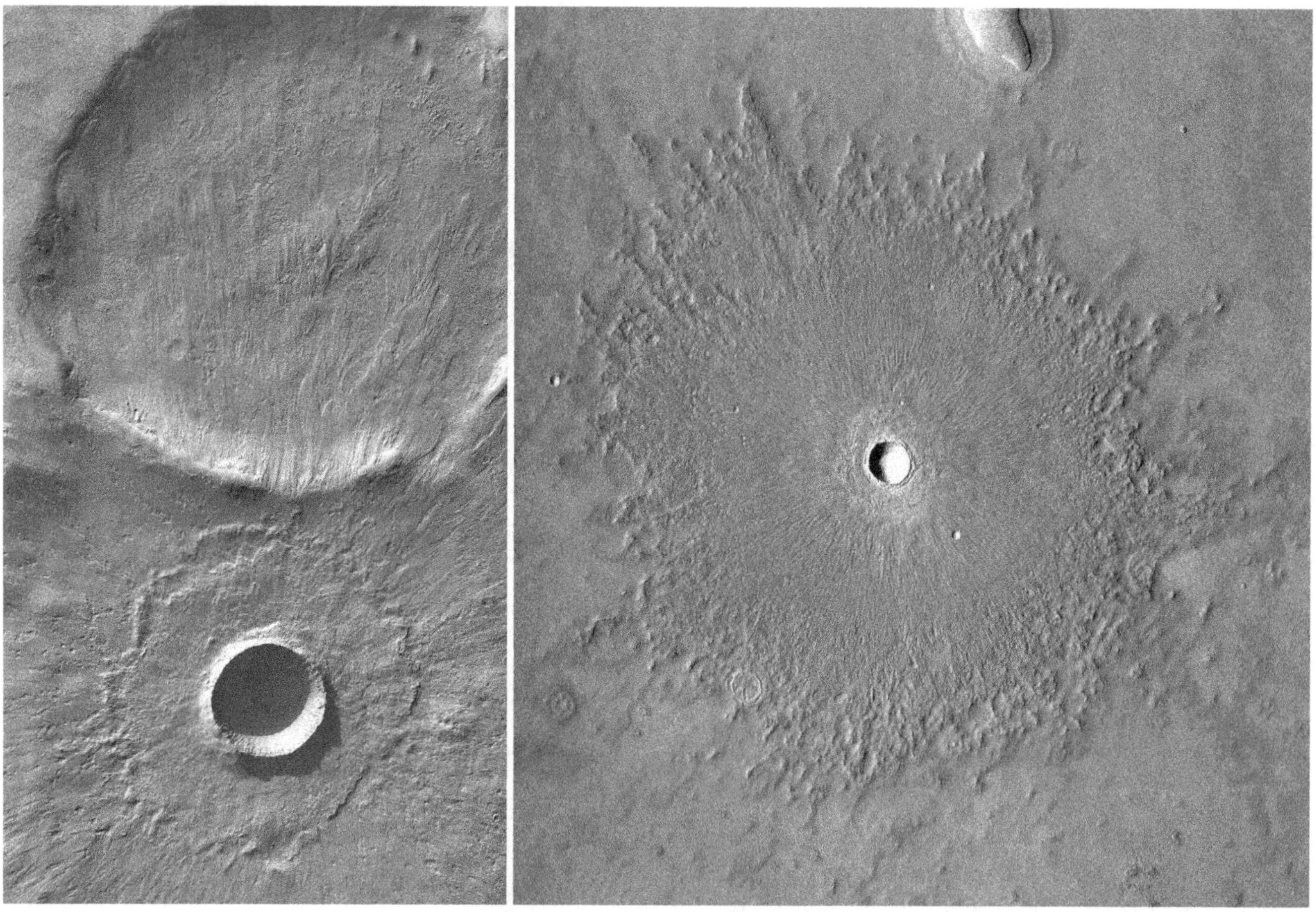

Figure 7.5 Two examples of LARLE craters, showing the unusually extensive ejecta deposits around these craters. Compare these examples to Figure 6.1a and Figure 6.1b. At left, Farim crater is a 4.2 km diameter LARLE crater located at 44.31°S, 139.28°E. Notice that this crater appears to be an MLE crater with multiple ejecta layers and distal ramparts, but that these ejecta lobes are confined to a short radial distance beyond the crater rim when compared to the radial extent of the LARLE material. Some of the LARLE material appears partially eroded at the greatest radial range. CTX image B19_016929_1364. At right, the ejecta blanket of a 1.6 km diameter crater (40.53°N, 175.68°E) appears to be surrounded by material that has been more resistant to erosion than the surrounding terrain, hinting that the original ejecta blanket extended far beyond that which is seen here. CTX image G23_027173_2212.

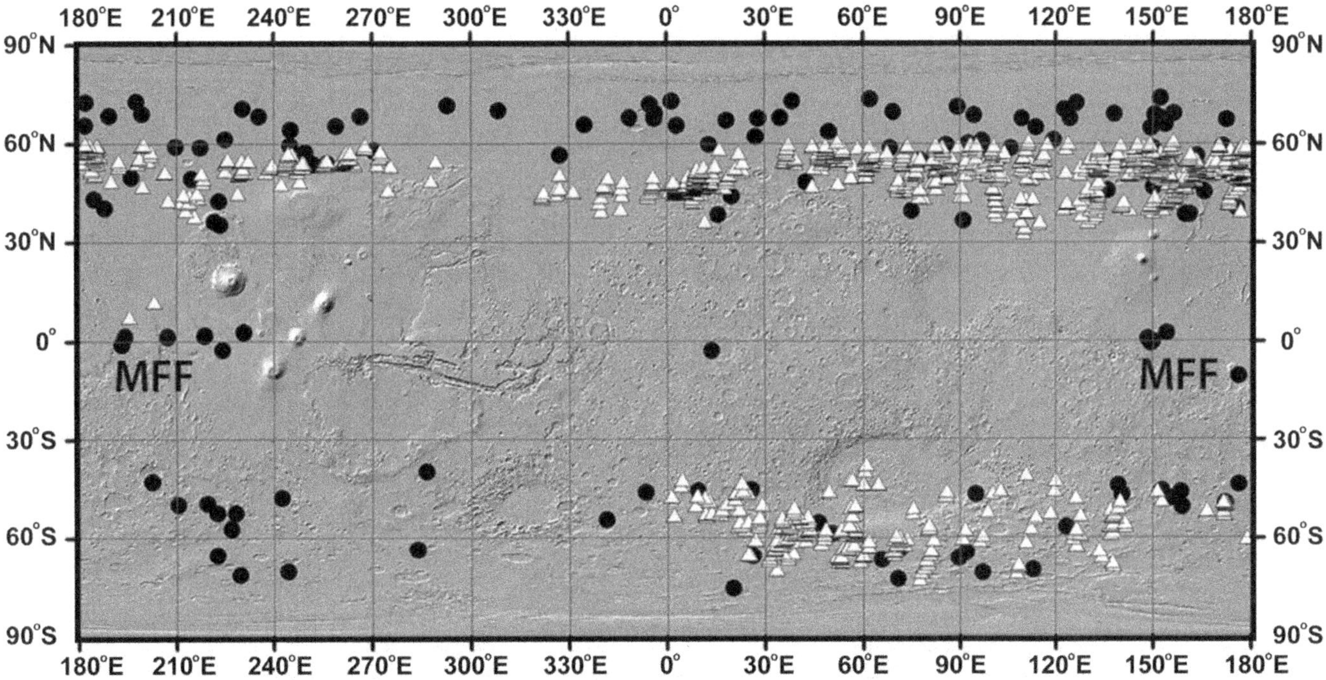

Figure 7.6 Distribution of 140 LARLE craters (black circles) identified by Barlow et al. (2014) and 2,696 pedestal craters (white triangles) identified by Kadish et al. (2008). Note that most of these two crater types occur in the same regions, namely at high polar latitudes. The few examples outside these areas are largely found in the thick fine-grained deposits of the Medusae Fossae Formation (MFF). The base image is a MOLA shaded relief image. Reproduced from Barlow et al. (2014). "Martian Low-Aspect-Ratio Layered Ejecta (LARLE) craters: Distribution, characteristics, and relationship to pedestal craters," *Icarus*, vol. 239, pp. 186–200, Figure 16. With permission from John Wiley and Sons.

layer that now protects the deposit from subsequent removal by eolian processes (Barlow et al., 2014).

7.5 PEDESTAL CRATERS

Pedestal craters (Figure 7.7) are another unusual crater type on Mars. At such craters the surrounding ejecta blanket has a volume greater than the interior of the crater below the rim crest. A plausible formation mechanism involves the presence of a thick ice-rich surface unit at the time of impact and the preservation of the material underlying the ejecta as the surrounding ice-rich unit disappears. Kadish et al. (2008) identified 1,027 pedestal craters within the 0–60°N latitude zone, with most located north of 35°N (Figure 7.8). The pedestals are believed to result from a climate-related mechanism and most likely result from impacts into ice-rich materials at mid-latitudes during periods of high obliquity (Kadish et al., 2010). Pedestal craters indicate that a considerable thickness of removed pedestals can reach up to ~250 m in height but are typically <60 m. Neither the northern nor southern hemispheres show any significant variation in the average pedestal height as a function of latitude. The removal of the surrounding material is therefore believed to be more rapid on the plains than on the ejecta blankets (Meresse et al., 2006). In some cases, the large ejecta volumes, some exceeding the volume of the cavity, suggest that the northern lowlands have experienced one or more episodes of resurfacing that involved both deposition and erosion (Meresse et al., 2006). In addition, the perimeter of the pedestal can be surrounded by pits

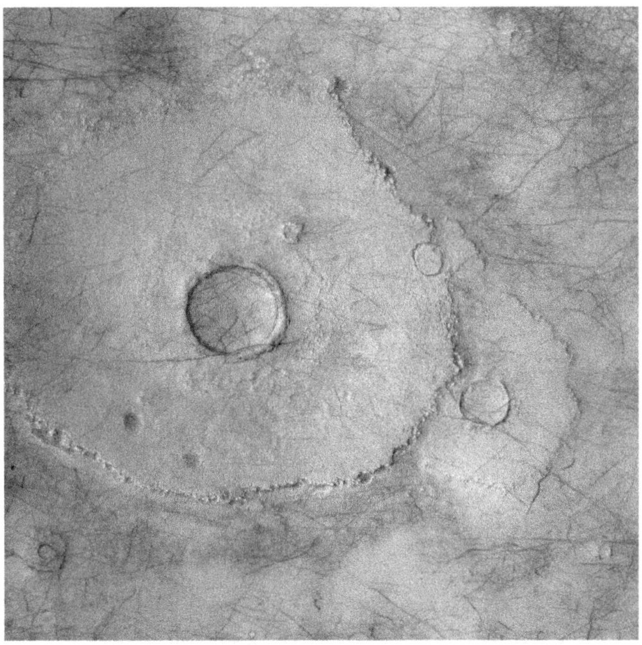

Figure 7.7 Pair of pedestal craters within Utopia Planitia (56.09°N, 107.96°E), the largest of which is ~2.5 km in diameter. The black streaks that cross this landscape were formed by dust devils and are unrelated to the craters. CTX image D04_028626_2361.

(Figure 7.8) indicating that larger volumes of ice may have been trapped near the point of impact and that this ice has subsequently sublimed away (Kadish et al., 2008).

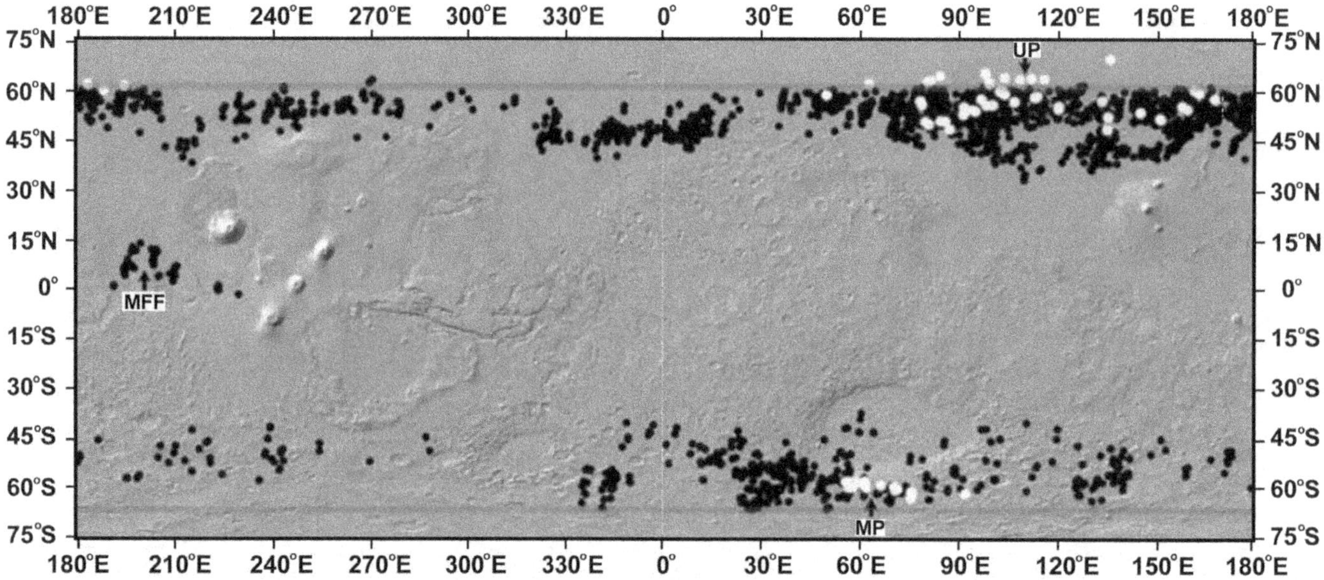

Figure 7.8 Geographic distribution of nonpitted pedestal craters (black circles) and pedestal craters with marginal pits (white circles) between ~60°N and 65°S. The Medusae Fossae Formation (MFF) contains the only population of equatorial pedestal craters. Pedestal craters with marginal pits were found in Utopia Planitia (UP) and Malea Planum (MP). Reproduced from Kadish et al. (2009), "Latitude dependence of martian pedestal craters: Evidence for a sublimation-driven formation mechanism," *Journal of Geophysical Research: Planets*, vol. 114(E10), doi:10.1029/2008JE0023318, Figure 1. With permission from John Wiley and Sons.

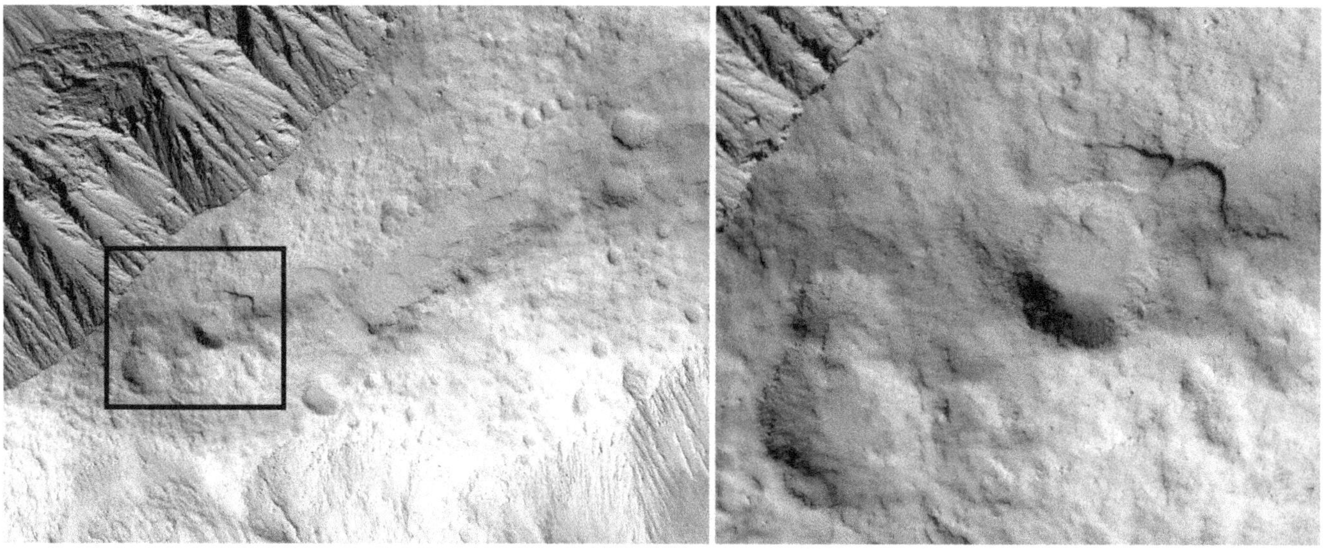

Figure 7.9 A relatively high-density area of small craters on the southern rim of Tooting crater. The image at right is an enlargement of the area shown at left. Note the lobate material that appears to have flowed down into and ponded within the two largest craters. Left image is ~1.4 km across and is centered at 22.96°N, 207.74°E. Both images are parts of HiRISE frame ESP_014157_2035.

7.6 SELF-SECONDARY CRATERS

Morris et al. (2010) identified small areas (<1 km^2) on the rim of Tooting crater (23°N, 152°E) containing anomalously high numbers of small (~20–70 m diameter) nearly circular craters (Figure 7.9). These areas have been attributed to the fallback of very high-ejection angle debris from the excavation of that crater. This interpretation has important implications for crater density-based age estimates of individual large impact craters (Hartmann et al., 2010), as well as for impact cratering mechanics (Melosh, 1989). These heavily cratered areas have several characteristics, most important of which are: (1) they represent a significantly more densely cratered area than other parts of the crater or the surrounding terrain, (2) these areas are restricted to near-rim locations (rim outward to <~5 km) on the parts of the rim where major terrace formation has not occurred; and (3) they formed on the up-range side of the crater when one considers the direction of impact of the crater-forming projectile, as inferred from the asymmetric distribution of the more widely distributed secondary craters (Mouginis-Mark and Boyce, 2012).

A preliminary survey of other fresh Martian craters (>~5 km diameter) by Boyce and Mouginis-Mark (2020) showed no such high-density crater areas on the rims. Because Tooting is a very fresh, unmodified crater, the location of the densely cratered areas on the rim may provide insight into the origin of these craters and their effects on crater density age measurements of the crater. For example, if these densely cratered areas are the result of ejecta fallback, as has been suggested for similar lunar examples (Zanetti et al., 2017; Plescia and Robinson, 2019), then such areas should occur around the entire crater rim, but they do not. But these areas are found only in places along the rim of Tooting where major rim collapse has not occurred. This may indicate that other heavily cratered areas of the rim formed but were subsequently destroyed by terrace formation. Alternately, this distribution could be, in some way, the result of the oblique impact that formed Tooting crater (Mouginis-Mark and Boyce, 2012; Lagain et al., 2021). Furthermore, if these areas were produced by a fallback, then this material must have been ejected on narrowly confined, nearly vertical, trajectories landing in a zone restricted to near the rim. The flight time of the ejecta must have been relatively short for the rotational speed of Mars to not disperse them. Indeed, impact melt on the rim was still molten after the ejecta landed as some impact melt flowed into these self-secondaries.

Figure 7.10 illustrates the places on the rim of Tooting crater where these high crater density areas occur, which are located on the parts of the crater rim that have not subsequently collapsed (Mouginis-Mark, 2015). A plot of the cumulative size frequency distribution of craters in the high crater density area on the southeast rim of the crater (i.e., the fallback craters), as well as the ejecta blanket and the adjacent area of Amazonis Planitia, shows that the densely cratered areas are significantly more cratered than either Tooting ejecta or the surrounding area of Amazonis Planitia. In "normal" circumstances, cumulative size-frequency distribution curves would indicate that these more heavily cratered parts of the rim are much older than the rest of the crater, but obviously this cannot be the case. Self-secondary cratering is thus the leading explanation for the origin of these high crater density areas. Similar self-secondaries have also been reported for Copernican-age craters on the Moon (Zanetti et al., 2017).

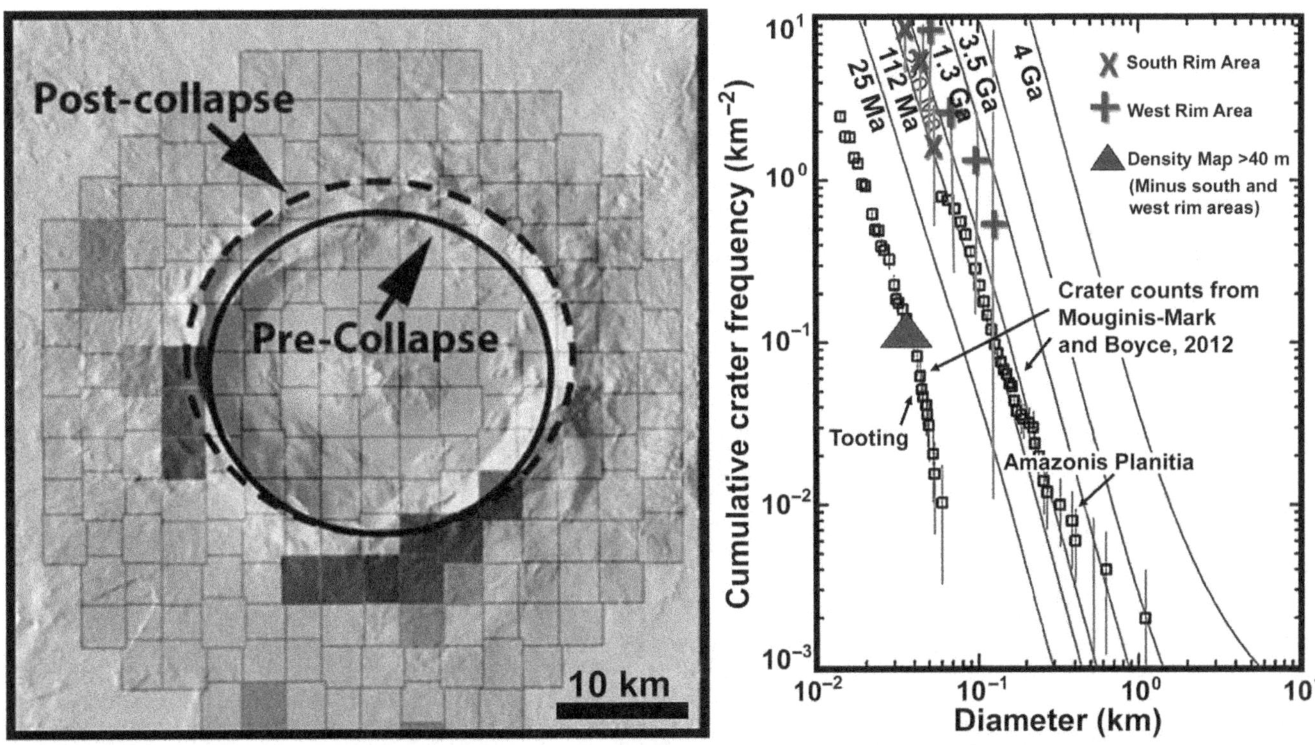

Figure 7.10 (a): Number density of craters >40 m diameter on the rim and outer wall of Tooting crater out to ~0.5 R from the crater rim. Each cell is ~3 km × 3 km, with the color of each cell a function of the number of craters in that cell. Light blue cells contain 2 ± 2 craters/cell, light green from 6 ± 1 crater/cell, burgundy from 9 ± 1 craters/cell, and red >11 craters/cell. The solid black circle is the approximate location of the rim before wall collapse and the formation of terraces. The dashed circle is the approximate location of the current rim. (b): Cumulative size frequency distribution of the craters on Tooting ejecta, the Amazonian plains surrounding Tooting, and the densely cratered areas on the south (red X) and west (red +) rim of Tooting. The cumulative number of craters >40 m diameter (normalized to 1 km^2) in the light blue and green cells is consistent with the density of craters on Tooting ejecta. The cumulative number of craters in the burgundy and red cells is consistent with the density of the fallback craters in this figure. Note: A black and white version of this figure will appear in some formats. For the color version, please refer to the plate section.

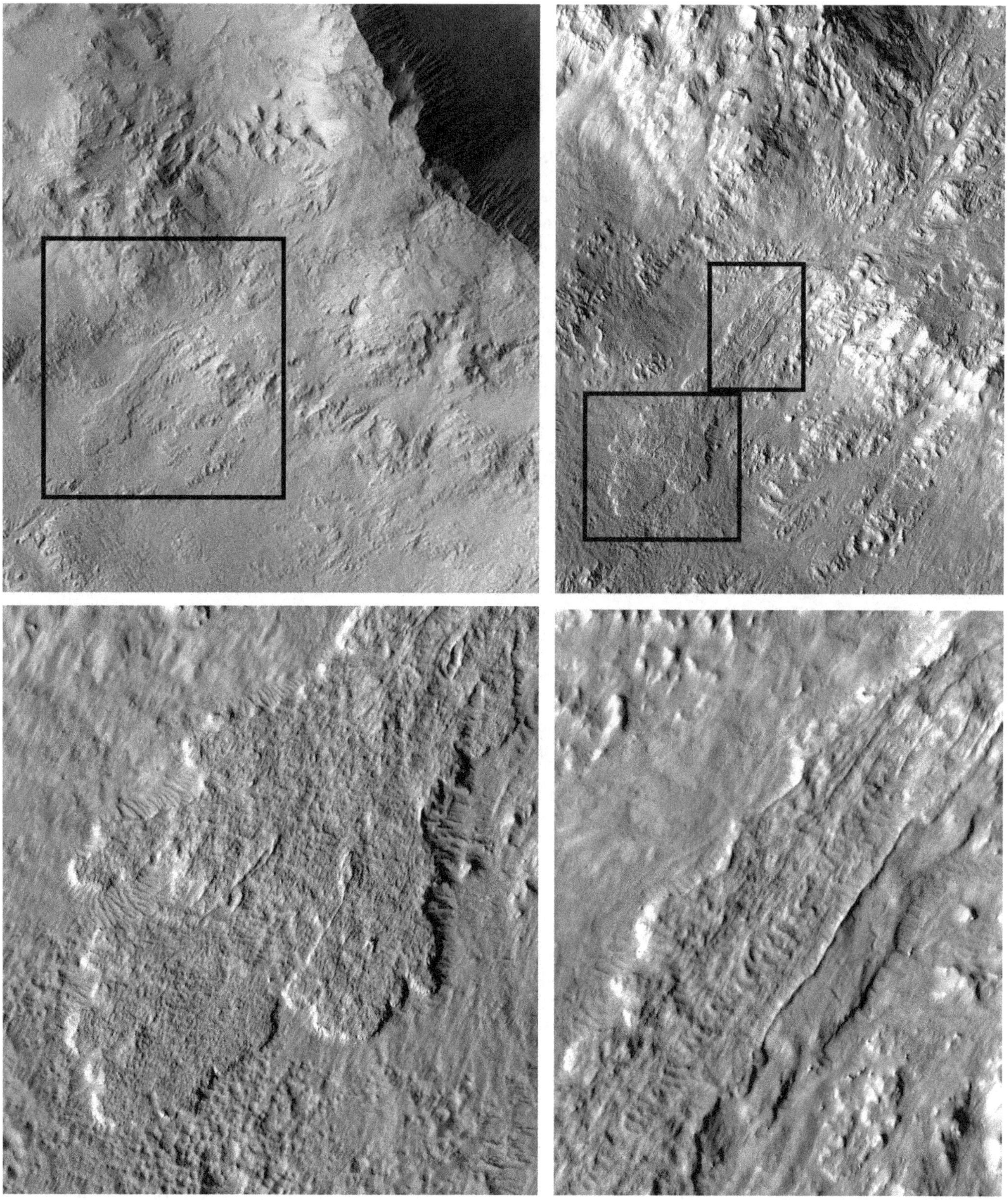

Figure 7.11 Location of a flow at Tooting crater identified by Morris et al. (2010). Width of image is ~1.5 km. The box denotes the area displayed at the top right, which in turn identifies the areas shown in the lower two images. All images are portions of HiRISE image PSP_002646_2035.

7.7 LOBATE FLOWS

Lobate flows have been identified both on the interior walls (see Section 5.3.2) and the exterior rim units of Tooting crater (Morris et al., 2010; Mouginis-Mark and Boyce, 2012). Morris et al. (2010) identified two types of exterior flows that are interpreted to have formed as sediment-rich flows. (1) On the southwest crater rim, a ridged region adjacent to the rim crest displays a 1.9 km long, 75–310 m wide flow (Figure 7.11).

Figure 7.12 Another type of sediment flow, also at Tooting crater, was identified by Morris et al. (2010). This one broke out onto the surface of the ejecta blanket from material emplaced higher up-slope. Width of image is ~4.0 km. Box denotes the area of Figure 7.13. Part of HiRISE image PSP_007406_2035.

(2) On the southern ejecta blanket, a large flow emanates from the collapse terrain to form a flow ~1.25 km in length and 225 m wide (Figure 7.12). This flow traverses a slope of ~4.5° and is interpreted to be a sediment-rich flow with the central channel carved by a water-rich flow.

The largest of these flows is 1,260 m long and 225 m wide, hummocky, and appears to emanate directly from collapsed terrain (Figures 7.12 and 7.13). There are levees of rubbly material, each 60–70 m across, that enclose a lineated channel region that is 86 m across at its widest. The channel is 490 m long and is characterized by linear features parallel to the direction of flow. The channel also contains several isolated, elongated features (longitudinal bars) that are similar in morphology to the levees. The outer margin of this flow is characterized by individual lobes on the scale of tens of meters or less in horizontal extent (parallel to the flow direction). The flow overlies material that is interpreted to be undivided sediments as well as regions of fractured and blocky material.

So, what does the occurrence of different types of sediment flow at Tooting crater imply for other, older craters? The preservation of the flows of sediment, likely lubricated by water released during or closely after (perhaps by only a few days?) an impact event, could be unique to Tooting crater. Our inspection of craters imaged by HiRISE reveals that evidence of similar sediment flows at slightly older craters on Mars is lacking and implies that if such flows ever existed, they most likely have been destroyed by subsequent crater modification. The flows at Tooting crater do indicate that a volatile-rich ejecta blanket might experience "dewatering" soon after emplacement. Such dewatering has been suggested for some of the smooth materials surrounding Bamberg crater (Mouginis-Mark, 1979b). The walls and interiors of young craters may also experience landslides that bring material high on the crater wall onto the crater floor. This may have relevance to the formation of the pitted material that is seen within young craters, as both landforms require the presence of water at the surface.

7.8 OBLIQUE IMPACTS

Natural impacts in which the projectile hits the target exactly vertically are almost nonexistent (Pierazzo and Melosh, 2000). Elongated craters with asymmetric ejecta patterns are clear demonstrations that many impacts took place at an oblique angle (Figures 7.14 and 7.15). Even some almost circular craters such as Tooting (Mouginis-Mark and Boyce, 2012) and Hale (Schultz and Wrobel, 2012) are interpreted to have formed by oblique impacts. As the angle of impact decreases from near-vertical to a grazing angle, the along-track crater diameter becomes larger than the cross-track dimension. Gault and Wedekind (1978) found that for angles <~30°, steeper interior slopes formed on the up-range walls of the crater. The ejecta deposits exhibit axial symmetry for impacts down to ~45°. As the angle of impact decreases to less than 45°, the ejecta becomes asymmetric. A "forbidden

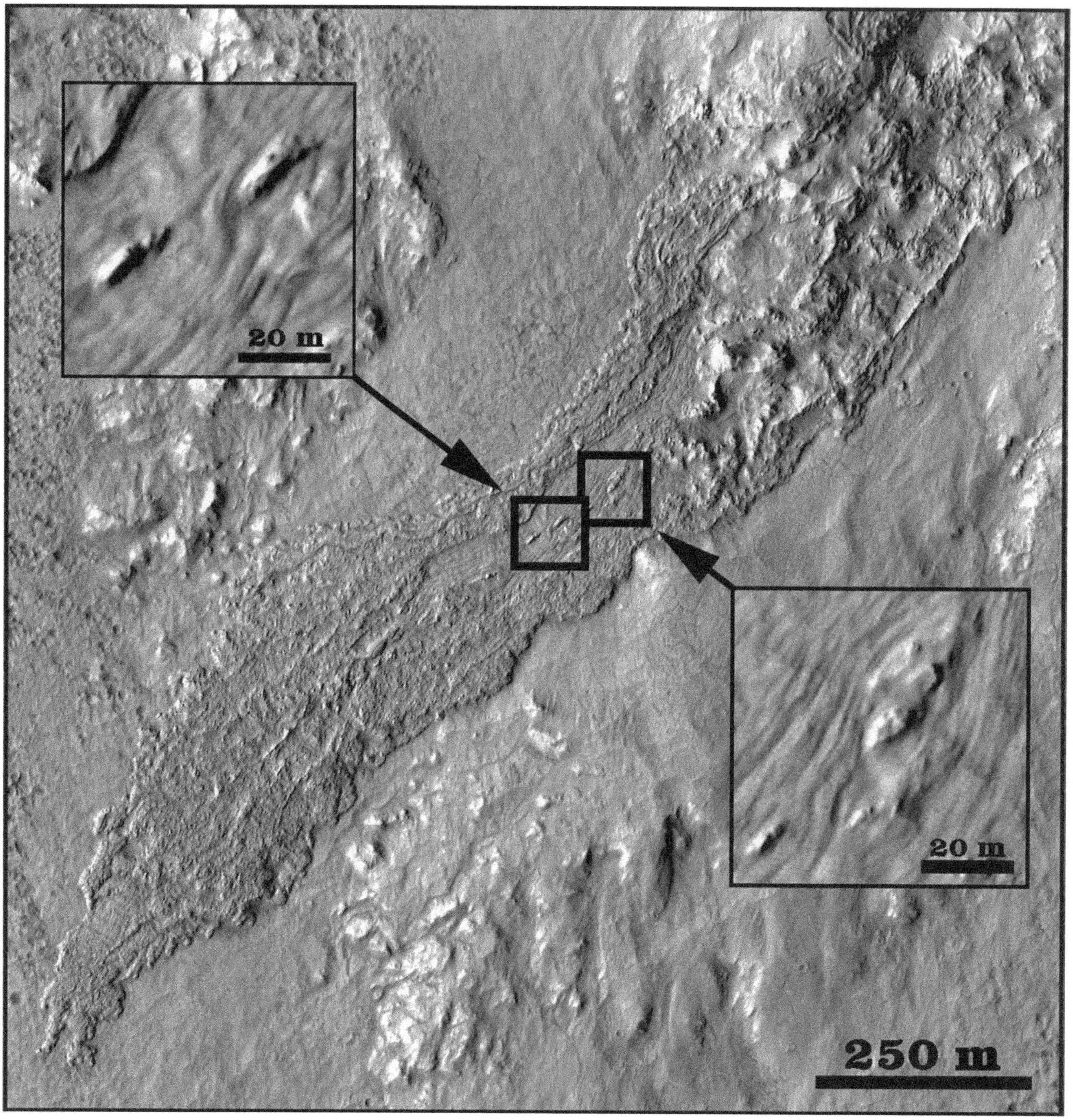

Figure 7.13 This type of flow appears to have broken out of the ejecta blanket of Tooting crater as a new surface flow. Enlarged images reveal that a central channel evidently formed where the flow was quickest, producing streamlines around blocks (bedrock) in the floor of the channel. See Figure 7.12 for location. Part of HiRISE image PSP_007406_2035.

zone" appears first up-range and then down-range of the crater for impact angles <~30° (Pierazzo and Melosh, 2000).

Oblique impacts have often been hypothesized to be responsible for meteorites being launched off Mars *en route* to Earth (Nyquist, 1983; O'Keefe and Ahrens, 1986; Lagain et al., 2021). When coupled with surface age estimates and the radial distribution of secondary craters, several young impact craters can be identified as candidates for being the source craters of many of the Martian meteorites (Lagain et al., 2021; Herd et al., 2024). The impact trajectory has also been linked to non-radial deformation of the floor of Martian craters (Wulf et al., 2012). As discussed in Section 3.5.2, the orientation of oblique impacts with respect to the Equator may even record different obliquity cycles, as many of these impactors may originally have been transient moons of Mars in equatorial orbit (Schultz and Lutz-Garihan, 1982; Chappelow and Herrick, 2008).

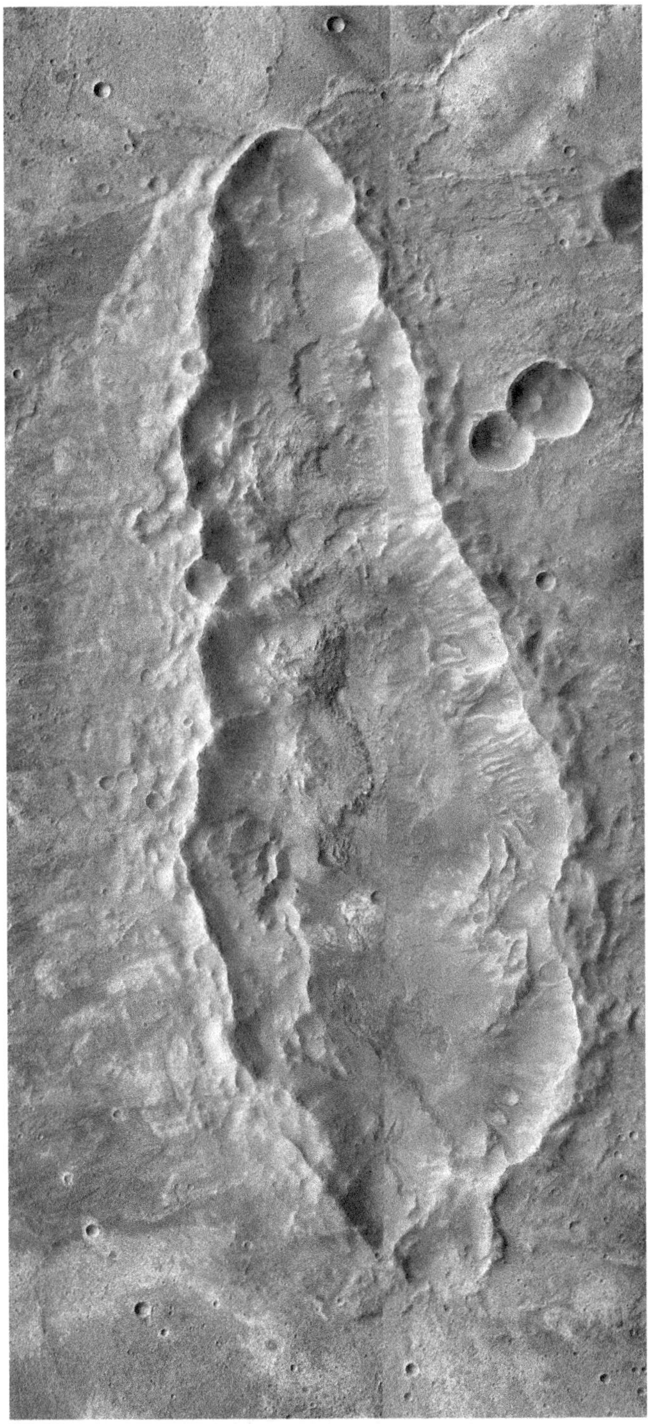

Figure 7.14 Unnamed highly elongated impact crater 77.6 × 22.6 km diameter, located at 21.14°S, 54.70°E. The projectile is inferred to have impacted at an oblique angle from the top of the image. Mosaic of CTX images B11_013965_1593 and CTX K13_058285_1586.

7.9 CRATER RAYS

Several of the morphologically freshest Martian craters have bright rays in addition to many secondary craters (Tornabene et al., 2006). Dilly crater is one of the very youngest craters and displays a clear ray system, while rays of bright material have also been traced for hundreds of kilometers from Zunil crater (McEwen et al., 2005). Comparison of MOC images and THEMIS nighttime TIR data reveal a 1:1 correspondence between bright streaks and concentrations of small bright-ejecta craters (McEwen et al., 2005). New craters known to have formed within the last few years (Daubar et al., 2014, 2022) provide some of the best examples of ray patterns from Martian craters as the fresh material provides a strong contrast to the surrounding weathered surface (Figure 7.16). Indeed, the fresh crater is often surrounded by a "blast zone" composed of diffuse low-albedo material (Figure 7.17). This diffuse material may be comparable to the extended LARLE-like deposits of larger craters and indicates that the effects of a cratering event often extend way beyond the visible boundary of the ejecta blanket.

7.10 HIGHLY FLUID EJECTA

El-Maarry et al. (2013) and Grant and Wilson (2018) drew attention to some unusual flow features ~450 km north of Hale crater, which is ~125 km in diameter. Hale crater (35.7°S, 323.6°E) formed in the Early to Middle Amazonian and is one of the best-preserved large craters on Mars. The impactor that created Hale crater hit directly into Uzboi Vallis, a very large channel thought to have periodically transported hundreds of thousands of cubic meters of water per second from Argyre Basin (Jones et al., 2011). Flow features associated with the ejecta from Hale crater (Figures 7.18 and 7.19) are typically tens of meters thick, are typically smooth at scales of tens to hundreds of meters, are relatively dark-toned, and are devoid of eolian bedforms. These features appear to be emplaced at low velocity because they tend to flow around, not over, low-relief pre-existing features and spread to form broad deposits in flat areas.

Evidence for ejecta flow resulting from layered ice-dust-rich deposits emplaced during a recent glacial epoch has also been identified at Gasa crater (Schon and Head, 2012). Channels on the lower crater wall transported material that ponded in a topographic low near the rim crest of Gasa crater (Figure 7.20). The interpretation is that the Gasa impact event deposited hot ejecta on top of a debris-covered glacier on the northern crater floor, and melting ice buried at shallow depth. This process resulted in the formation of a slurry that drained in directions related to local topography.

7.11 PITS ON THE EJECTA LAYERS

Hundreds of small pits are ubiquitous on the ejecta layers of the Kotka crater (Figure 7.21), and several other large fresh craters display a similar morphology. At Kotka crater, these pits are in close association with secondary craters but appear to have formed after the secondaries. The small pits appear to have a different origin as they have no ejecta blankets. The pits are like secondary craters but formed on top of the ejecta layer after this layer was emplaced, so in this case pit formation was a relatively late-stage component of the cratering process. Pits often occur in chains extending several kilometers in a radial

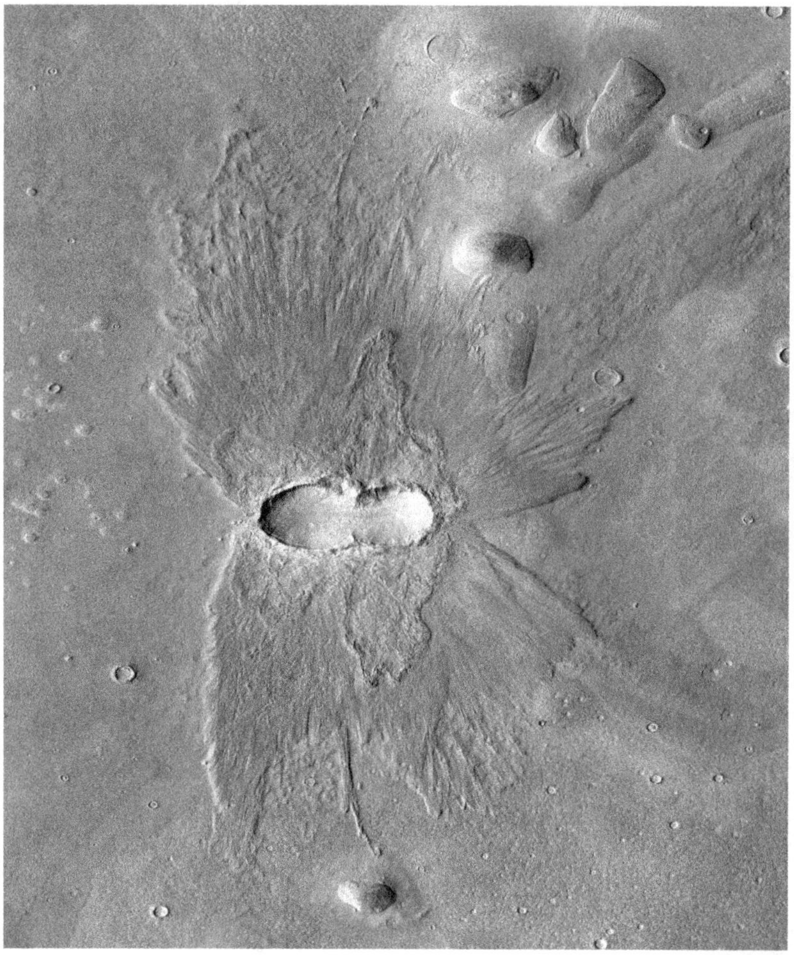

Figure 7.15 A spectacular double crater (~2 km by ~4.4 km) formed by an oblique impact shows how ejecta from such impacts are distributed perpendicular to the direction of impact, which can be determined from the greater elongation of the craters in the east-west direction. CTX image B21_017903_2196, located at 39.32°N, 174.85°E.

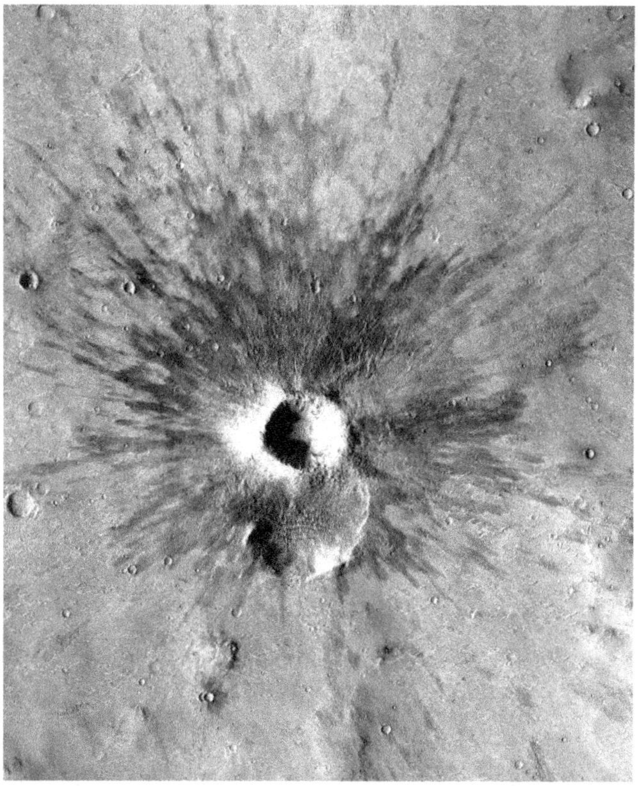

Figure 7.16 Very recent small impact crater ~400 m in diameter, with a prominent ray pattern. Located at 3.65°S, 347.48°E. HiRISE image ESP_067809_1765.

Figure 7.17 A "blast zone" exists around a fresh impact crater at 13.24°S, 200.72°E. In addition to prominent rays, the crater has a complex diffuse low-albedo halo (Dauber et al., 2016). Width of the image is ~650 m. HiRISE image ESP_034372_1665.

Figure 7.18 Ejecta from Hale crater most likely had a rheology like mud. The rim crest of Hale is ~105 km to the south of the flow. Direction of flow was from the bottom of the image. Width of image is ~20 km. CTX image B22_018188_1472, centered at 32.46°S, 322.90°E.

direction from the crater rim. These chains also cross onto the surrounding terrain. No models have been developed to explain the formation of these pits, but they are clearly associated with a late stage of the impact event.

7.12 "DEWATERING" OF EJECTA BLANKETS

When fluidized ejecta layers contain at least some proportion of water or ice during their formation, it is not unexpected that at least some ejecta blankets exhibit signs of some of the water being released onto the surface after ejecta emplacement. Although it seems likely that most volatiles contained within an ejecta blanket would either seep into the underlying substrate or sublime at the surface, some examples of channel networks produced by surface water flow have been found on fresh ejecta blankets. This phenomenon was first observed at Bamberg crater (Figure 7.22), where numerous examples of dendritic channels feeding into topographic lows appear to be partially filled with sediments (Mouginis-Mark, 1979b).

Parts of the Hale crater's ejecta blanket also display abundant evidence for copious amounts of water within the flowing ejecta; Jones et al. (2011) identified a well-developed channel system to the west of the crater (Figure 7.23). Pre-existing craters close to the Hale and other impact craters were filled with fluid ejecta that has now been partially eroded, perhaps by sublimation of the volatile-rich ejecta (Figure 7.24). In addition, the interior of the crater was evidently water-rich, as there are numerous gullies on the flanks of the central peak.

The formation of Hale crater is relevant to the broader analysis of older craters on Mars, particularly those with diameters >~100 km. Excavation of material from greater depths would have been common. Any deep ancient aquifer could have been more easily tapped by these larger impacts. It is therefore relevant to think of Hale crater as serving as an analog to some of the larger older craters and basins, such as the 467 km diameter Huygens crater (13.77°S, 55.52°E), where an extensive drainage network exists (Mukherjee et al., 2020). Lyot crater (50.47°N, 29.34°E) is one of the youngest impact basins >200 km in diameter and includes an extensive "channeled scabland" extending to the north, west, and east of the crater that could well be the product of impact-induced fluid flow (Harrison et al., 2010). On an even larger scale, impacts such as those that created the Argyre, Isidis, and Hellas Basins may well have produced large volumes of fluid ejecta that helped produce the drainage networks seen in the Southern Highlands (Irwin et al., 2013).

7.13 EJECTA FROM LARGE CRATERS

The study of ejecta blankets of large impact structures provides insights into crustal structure by virtue of the redistribution of the target material on the surrounding terrain. Unfortunately, there are few craters larger than ~150 km that are sufficiently fresh for the ejecta blankets to be mapped. Bakhuysen crater (Figure 7.25) is one such example (Caudill et al., 2018). Bakhuysen is a Late Noachian to Early Hesperian age ~150 km-diameter crater with two distinct ejecta units: a lowermost unit having features consistent with ejecta emplacement

Figure 7.19 A central channel and lobate flows are visible in the distal portion of one of the ejecta flow lobes from Hale crater at 32.52°S, 322.90°E. Direction of flow was towards the top of the image. Width of the image is ~5 km. HiRISE image PSP_008259_1470.

Figure 7.20 A broad, flat deposit ("A") could be a mud flow or a low-velocity debris flow resulting from remobilized ejecta from Gasa crater (35.72°S, 129.43°E), this one identified by Schon and Head (2012). Width of image ~600 m. Part of HiRISE image PSP_009901_1440.

via ballistic sedimentation and an upper discontinuous unit with a distinct ponded morphology with a pitted texture as well as lobate flow deposits. One of the most prominent attributes is a chain of secondary craters that extends almost 250 km from the crater rim. This chain must have been produced by a series of ejecta blocks thrown in a specific orientation from the opening crater cavity.

A few other large craters are noteworthy here. The *Viking Lander 2* (*VL2*) spacecraft landed within ~20 km of an ejecta lobe from the 100 km diameter Mie crater (Shorthill et al., 1976). The density of craters superposed on Mie crater ejecta is consistent with a Late Hesperian age. The number of rocks on the surface observed by the lander (Figure 7.26) was surprising to mission planners as the landing site was within the northern plains where the Vastitas Borealis Formation (VBF) is pervasive (Mutch et al., 1977). A smooth, rock-free landing site was expected. Rock abundances observed at *VL2* are consistent with the addition of impact-emplaced material on top of a thick mantle layer that has subsequently been eroded away (Thomson and Schultz, 2007). It seems most likely that the VL-2 rocks were ballistically emplaced, although due to the erosion of the area, a surface-flow origin cannot be ruled out.

Small basins on Mars can also provide some information about the ejecta emplacement process, even if most of the fine-scale materials have been eroded. Schiaparelli Basin (~410 × 440 km in diameter; 2.63°S, 16.70°E) displays radial textures comparable to the Imbrium sculpture of the Moon (Head, 1976) and the radial sculpture around the Caloris Basin on Mercury (Fassett et al., 2009). Mouginis-Mark et al.

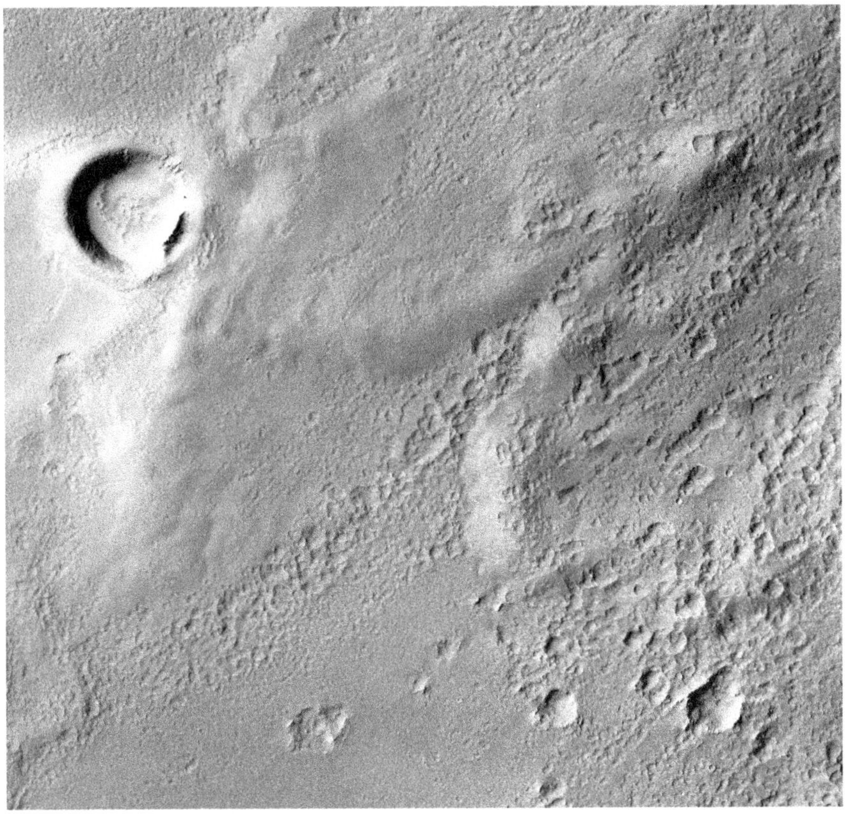

Figure 7.21 Many secondary craters can be observed on the ejecta layers of the Kotka crater, which lies to the right of this image. Image centered at 18.35°N, 169.13°E. On the left side of this image, some of the strings of pit craters extend into the surrounding terrain. The largest crater in this view is ~2 km in diameter. CTX image B01_010018_1993.

Figure 7.22 Part of the southwest ejecta blanket of Bamberg crater has multiple dendritic channels that appear to be drainages associated with the dewatering of the ejecta blanket after emplacement. Bamberg crater lies at the top right of this image, and the flow direction was towards the bottom. The small (2.66 km diameter) crater at the top of this image is Gaan crater (38.69°N, 356.60°E). CTX image J01_045298_2193.

Figure 7.23 Channels within the ejecta flow to the west of Hale crater. Image width ~4 km and is centered at 35.10°S, 321.19°E. HiRISE image PSP_007033_1445.

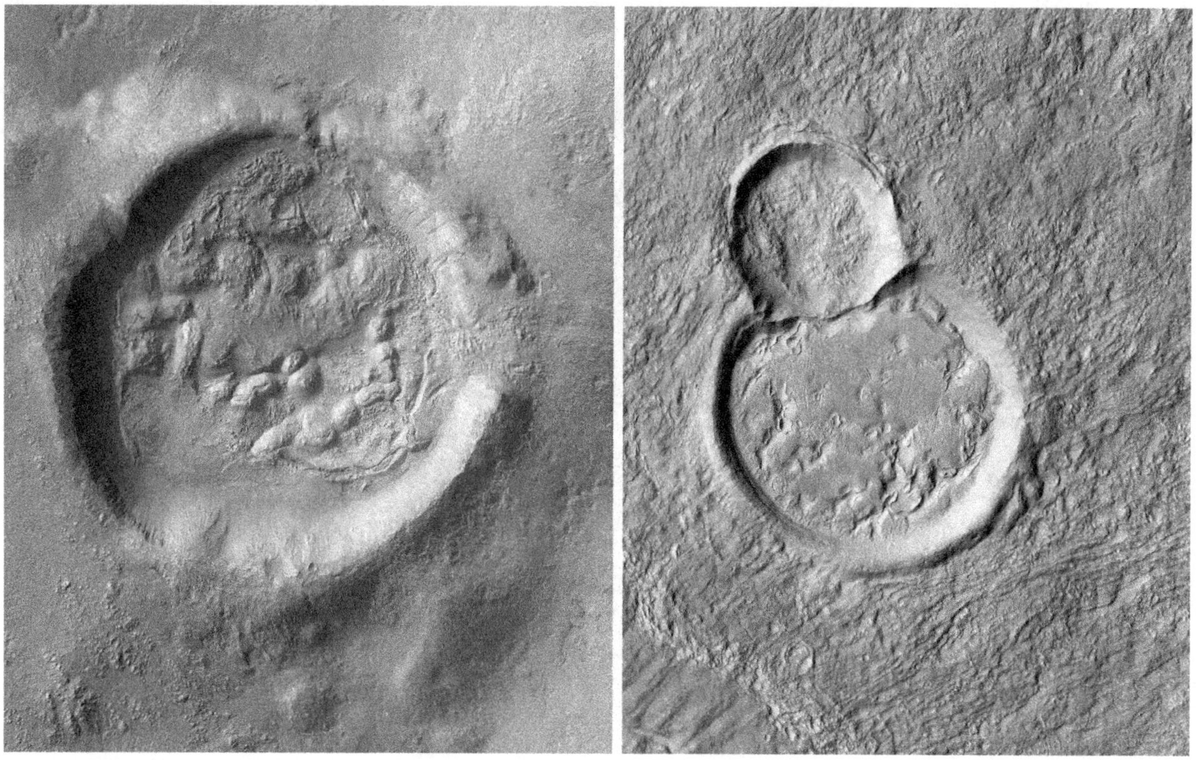

Figure 7.24 Examples of pre-existing craters partially filled with fluid ejecta. Notice that both craters display hummocky surfaces suggestive of partial collapse of the ejecta layer after emplacement. Left: Close to the eastern rim of Hale crater, the ejecta ponded inside an older 9.4 km diameter crater (36.69°S, 325.10°E) and has subsequently deformed. CTX image B03_010896_1425. Right: Pair of craters, the largest of which is ~6.2 km in diameter, to the northeast of Bacolor crater (33.33°N, 118.90°E). These craters lie within the outer layer of the Type-1 DLE crater. CTX image P21_009044_2132.

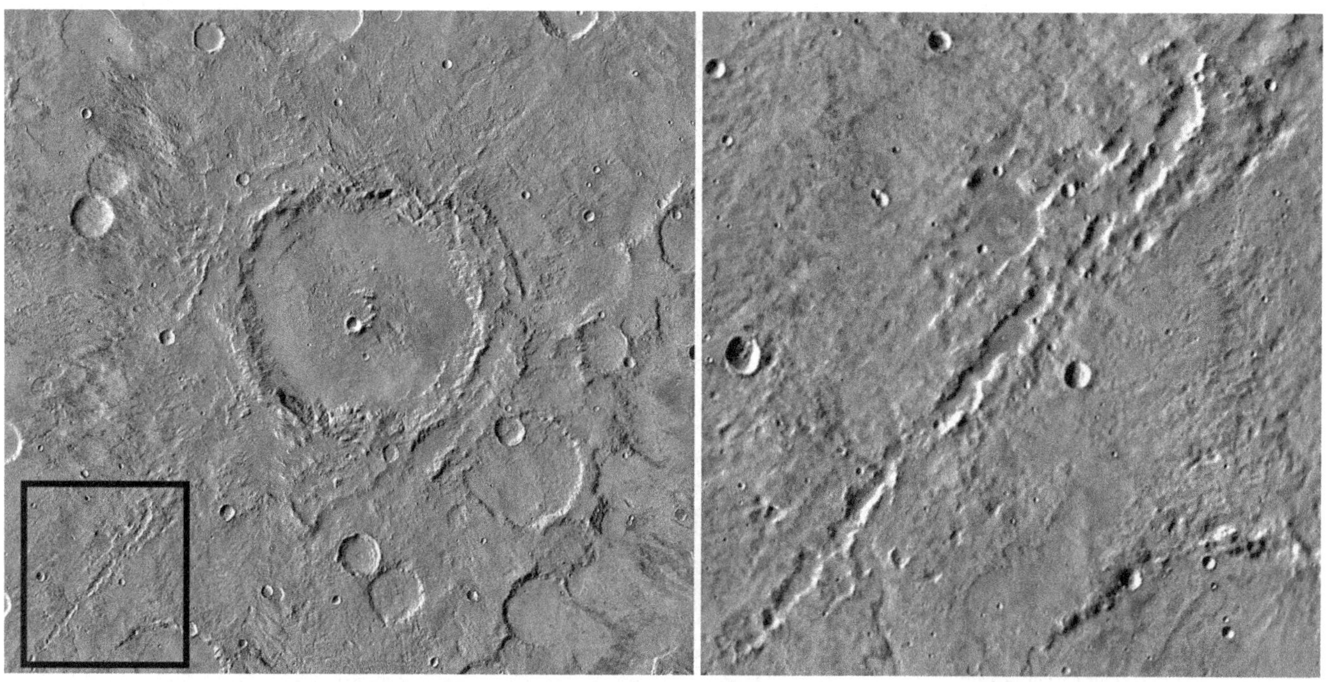

Figure 7.25 Bakhuysen crater (~155 km in diameter, 22.88°S, 15.71°E) is one of the freshest large craters on Mars, preserving several different ejecta morphologies, including a prominent secondary crater chain to the southwest of the crater. The box in the left-hand image shows the location of the image shown at right. Both images are part of the global THEMIS daytime TIR mosaic.

Figure 7.26 Afternoon mosaic of Utopia Planitia (47.84°N, 134.43°E) from Viking Lander 2. The large number of rocks on the surface is believed to originate from the secondary impacts of ejecta from the Mie crater. Viking Lander 2 image mosaic L2.C2.PM.Q1+Q2.

Figure 7.27 This 14.4 km unnamed crater contains heavily eroded, nearly horizontal layers of the Vastitas Borealis Formation (VBF). The smooth area outside the crater rim is also most likely the VBF and appears to be less than ~100–300 thick (Boyce et al., 2005). CTX image P16_007199_2216, located at 40.97°N, 98.25°E.

Figure 7.28 Possible remobilized ejecta on the southern rim of the crater Irharen (6.2 km diameter, 34.50°N, 140.81°E). This part of the ejecta blanket is partially covered by materials originating from Hrad Vallis, which has been postulated to be a mudflow (Hamilton et al., 2018). HiRISE image ESP_023337_2150.

(1981) identified both ballistic and ground-flow ejecta deposits beyond the rim of Schiaparelli Basin.

7.14 EJECTA PARTIALLY BURIED BY THE VASTITAS BOREALIS FORMATION

The VBF is a unique geologic unit on Mars (see Figure 3.5) that extends over much of the Northern Plains (Tanaka et al., 2014). Thus, it is reasonable to expect that the craters that formed in this material may have unusual characteristics. Boyce et al. (2005) demonstrated that craters in this area fall into two major populations: those craters that formed in the VBF and those craters that do not. The deepest craters comprise a Late Hesperian/Early Amazonian-age population. More subdued craters of probable Hesperian age are partially buried by VBF. This modification is not only clear from the interior of the craters containing VBF (Figure 5.7) but also on the exterior ejecta blankets such as the example illustrated in Figure 7.27. In this example, the inner ejecta layer of a Type-1 DLE crater has been mantled by the VBF and only where this mantle has been eroded away can the morphology of the ejecta blanket be observed. Other mantle deposits, possibly associated with mud flows (Hamilton et al., 2018), can be identified at Irharen crater to the west of the outflow channel Hrad Vallis (Figure 7.28).

7.15 SUMMARY

This chapter has shown that there is much diversity in the morphology of ejecta blankets beyond the three principal ejecta types described in Chapter 6. Small crater diameters (say, those less than 3 km) or the geographic setting of the crater may also influence ejecta morphology. Unusual properties of the target at the time of impact may also produce unusual ejecta, as in the case with Hale crater. Similarly, the unique attributes of the Vastitas Boreas Formation and the sequence of deposition and erosion events act to modify an ejecta blanket.

We have not yet dealt with some of the smaller scale features associated with ejecta layers. In particular, morphologically fresh craters may provide useful insights into the flow properties of the ejecta, enabling older examples to be more fully understood. Thus, in Chapter 8, we take a closer look at some fresh ejecta flow features.

8

Ejecta Flow Features

8.1 INTRODUCTION

In Chapters 6 and 7, we reviewed the primary types of ejecta surrounding different types of craters. The single-layered ejecta (SLE), Type-1 double-layered ejecta (Type-1 DLE), and multi-layered ejecta (MLE) were identified in Chapter 6 as the main crater types, but inevitably there are considerable differences within a single crater type, based on the morphology of the ejecta layers. This is also true for the other ejecta types, as described in Chapter 7. Here we delve more deeply into these differences and show some of the rare features and characteristics associated with the freshest examples of craters. When trying to understand the flow processes displayed by the ejecta, these features no doubt provide additional details on the emplacement process as well as illustrate the potential variability across the planet as a function of geographic location.

8.2 THERMAL PROPERTIES OF EJECTA?

The thermal properties of any ejecta blanket at the instant of emplacement are hard to determine. Presumably some of the ejecta were at quite high temperatures, but particularly the finer materials would have been cooled quite rapidly due to their intimate contact with the cooler ejecta that originated further from the point of impact. This cooling may have been particularly important during the surface flow of the fluidized ejecta, but the ejecta may still have come to rest at an elevated temperature. Kieffer and Simonds (1980) considered the role of volatiles and lithology during the production of impact melt in Martian craters, but they did not investigate the probable temperature distribution within the ejecta layers after they were emplaced. Similarly, models of impact-induced hydrothermal activity at a crater have previously only considered the crater interior (Abramov and Kring, 2005). Few terrestrial craters show evidence for hydrothermal activity within the ejecta blanket and so offer little in the way of a planetary analog. One good example was documented by Sapers et al. (2017) who found mineralogical evidence for localized hydrothermal activity in the ejecta beyond the rim of the Ries crater, Germany, implying that these ejecta were at a relatively high temperature. Unfortunately, determining the temperature distribution of an ejecta layer on Mars at the time of emplacement is beyond the available data, as the volume of ice within the target, as well as the stratigraphy of target rocks, would conspire to make most thermal regimes plausible. But a search for morphologic signs of hot ejecta on Mars at the time of their deposition may provide us with some clues as to the thermal regime.

Figure 8.1 Does this unnamed 830 m diameter crater (48.76°N, 127.45°E) show evidence that the ejecta were emplaced at an elevated temperature, and therefore able to sink into a substrate containing ice? The illumination direction is from the left, so the ejecta blanket is topographically lower than the surrounding terrain. HiRISE image ESP_018709_2290.

One possible indicator that at least some ejecta layers were emplaced warm (or hot) comes from the Northern Plains (Figure 8.1). Here a small (830 m diameter) crater has apparently formed on target materials that were volatile-rich. After emplacement, all parts of the ejecta blanket subsided, most likely because the hot ejecta heated the substrate causing the ejecta to sink into the underlying layers. There is no clear difference here in azimuth from the crater cavity for the depth to which the ejecta sank, as might be expected if hotter ejecta were to land at greater radial distances because they were ejected from closer to the point of impact. Potentially, this might suggest that the flow process was sufficiently turbulent that the hot ejecta were mixed with cooler material in a fairly uniform manner.

8.3 CHARACTERISTICS OF DISTAL EJECTA

The prominent ejecta layers that characterize SLE and MLE craters were evidently not the only material distributed around the primary during a cratering event. Craters such as Gratteri

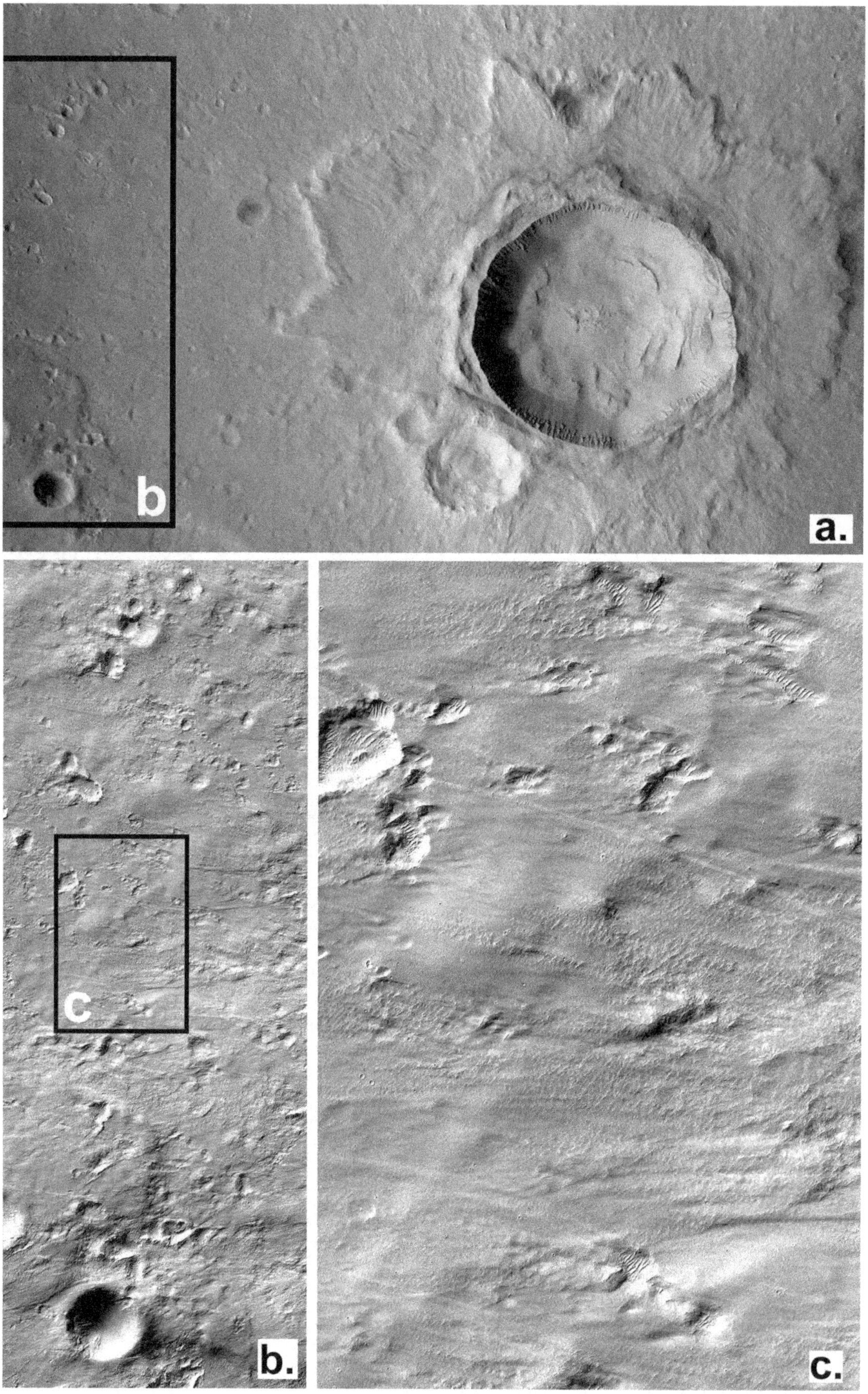

Figure 8.2 The fresh SLE crater Gratteri (7 km in diameter, located at 17.70°S, 199.93°E) shows that craters have a considerable influence on the terrain surrounding the visible ejecta layers. In subscene (b), which is ~10 km from the crater rim, smooth ejecta flows are visible. At higher resolution (subscene c), these are clearly flows of material but not necessarily of material ejected from the cavity. Potentially, this material may be reworked sediments that were on the surface at the time of the impact event and were subsequently moved by the radial flow of material from the impact site. Top and lower left: CTX image P07_003820_1628. Lower right: HiRISE image ESP_025946_1620.

(Figure 8.2) display flow materials at greater radial distances than were reached by the ejecta layers. At a radial distance of ~3 crater radii (~10 km), a considerable amount of material was displaced to produce flow-like features not associated with secondary craters. Potentially this material may be similar to the "blast zone" deposits observed by Daubar et al. (2016) around small craters formed within the satellite era (Figure 7.17). This material may comprise reworked sediments that were on the surface before the impact, or the deposit might comprise fine particles ejected to a greater radial distance by atmospheric processes generated by the impact itself. A second crater, this time ~4 km in diameter, further illustrates the subdued nature of this distant material (Figure 8.3). At this crater, smooth material that lacks individual boulders is at comparable distances to much more rugged deposits with abundant boulders on the surface.

These bland ejecta deposits at relatively large radial distances may be less eroded examples of the LARLE ejecta blankets that were discussed in Section 7.4. A rare example of a Type-1 DLE crater with subdued ejecta that extends more than 12 crater radii from the rim crest of the parent crater is shown in Figure 8.4. Unlike the LARLE craters illustrated in Chapter 7, this example appears to be morphologically fresh with numerous secondary craters between 18 and 34 km from the rim. This raises the intriguing possibility that more Type-1 DLE craters had secondary craters when they formed; it is only due to the erosion of the distal ejecta that there is the impression that these craters lack secondaries (e.g., Boyce and Mouginis-Mark, 2006).

8.4 CLUES TO EJECTA LAYER EMPLACEMENT

The timing of ejecta emplacement was evidently quite complex, with different ejecta layers potentially moving at different speeds or, alternatively, layers were formed at slightly different times. Detailed mapping of Tooting crater (Mouginis-Mark, 2015) was able to deduce the general sequence of events. This sequence may have unfolded in seconds to tens of seconds, as ejecta from different parts of the crater cavity reimpacted the surface. The complexity in timing is illustrated in Figure 8.5 where three ejecta layers, each with their own distal ramparts, interacted with each other. Ejecta layers may also have become "disconnected" or split into several individual parts leaving distal ramparts separated from the original ejecta layer (Figure 8.6). This split in the layers may be a result of the emplacement process or it may simply indicate that the thickness of the layer proximal to the rampart was so thin that eolian activity has buried the thinnest parts.

Boyce and Mouginis-Mark (2025) proposed a conceptual model for the cause of differences in rampart relief and as such may provide further information on the timing of ejecta layer emplacement. The disconnected rampart in Figure 8.6 illustrates the thin nature of the ejecta layer upslope toward the crater rim. The key attribute associated with rampart formation is likely to be that they were produced by flows that initially increased in velocity progressively outward from the crater. As illustrated in Figure 1.9, ballistic ejecta from the crater cavity would re-impact the ground at higher velocities as the radial range increased, thereby accelerating the flow and adding new material as the radial distance increased (Boyce and Mouginis-Mark, 2025). If this model is correct,

Figure 8.3 (a) The secondary craters from an unnamed 4.2 km diameter crater (located at 15.62°S, 121.72°E) display herringbone textures reminiscent of the lunar craters (see Figure 2.19). These Martian examples may also have had an effect on the target material close to the primary crater and reworked fines on the surface in a manner comparable to the "blast zone" observed around some of the freshest small craters. HiRISE image PSP_010033_1640. The boxed area denotes the coverage of Figure 8.3b. (b, c, d) Details of the ejecta pattern illustrated in Figure 8.3a. In part "b," a secondary crater chain and (at right of the image) a pair of large secondary craters have created topographically rugged terrain. In parts "c" and "d," notice that the ejecta with a subdued morphology appears to overlie the rugged terrain associated with the secondary craters. The subdued area is almost devoid of boulders on the surface, while the adjacent deposits display abundant boulders. This suggests a late-stage process formed a mantle on top of the earlier components of the ejecta blanket. All subscenes are parts of HiRISE image PSP_010033_1640.

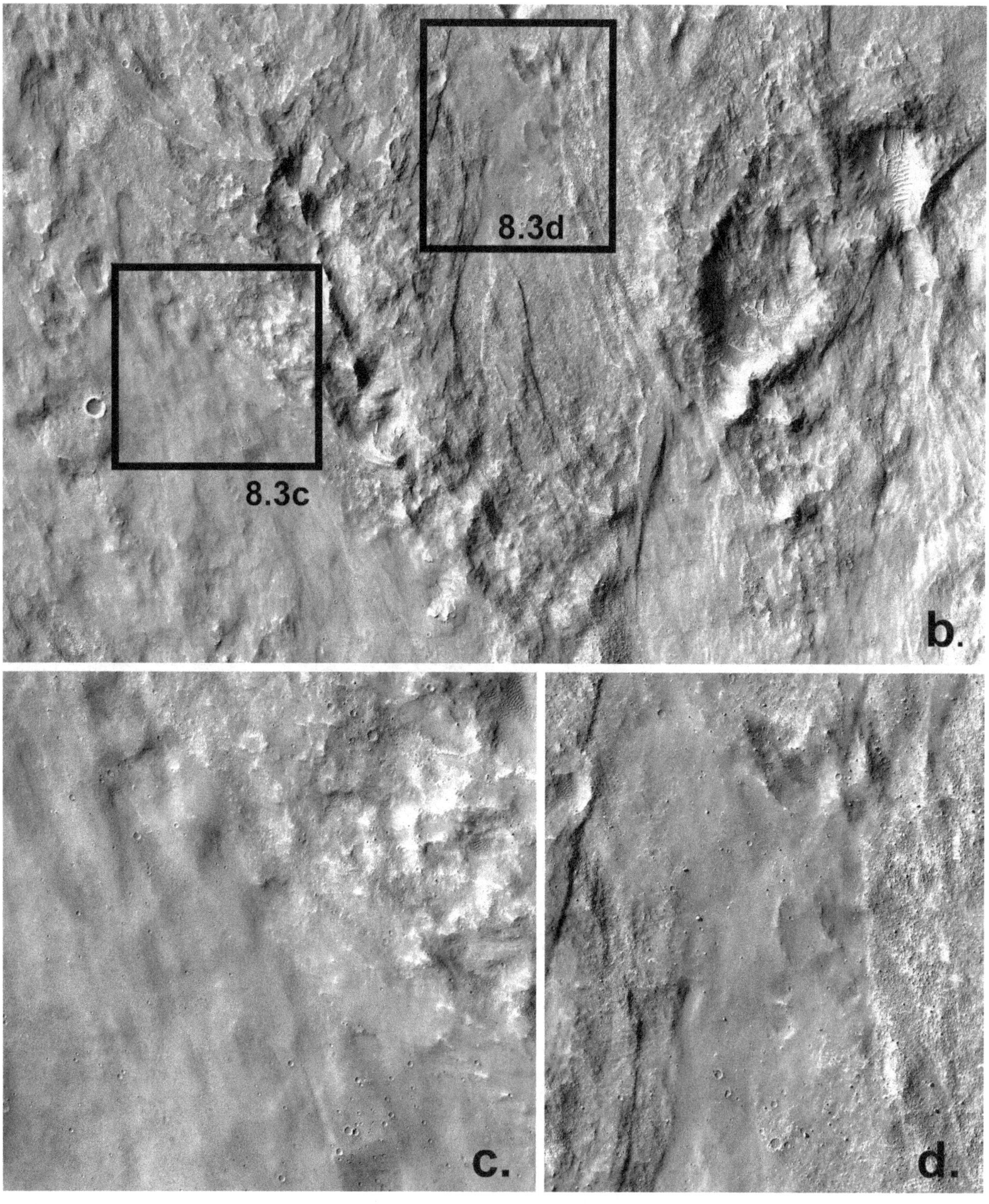

Figure 8.3 (Cont.)

Figure 8.4 Lonar crater is a very fresh 11 km diameter Type-1 DLE crater at a high latitude (72.99°N, 38.29°E) and shows the extensive effects of cratering at these latitudes. While both the inner and outer ejecta layers are reminiscent of lower latitude craters, additional features can be seen such as deposits extending far beyond the edge of the outer layer that are more characteristic of LARLE craters (see Section 7.4). Dunes and swirl-like features can be identified out to ~50 km from the edge of the outer ejecta layer and ~65 km from the parent crater rim. The dark stripes running across the ejecta blanket are eolian features unassociated with the crater. CTX image P22_009798_2530.

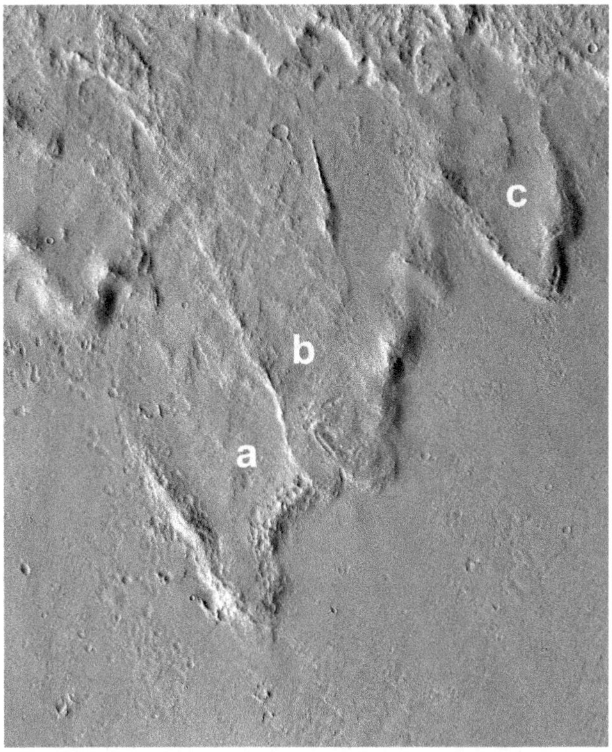

Figure 8.5 Ejecta flow patterns in the ejecta of an unnamed 22 km diameter MLE crater located at 18.02°N, 249.06°E. Here we see that the fluidized layers came to rest at slightly different times. At a radial distance of ~24 km from the rim, flow "b" over-ran the previously emplaced flow "a." In turn, flow "b" may have been partially over-run by flow "c." Notice the abundance of pits along the crest of the rampart of flow "a." Direction of ejecta flow was toward the bottom of the image. Image width is ~5 km. CTX image G04_019786_1985.

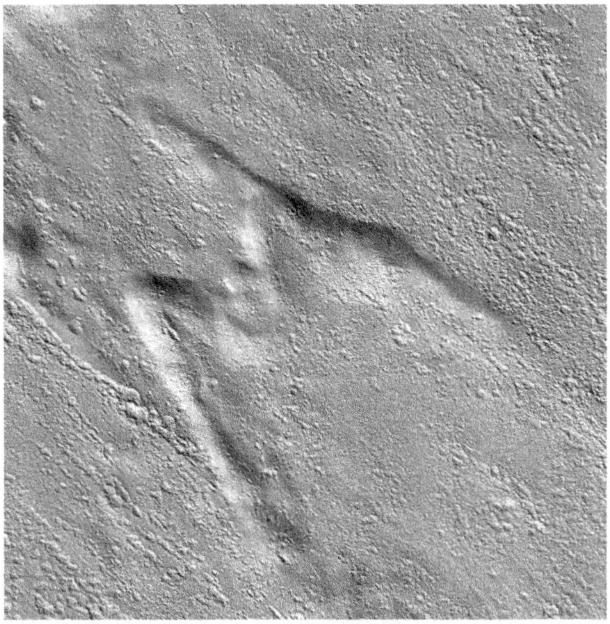

Figure 8.6 Sometimes the distal rampart became "disconnected" from the main ejecta layer. This "V-shaped" rampart is associated with Tooting crater and is ~42 km (3 crater radii) from the rim of the crater, which lies to the bottom right in this image. Image centered at 23.91°N, 207.02°E. Further examples of these disconnected ramparts can be found up to 56 km from the rim. Notice the myriads of small secondary craters on the rampart as well as behind and in front of the topographic high. Image width is ~5.6 km. HiRISE image ESP_013089_2040.

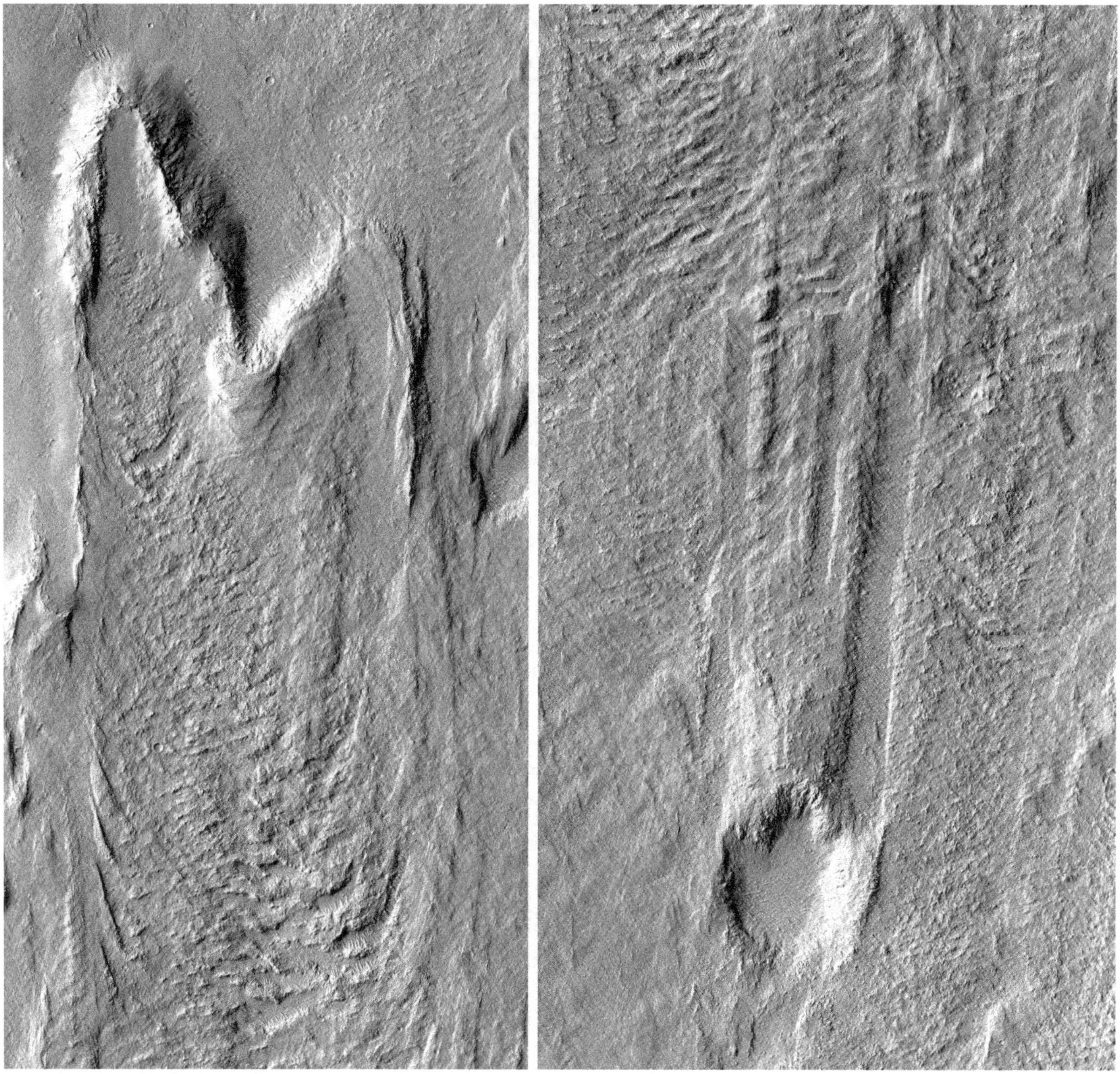

Figure 8.7 Left: Hummocky facies (23.70°N, 207.80°E) associated with an ejecta layer ~14 km from the northern rim of Tooting crater (Mouginis-Mark, 2015). This morphology may have resulted from the movement of the ejecta layer over undulating topography. Direction of flow was toward the top of the image. Image width is ~2.7 km. Right: Grooves extend away from a secondary crater 22 km north of the rim of Tooting crater (23.83°N, 207.81°E). This implies that the secondary crater formed prior to the arrival of the ejecta layer, which is inferred to be ground-hugging. Direction of flow was toward the top of the image. Image width is ~1.2 km. Both images are parts of HiRISE frame ESP_018258_2040.

then the multiple ejecta layers displayed by MLE craters may well have had their own velocity linked directly to the arrival of the ejecta from the still-forming crater cavity. Such timing might be controlled by the depth of ejection of the material as well as the angle and velocity at which the ejecta left the cavity. It is therefore not surprising that there is an observed transition between small SLE craters and larger MLE craters as their ejecta velocities were no doubt influenced by the depth of cavity excavation.

Within fresh ejecta layers, up to four different ejecta facies can be identified (Mouginis-Mark and Boyce, 2012; Mouginis-Mark, 2015). Such layers can be well defined, with variations in surface texture identified within individual layers. The near radial symmetry of these layers suggests individual pulses of material sweeping out from the rim of Tooting crater. A "smooth facies" was interpreted to indicate uniform flow with no relative velocity gradient within the layer. A "hummocky facies" (Figure 8.7a) was interpreted to characterize areas where ejecta

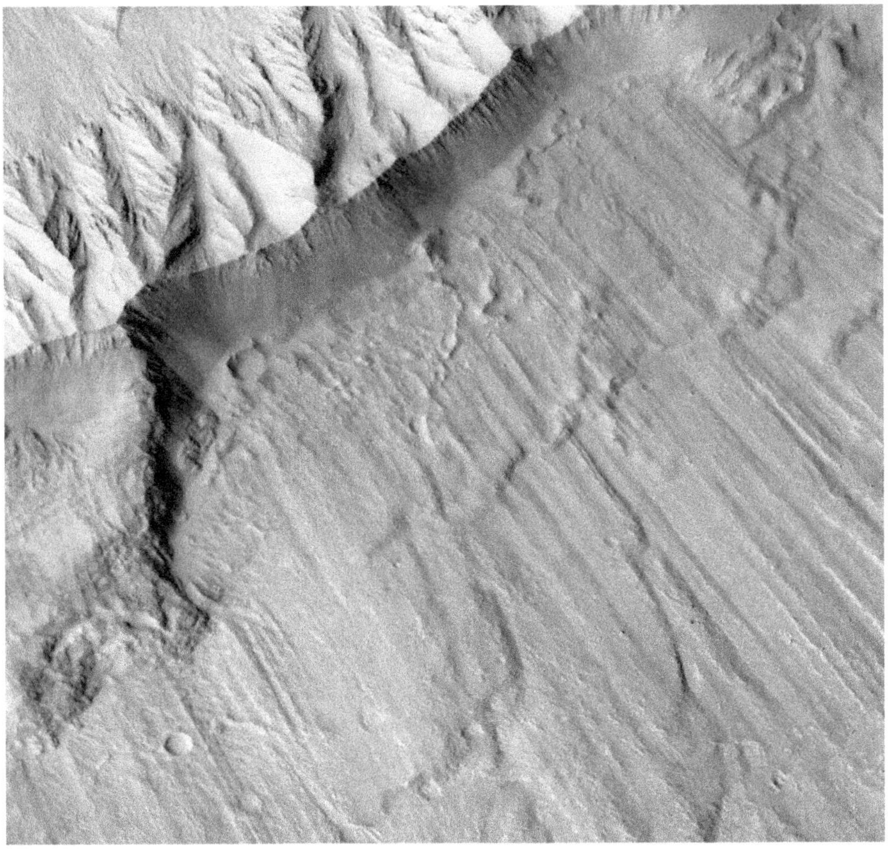

Figure 8.8 Oblique view looking north at the southern rim of Bacolor crater (33.01°N, 118.58°E) showing the steep, high-relief scarp, which was probably produced when a portion of the outer rim collapsed. Note that there is no trace of a deposit produced by the collapse. The absence of this material suggests that it was most likely transported away during the radial flow of ejecta and incorporated into the material that now forms the inner ejecta layer. Also, the inner layer grooves are superimposed on the debris where the collapse took place indicating that the grooves must have formed after the rim collapse. Image width is ~8.0 km. CTX image P16_007462_2133. The digital elevation model was derived from CTX frames P15_006750_2133 and P16_007462_2133.

Figure 8.9 Clear evidence of the erosional process that carved the groves on the inner layer of Type-1 DLE craters can be seen in the walls of smaller impact craters. The larger crater in this image is ~1 km in diameter and shows deep incisions on both the up-slope ("A") and down-slope ("B") inner walls where grooves have been cut into the pre-existing terrain. The direction of ejecta flow was from the top of the image. The parent crater (Jaisalmer crater; 33.52°N, 84.11°E) is ~15 km in diameter, and the rim crest is ~5 km from this scene. HiRISE frame ESP_044675_2135.

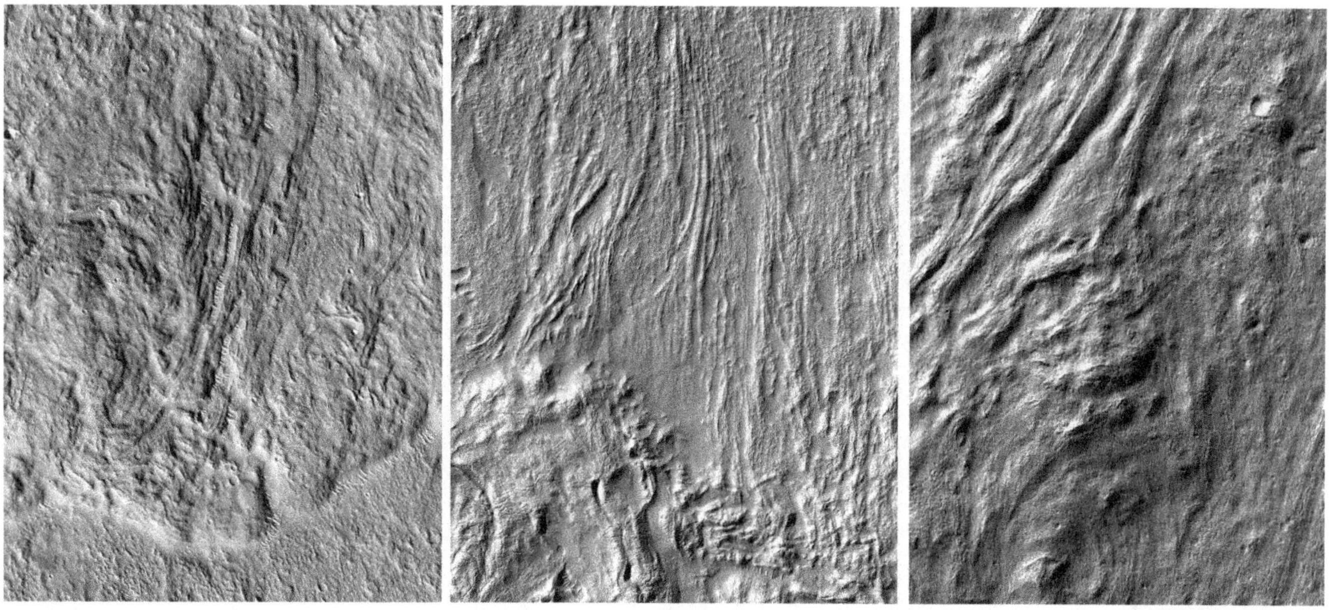

Figure 8.10 The pronounced meandering grooves formed on the outer ejecta blankets of Type-1 DLE craters are very similar even when there is a considerable difference in the elevation of the crater. This suggests that the material from each crater had similar flow characteristics. These examples were formed on terrains with an elevation difference of more than 5 km. On the left we see the outer layer of Bacolor crater (33.01°N, 118.58°E). Center image shows a relatively straight groove from a 16 km diameter crater (38.00°S, 16.43°E). Illustrated at right is the inner layer of a 13 km diameter crater at 41.12°S, 196.55°E. The crater rim is at the top of the image. The direction of ejecta flow was toward the bottom of the image. At left, HiRISE image ESP_020305_2125, center HiRISE image ESP_028010_1415. At right, HiRISE image ESP_044658_1380. Width of each image is ~4.5 km.

Figure 8.11 The distal margin of the outer layer of Type-1 DLE craters can have a pronounced outline. Here we see ridges that curve with the outline of the margin from a 14.3 km crater at 41.12°S, 196.55°E. Notice the convoluted "swirls" of material at this leading edge. This lobe traveled ~23 km from the rim crest of the crater, which is located at the top of the image. Image width is ~4.5 km. HiRISE image ESP_044658_1380.

flow overrode pre-existing topography, most likely created by earlier components of the ejecta. A more "crenulated facies" was identified close to the perimeter of the flow where the ejecta. A "radial facies" was identified down-range of discrete topographic obstacles such as large, partially buried secondary craters (Figure 8.7b).

Further details of the timing of ejecta emplacement are often hard to confidently identify, but one fine example where this

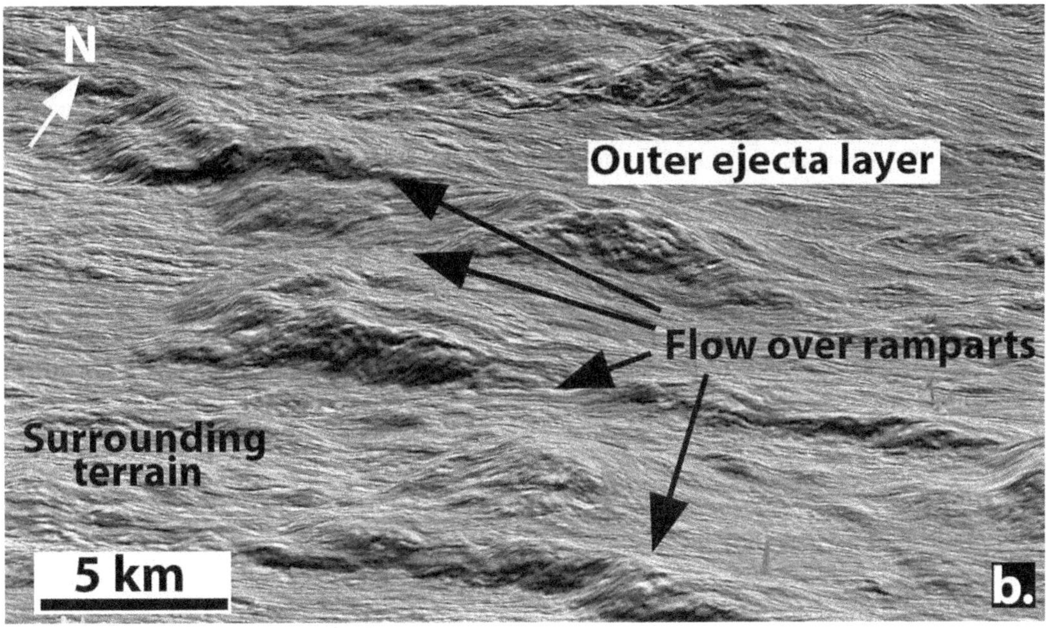

Figure 8.12 Oblique view of the inner and outer ejecta layers of Jaisalmer crater (15 km diameter, 33.52°N, 84.11°E), showing some of the clearest evidence on late-stage flow of the outer ejecta layer of a Type-1 DLE crater as it crosses the inner ejecta layer. The direction of ejecta flow was toward the left of the image. Digital elevation model derived from CTX images F23_044886_2136 and J01_045229_2136.

is possible is the southern rim of Bacolor crater (Boyce and Mouginis-Mark, 2021). Bacolor is a 20 km diameter Type-1 DLE crater that formed on the eastern margin of Utopia Planitia. At this crater the outer southern rim has collapsed during ejecta layer emplacement (Figure 8.8). Almost the entire rim collapsed along a 3.5 km segment of the outer wall and the resultant material was transported radially away from the crater by the ejecta layers. No trace of this collapsed material can be seen within the ejecta blanket. Most noteworthy is that at the scar base where the rim once stood, grooves cut into the inner ejecta layer are continuous (Figure 8.8). It seems likely that were this collapse to have occurred in coherent, blocky material, some evidence of the displaced mass would be visible in the ejecta layer, but it is not. The implication is that the rim of Bacolor crater is composed of fine material that could be easily reworked during lateral transport.

This absence of material that once formed the rim of Bacolor crater must have first been uplifted during the early stages of cavity formation, and then underwent a catastrophic collapse before the grooves were formed. Mouginis-Mark (1981) and Boyce and Mouginis-Mark (2006) used this observation to conclude that the outer layer of the ejecta blanket was emplaced after the inner layer because the grooves cross continuously from the inner layer to the outer layer. Some layering can be identified within the collapse scar on the rim, so that the material from which the slide was made was, temporarily, in a coherent form. Clear evidence exists to show that the formation of the grooves on Type-1 DLE craters was quite erosive (Pietrek et al., 2020). For example, the erosive character of the inner ejecta layer on the pre-existing surface can be seen in Figure 8.9. Here the ejecta layer has crossed over a smaller pre-existing crater and scoured a radial groove into the smaller crater's rim crest. It is apparent that not only the up-slope and down-slope crest of the crater rim was scoured but also the erosion continued onto the floor of the crater. The grooves can be traced across the crater floor and continue downrange with a comparable spacing and shape.

Most Type-1 DLE craters such as Bacolor are found in the Northern Plains, but as we have seen in Chapter 6 rare examples can also be found in the southern hemisphere (see Figure 6.3). There is a difference in elevation of >~5 km between these two localities and so some fundamental differences in emplacement might be expected due to differences in the rheology of the ejecta. However, this does not appear to be the case. Two of the freshest Type-1 DLE craters in the southern hemisphere show the same type of meandering groove morphology that can be found at Bacolor crater (Figure 8.10). The meandering grooves have lateral ridges, more suggestive of leveed channels than depressions carved into the substrate, and the distal margin of the outer layer of Type-1 DLE craters can have a distinctive outline (Figure 8.11). The outer edge of the ejecta layer has a pronounced curvilinear edge, and the flow appears to have stopped incrementally along the margin. Jaisalmer crater displays one of the clearest examples of the emplacement of ejecta layers for a Type-1 DLE crater (Figure 8.12).

Salient features of the ejecta layers of layered ejecta from different crater types and sizes allow the general characteristics to be summarized (Table 8.1). To a first order, the flow features on SLE craters are like those of MLE craters, but both are different than those on Type-1 DLE craters. All ejecta layers are crossed by radial grooves and ridges that differ with crater type and layer. The grooves of the inner layers of Type-1 DLE craters are relatively wide and form straight troughs that appear to be insensitive to most obstacles. These grooves also show no fanning at lobe edges. The radial grooves on MLE and SLE ejecta layers are generally straight but can be deflected around topographic obstacles. Some grooves fan at the edges of lobes. In contrast, the radial grooves of the outer ejecta layers of Type-1 DLE craters are typically long and sinuous. These features can be deflected by pre-existing topography. All ejecta layers, except the outer ejecta layer layers of Type-1 DLE craters, have closely spaced sets of transverse ridges and troughs. The ridges and troughs have been interpreted to

Table 8.1 *Morphologic characteristics of flow features on layered ejecta.*

Inner ejecta layer	DLE	MLE and SLE
Straight radial grooves and ridges	Broad undeflected by topography. Terminates abruptly, no curving of fanning at layer edge	Deflected by topography. Commonly terminates abruptly at layer edge, but can curve and fan
Transverse troughs and ridges	Sets in wave-like patterns	Same as DLE craters
Transverse graben-like troughs and scarps	Frequently segments intersect in V-shaped patterns	Same as DLE craters
Layer termination	Relatively broad rampart	Relatively narrow rampart
Smooth pitted material	No	Small patches on MLE ejecta, and mostly beyond rampart on SLEs
Outer ejecta layer	**DLE**	**MLE**
Straight radial grooves and ridges	No	Yes, can fan out at ramparts and be deflected by topography
Radial sinuous grooves and ridges	Yes, deflected by topography	No
Transverse troughs and ridges	No	Sets in wave-like patterns
Layer termination	Low, relatively broad rampart	Relatively narrow rampart
Smooth pitted material	No	Mostly blanketing outer ejecta layer and surrounding surface

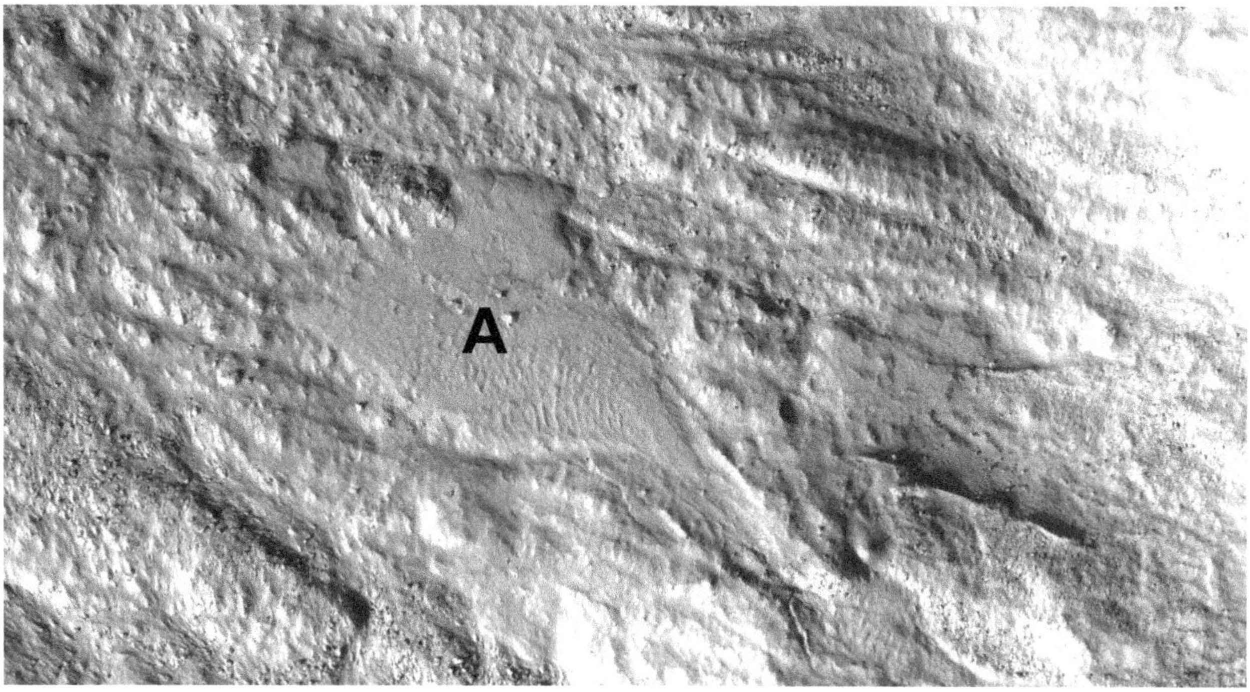

Figure 8.13 Once coming to rest, the ejecta layers rarely seem to have become remobilized. Here we see a rare case where a lobe of material ("A") has flowed after the inner layer of a Type-1 DLE crater was deposited. The crater is ~6.3 km in diameter at 43.26°N, 225.77°E. The source of this flow (to the right of the image) is ~200 m from the rim of the crater, and the distal part of the flow is 930 m from the rim crest. Width of image ~600 m. HiRISE image ESP_020512_2235.

be roll waves (Boyce and Mouginis-Mark, 2021). Extensive smooth pitted deposits are common on and around fresh MLE and SLE craters. Smooth pitted deposits are found on, and beyond the outer layer of MLE craters. For SLE craters, pits are seen beyond the ejecta layer and appear to be like those on the floors of fresh craters.

8.5 EARLY PHASES OF MODIFICATION OF DLE EJECTA

Most of the Type-1 DLE craters in the northern hemisphere have been modified by erosion from their pristine form. However, there are a few examples where the ejecta layer appears to have

Figure 8.14 There are a few clear examples of deformation of the inner ejecta layer of a Type-1 DLE crater. More chaotic terrain appears to have been remobilized after initial emplacement. Jaisalmer, the parent crater (33.52°N, 84.11°E), is ~15 km in diameter, and the rim crest (to the top of the image) is ~5 km from this scene. Width of the image is ~2.6 km. HiRISE image ESP_044675_2135.

preserved modifications from soon after crater formation. In some instances, the flow of remobilized material after the ejecta came to rest can be identified (Figure 8.13). Evidence of such remobilization is quite rare but, in this case, it hints at the existence of a considerable amount of fluid being contained within the ejecta layer. What the cause of the remobilization is remains unclear. The transient creation of local topographic highs and the subsequent slumping of material appear to be reasonable. Other forms of early modification of the ejecta layer can also be found. While most ejecta layers have remained structurally unmodified, slumping of the ejecta where it came to rest on local topographic highs appears to have been possible (Figure 8.14).

8.6 SUMMARY

In this chapter we have illustrated some of the smaller-scale details of the ejecta layers, drawing upon some of the morphologically freshest craters. From these examples, the kilometer-scale landscape provides useful information to further interpret the emplacement process. But even with this higher resolution, numerous questions remain concerning crater formation, particularly when one considers differences in target material, location on the planet, and the age of the target. Throw in the possibility of obliquity changes producing very different climate conditions even for craters formed adjacent to each other, and the puzzle becomes even more challenging. So, in Chapter 9, we try to resolve some of the puzzles and make suggestions for how future research might bring greater clarity to the topic.

9

Remaining Puzzles Concerning Martian Craters

9.1 INTRODUCTION

In this final chapter, we consider some of the previously discussed hypotheses concerning the formation of impact craters on Mars. One is the hypothesis that the degree of fluidization of crater ejecta is related to volatiles within the target at the time of crater formation, rather than being an atmospheric effect. Soon after Martian craters were first seen in *Viking Orbiter* images, the debate focused on the relative importance of volatiles within the target compared to the fluidization of ejecta because of the inclusion of the atmosphere within the ejecta (Schultz and Gault, 1979; Schultz, 1992). The importance of subsurface volatiles rather than an atmosphere is supported by the observation that fluidized ejecta from craters of varying size are also found on airless bodies (Figures 2.26 and 2.27) (Boyce et al., 2010; Robbins et al., 2018b).

Regional differences in the target elevation must also be considered. Coupled with the increasingly higher image spatial resolution over the past few decades (e.g., Figure 3.1), there have been major steps taken in improving our knowledge of the topography of impact craters, and these data have enabled the quantification of crater geometry to the extent that the role of volatiles can be assessed. From the Earth-based radar measurements made by Roth et al. (1989), to the laser altimetry provided by MOLA (Smith et al., 2001), and the use of stereo-derived digital elevation models (DEMs) (Mouginis-Mark et al., 2018), we have gained over time a better understanding of crater morphology. Spectral measurements from orbit (Goudge et al., 2015; Christian et al., 2022) have further refined our knowledge of the geology of Mars. This improvement in data coverage and quality has led to the interpretation that the fluidized morphology is most likely the product of water or ice within the target material at the time of crater formation (e.g., Mouginis-Mark, 1987; Barlow and Perez, 2003; Barlow, 2004). So far in this book, the temporal and geographic distribution of these volatiles have not been addressed even though Mars has clearly experienced dramatic climate variations over its history (see Chapter 3). Similarly, we have not yet tried to correlate interior and exterior attributes of different craters.

Questions pertaining to the use of impact craters as probes of the Martian subsurface can still be addressed. Is it possible, for instance, to use the surface area of ejecta blankets around impact craters of similar size and age to infer changes in the distribution (either with depth or total abundance) as a function of time? Could the emplacement of crater ejecta have been controlled by the elevation of the target rock at the time of crater formation? As an important check on our interpretations, if no variation in ejecta blankets can be identified over a broad elevation range, then the atmosphere would appear to have a minor role (at best) in controlling the emplacement process. We may also ask if there are any anomalous craters that fail to fit any simple model for the formation of fluidized ejecta layers.

We will discover in this chapter that none of the earlier questions can be fully resolved with the available data. Despite five decades of analysis, many aspects of Mars crater morphology and evolution remain enigmatic, and it seems likely that new types of data will be needed to find the answers. As a final section in this chapter, we therefore offer new approaches to solving these questions. Finding the answers will require a new orbital data set. Our recommendation is for a new data set that is comparable to many that have been collected for other planets in the Solar System, and thus well within the capabilities of the National Air and Space Administration (NASA) and other international space agencies. If an orbital mission cannot provide the required answers, then the characteristics of a rover mission equivalent to those already operating on Mars may provide sufficient imperative to further study craters.

9.2 ARE SIMILAR-SIZED CRATERS CREATED EQUAL?

One of the underlying assumptions made by most investigators is that fluidized ejecta is due to water in the ejecta and that the "degree" of fluidization is due to the amount of water rather than the different physical properties of the target (e.g., dense lava flows or brecciated uplands appear to host comparable craters). To a certain extent, this can be tested by comparing similar-diameter craters formed on different rock types around the planet (Figure 9.1), as well as comparing comparable-sized craters that formed in close geographic proximity to each other (Figure 9.2). It is reasonable to expect that craters formed in the Southern Highlands, where weathering and protracted cratering have produced a fragmented stratigraphy, would appear different from craters formed in the coherent young lava flows of the Tharsis region. As can be seen from Figure 9.1, this does not appear to be the case. Variations in target properties do not appear to produce significantly different crater morphologies, as similar-sized craters have multi-layered ejecta (MLE) blankets with prominent distal ramparts irrespective of the rock in which the craters formed.

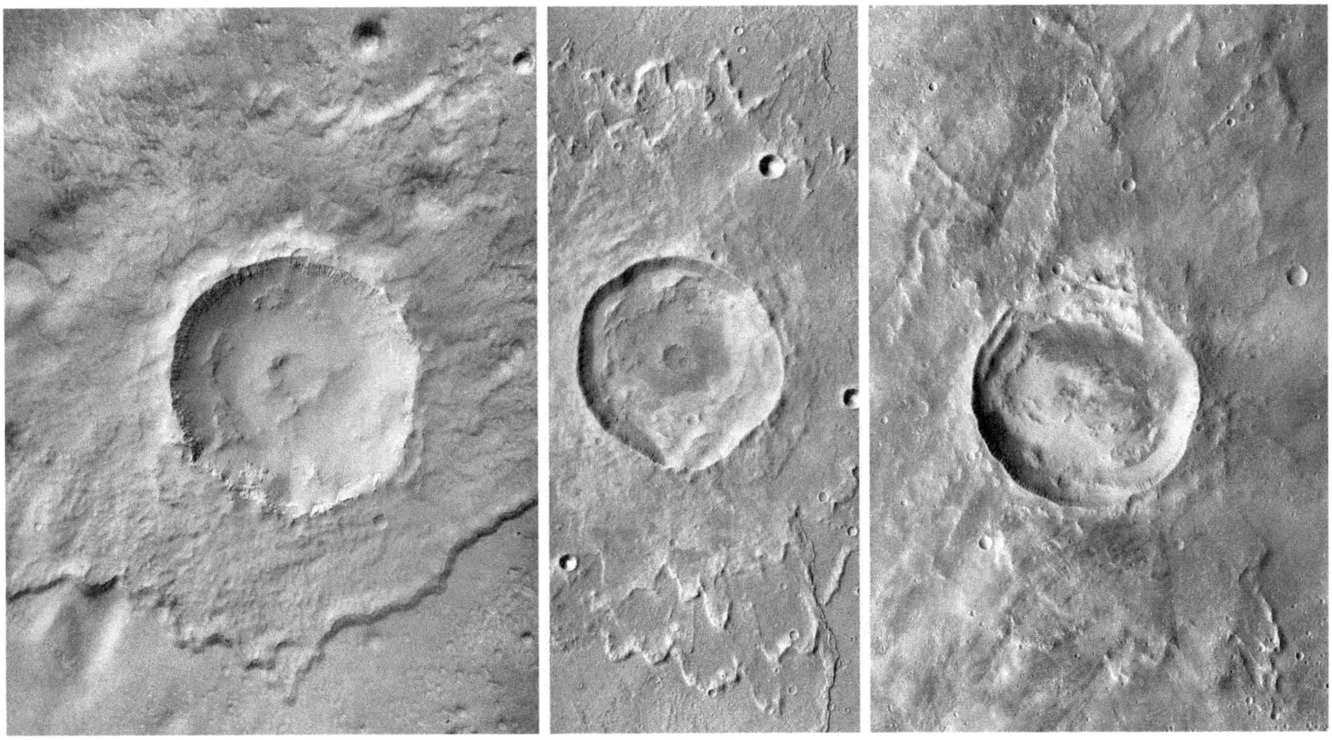

Figure 9.1 Morphologically similar craters with similar diameters have formed on diverse terrains on Mars, suggesting that rock type is not the determining factor for the formation of fluidized ejecta. At left: 12 km diameter crater located in the highlands at 0.40°N, 357.27°E. CTX image F01_036054_1803. Center: 15.9 km diameter formed on ridged plains at 6.83°N, 296.44°E. CTX image B02_010554_1869. Right: 14.1 km diameter crater formed on lava flows from Elysium Mons, located at 27.01°N, 147.70°E. Mosaic of CTX images P16_007250_2051 and P20_008832_2094. All three of these craters display distal ramparts on their ejecta blankets, as well as terraces and a probable central pit on the crater floor.

Three craters that formed on Lunae Planum (Figure 9.2) suggest a similarity of final crater form for comparable impact events. In this example, the diameter of each crater ranges from 18.6 to 19.1 km. The crater interiors all display flat floors with a subtle central pit, the wall units are comparable with terrace blocks, and the ejecta layers extend similar distances from the respective crater rims. While one cannot be certain that these three craters formed close in time (so that the subsurface conditions would have been very similar), the stratigraphy of the ejecta layers indicates that the southernmost crater formed before the middle crater, which in turn formed before the northernmost crater. Of course, we do not know if the masses or types of projectiles that formed the craters were similar. Nevertheless, the existence of these three craters so similar in size and morphology near each other provides some confidence in saying that comparable impact energies in the same target material can produce craters with a similar appearance.

9.3 IS THERE AN ELEVATION EFFECT FOR EJECTA FLUIDIZATION?

If the Martian atmosphere had a significant role in ejecta fluidization, it seems reasonable to expect that craters formed at low elevations (where the atmosphere is thicker) would systematically have more highly fluidized ejecta than craters that had formed at high elevations. Syria Planum (Figure 9.3) offers a place where this idea can be tested. The plateau is at an elevation of >7 km about Mars datum and has several large (>5 km diameter) impact craters. All these craters possess fluidized ejecta with prominent distal ramparts. The high elevation appears to have had no influence on the formation of the type of ejecta. Unfortunately, the data set of large craters at even higher elevations is very small so that the notion that high elevation precludes ejecta fluidization cannot be rigorously tested. Only Pangboche crater (elevation ~20.9 km; Figure 4.26) and Karzok crater (elevation ~18.8 km; Figure 4.16) are at higher elevations on the upper flanks of Olympus Mons volcano. Both Pangboche and Karzok craters lack fluidized ejecta, but obviously two craters do not define a general rule. The critical test would have been to search for craters at an elevation between 8 and 18 km above datum and look for any systematic trend in the amount of fluidized ejecta. But no such craters exist on the upper flanks of the volcanoes.

When studying smaller craters at lower elevations, it is possible that elevation did play a role (alternatively, some small craters may have excavated to sufficient depth to reach a volatile layer). Figure 9.4 illustrates two craters that are each ~2.1 km in diameter and that formed over an elevation range of >8 km. Dilly crater has been interpreted by Tornabene et al. (2006) to be one of the youngest craters on Mars by virtue of its prominent ray system. Dilly crater may therefore represent

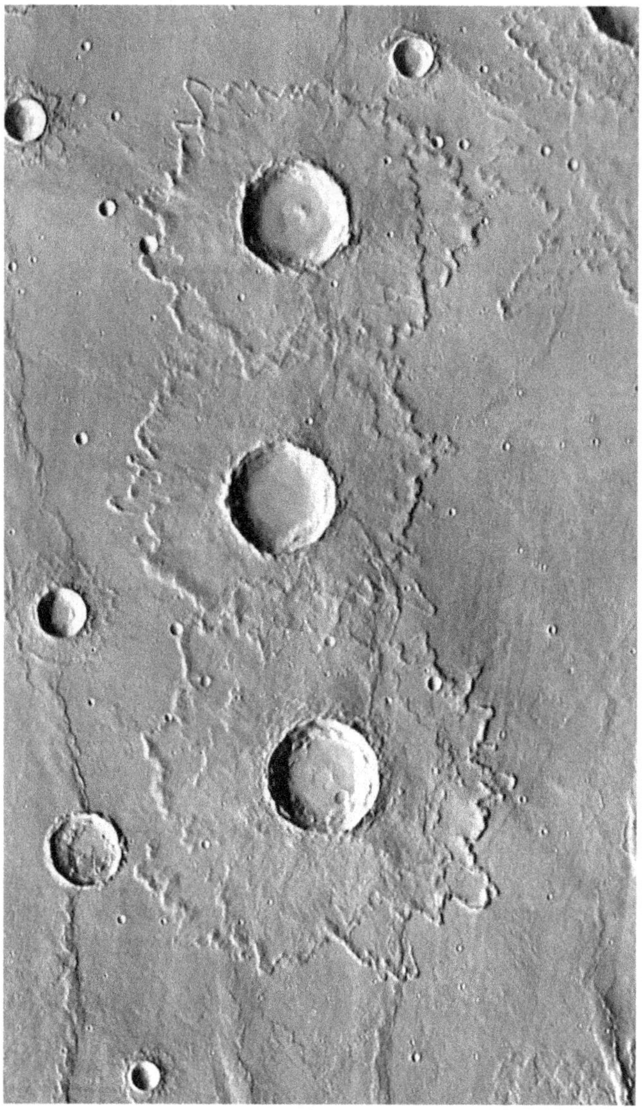

Figure 9.2 It seems likely that comparable-sized craters formed in the same material have similar interior and exterior morphologies. Of course, this may only be the case if the craters formed close in time to each other so that the climate conditions were similar. This triplet of craters formed on the ridged plains materials of Lunae Planum (the middle crater is located at 8.04°N, 287.68°E). The northernmost two craters (toward image top) are ~18.5 km in diameter and the southernmost is ~19.1 km in diameter. THEMIS daytime TIR mosaic.

near present-day conditions in terms of the distribution of volatiles on Mars. Both craters illustrated in Figure 9.4 have numerous secondary craters and yet Dilly crater more clearly shows lobate ejecta flows both north and south of the crater. Measuring the average ejecta mobility (EM) (see Section 6.3) of all craters of a given diameter and apparent age, and then correlating this parameter with elevation, may be a way to resolve if height above mean Mars datum is an important consideration. However, it does not at this time seem that elevation was a controlling factor in ejecta fluidization. Given this lack of pronounced correlation between elevation and ejecta fluidization, the role of the atmosphere is believed by most investigators to be minor.

9.4 CORRELATION BETWEEN INTERIOR AND EXTERIOR CRATER MORPHOLOGIES

In Chapter 4 we reviewed the landforms seen within morphologically fresh Mars craters and provided a similar review of the exterior features in Chapter 6. In those chapters we discussed a few relationships between the interior and exterior of individual craters. The Type-1 double-layered ejecta (Type-1 DLE) craters offer the clearest evidence that the defining interior and exterior morphologies are correlated. Figure 9.5 well illustrates the relationship between a crater interior with a small central peak, no terraces, and the double-layered ejecta blanket with prominent radial grooves. No such interior morphology can be found at either single-layered ejecta (SLE) or MLE craters; these two types of craters display larger central peaks and terraces in the interior of the crater and one or more ejecta layers with prominent distal ramparts. However, we note that some exceptions to this rule can be found, such as the Type-1 DLE crater Steinheim (Figure 6.1c) where the crater interior is comparable to that of an MLE crater.

The fact that most of the Type-1 DLE craters are found in the Northern Lowlands (Figure 6.3) suggests that the physical properties of the Vastitas Borealis Formation (VBF) have a prominent role in their formation. Geologic mapping indicates that the VBF most likely comprises sedimentary deposits (Tanaka et al., 2014), which would support this interpretation. Of course, not all Type-1 DLEs are found on VBF, while some SLE and MLE craters are located on this material. Some Type-1 DLE craters can be found on young lava flows (Figure 6.4), but they too lack interior terraces and large central structures (central peaks or central pits). One is left to ponder if the conditions at the specific sites where Type-1 DLE craters formed in the southern hemisphere were unusual. Ice-filled craters at a different time during the obliquity cycle might be one possibility, or the formation of the crater in lake deposits that were unusually fine-grained might also be plausible.

9.5 COULD FLUIDIZED EJECTA CRATERS INDICATE TIME-VARYING AMOUNTS OF VOLATILES?

Determining the evolution of subsurface ice is a key component to understanding Mars's climate and geologic history (Fanale et al., 1986; Clifford et al., 2010). Studying the amount of fluidized ejecta produced during the formation of layered ejecta craters, which likely formed by tapping subsurface ice, may provide constraints on this evolution when the relative timing of crater formation can be determined. Rare examples of craters of similar size but with very different morphologies do exist, such as the examples shown in Figure 3.6, but they are very unusual. Results from both regional and local analyses of the ejecta mobility (EM) (Barlow, 2004, 2015) suggest that concentrations of subsurface volatiles have remained approximately constant over time at the depths reached by these craters. The possibility that changes over time, perhaps due to the obliquity variations that Mars has experienced, is an intriguing issue to address through future research. Identifying what

Figure 9.3 Syria Planum (centered at 10.00°S, 257.50°E) should be a test of the idea that fluidized ejecta craters are less common at higher elevations due to a decrease in the volatile content of the target. At an elevation of >7.4 km, craters formed here are more than one scale height above the 6 mb atmospheric pressure at the datum, and yet eight craters ("A") in the diameter range 9–24 km all possess fluidized deposits. Furthermore, there are no examples of large fresh craters without fluidized ejecta at this elevation. Maybe elevation is not a key factor? THEMIS daytime TIR mosaic.

Figure 9.4 At left is an unnamed 2.1 km diameter crater at an elevation of +6.0 km at 3.2°S, 254.9°E. At right is Dilly crater, also ~2.1 km in diameter that formed at an elevation of −2.1 km (i.e., there is an 8 km difference in elevation of the two targets) at 13.28°N, 157.19°E. Left: CTX image B12_014340_1744. Right: CTX image D09_030866_1933.

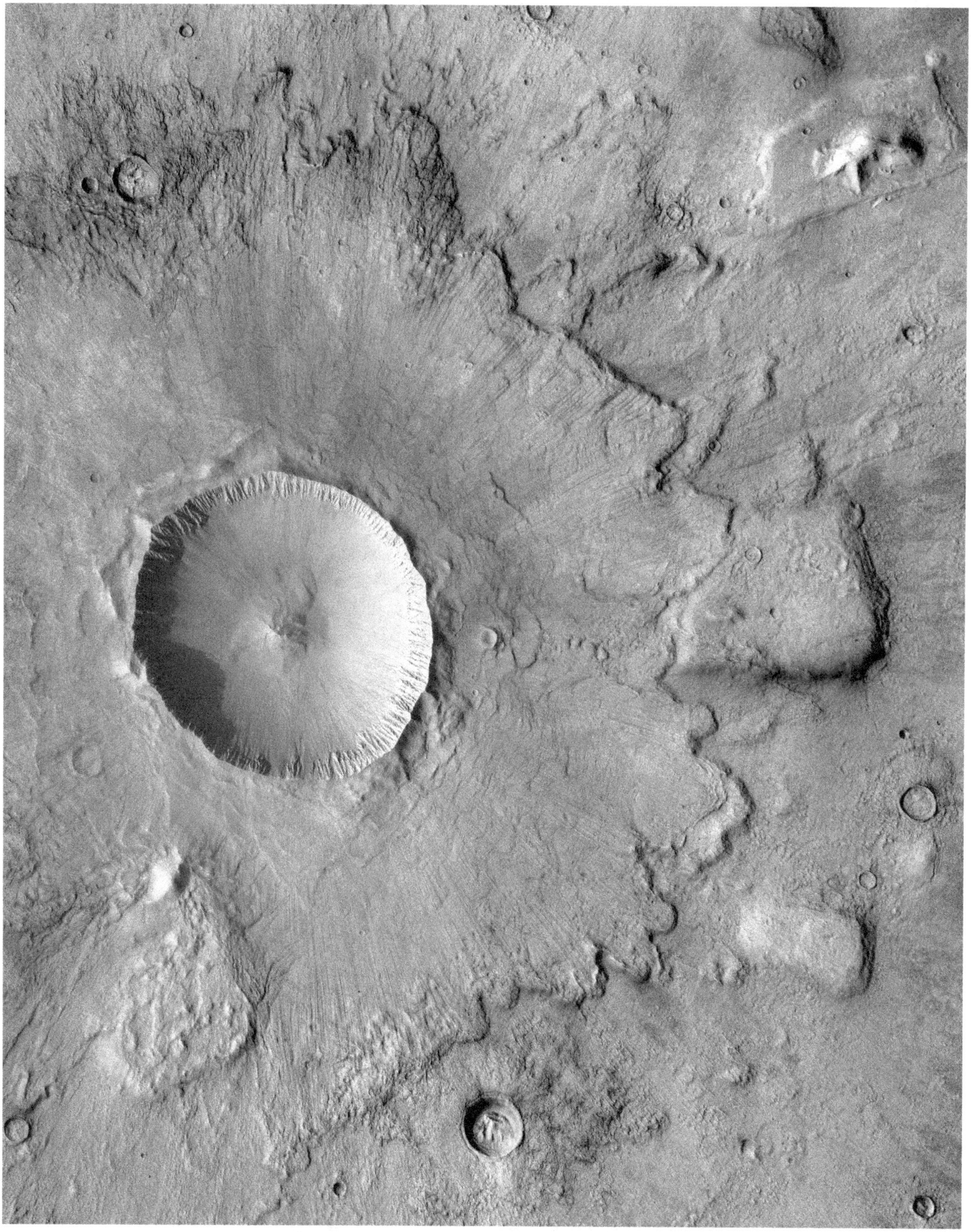

Figure 9.5 Pairing of Type-1 DLE ejecta and interior morphology seems to be common, with a small central peak, no terraces, and gullies on the inner wall. This unnamed crater is located at 40.87°N, 323.20°E and is ~8.2 km in diameter. CTX image D04_028961_2214.

controls the dynamics of ejecta emplacement, in terms of how much water or ice needs to be within the target and at what depth to provide the fluidizing medium, is an ongoing research topic (Sokolowska et al., 2024).

Measuring the runout distance of the ejecta is a logical way to assess variations in ejecta fluidization. Barlow (2004) documented the average EM value for SLE craters larger than 5 km in diameter in nine quadrants of the northern hemisphere. However, she did not calculate the azimuthal variations of EM for individual craters. Barlow's (2004) analysis only took the maximum extent of the ejecta compared to the crater radius, thereby ignoring the possibility of highly asymmetric ejecta deposits (such as those produced by oblique impacts) contaminating the statistical trends. Conversely, Mouginis-Mark

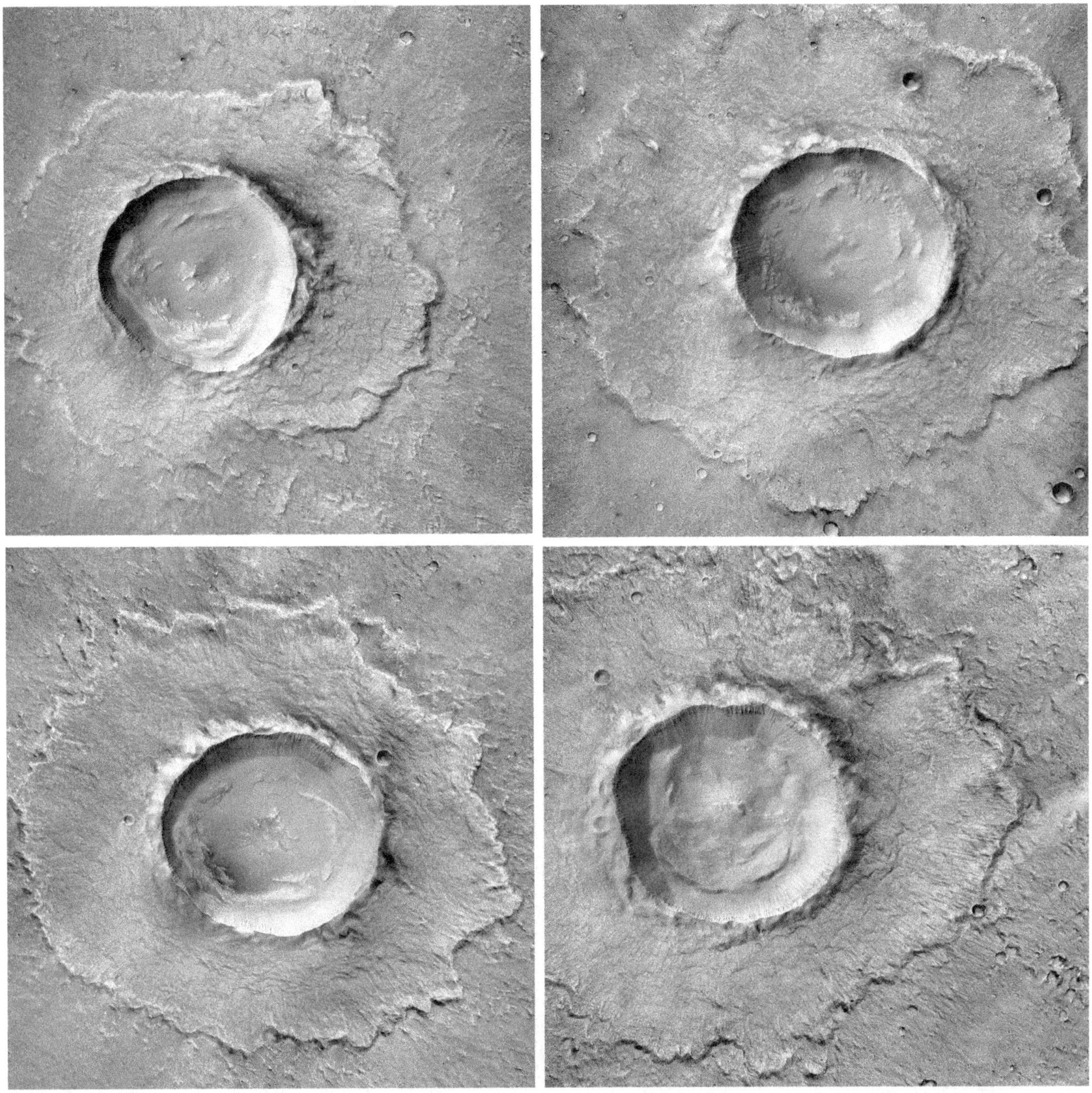

Figure 9.6 Here are four craters of comparable size (shown at the same scale) within Chryse Planitia that may help answer the question that temporal variations in the amount or distribution of volatiles within the target will allow craters with different ejecta mobilities to form. Top left: Belz crater (10.2 km diameter, 21.58°N, 316.76°E). CTX image B19_016909_2009. Top right: Puńsk crater (11.6 km diameter, 20.62°N, 318.87°E). CTX image B21_017766_2007. Lower left: Naar crater (11.4 km diameter, 22.93°N, 317.86°E). CTX image G19_025691_2031. Lower right: Grindavik crater (12.0 km diameter, 25.42°N, 321.00°E). CTX image P19_008562_2055.

and Carey (1980) did measure the surface area of 48 craters in Acidalia Planitia, but their sample size was too small to draw any regional conclusions. Most of the craters studied by Mouginis-Mark and Carey (1980) were Type-1 DLE craters in the diameter range of 4–40 km, so that regional trends could not be deduced from such a small data set. To our knowledge, no systematic analysis of ejecta areas as a function of crater size and/or crater location has been conducted since fluidized craters were first identified in the mid-1970s!

To infer changes in volatiles within a target as a function of time requires good relative ages, as a detailed chronology for the craters must be established. Manual counting of smaller craters on the ejecta blanket of SLE craters has been attempted by Niu et al. (2022). But to obtain adequate statistics over a wide geographic area, the automated system discussed by Lagain et al. (2021) would provide a better solution. Spatial variations in target properties must also be taken into consideration (Jones et al., 2016), as any variation in the material properties of the target (e.g., degree of cementation, mean particle size, amount of eolian deposits) might influence the maximum run-out distance of the ejecta layers, and hence the count area. In addition, surface processes (e.g., eolian, periglacial) have a substantial effect on the way features degrade. The challenge is therefore to find craters that are similar in size on the same target material and yet show signs that they were formed at slightly different times.

The SLE craters within Chryse Planitia may represent a candidate set of craters where such an investigation can be undertaken and produce reasonably accurate results (Salvatore et al., 2010; Jones and Osinski, 2015; Jones et al., 2016; Niu et al., 2022; Gou et al., 2024). Here can be found several craters of very similar diameters (10.2–12.0 km) that all possess similar ejecta morphologies (Figure 9.6), and the target material appears to be uniform over a large geographic area. Unfortunately, the number of these suitable craters is small, and their age spread derived from superposed impact craters is not ideal for detecting any variation in ejecta characteristics over the timespan of the floor of Chryse Planitia. Of course, selecting smaller diameter primary craters might be a solution, but this brings challenges in finding enough superposed craters in the size range that can be used to confidently obtain crater ages. Correlating any apparent changes in the degree of fluidization of the ejecta over geologic time will also bring uncertainties in terms of the possible subsurface stratigraphy of each target, and the unknown effects of obliquity changes throughout the period to be studied most likely introduce random "noise" into the analysis.

There are distinct challenges in estimating the relative chronology of such a small sample of craters. Figure 9.7 illustrates the northeast ejecta blankets of two Chryse Planitia craters (Naar and Belz) that have been investigated with this objective in mind (Niu et al., 2022; Gou et al., 2024). Even at HiRISE resolution (~25 cm/pixel), there is a paucity of small impact craters on each ejecta blanket and there is the necessary assumption that these are all primary (rather than secondary) craters. Making this task more challenging is the lack of HiRISE images for several of the craters of interest. Consequently, further research clearly needs to be done to ascertain if there is a time-varying signal contained within EM measurements for a specific area of Mars.

9.6 WHAT HAPPENS TO EJECTA AFTER EMPLACEMENT?

Arvidson et al. (1976) were among the first to propose that wind erosion was the dominant process in eroding crater ejecta deposits and surrounding materials. The remnants of eroded

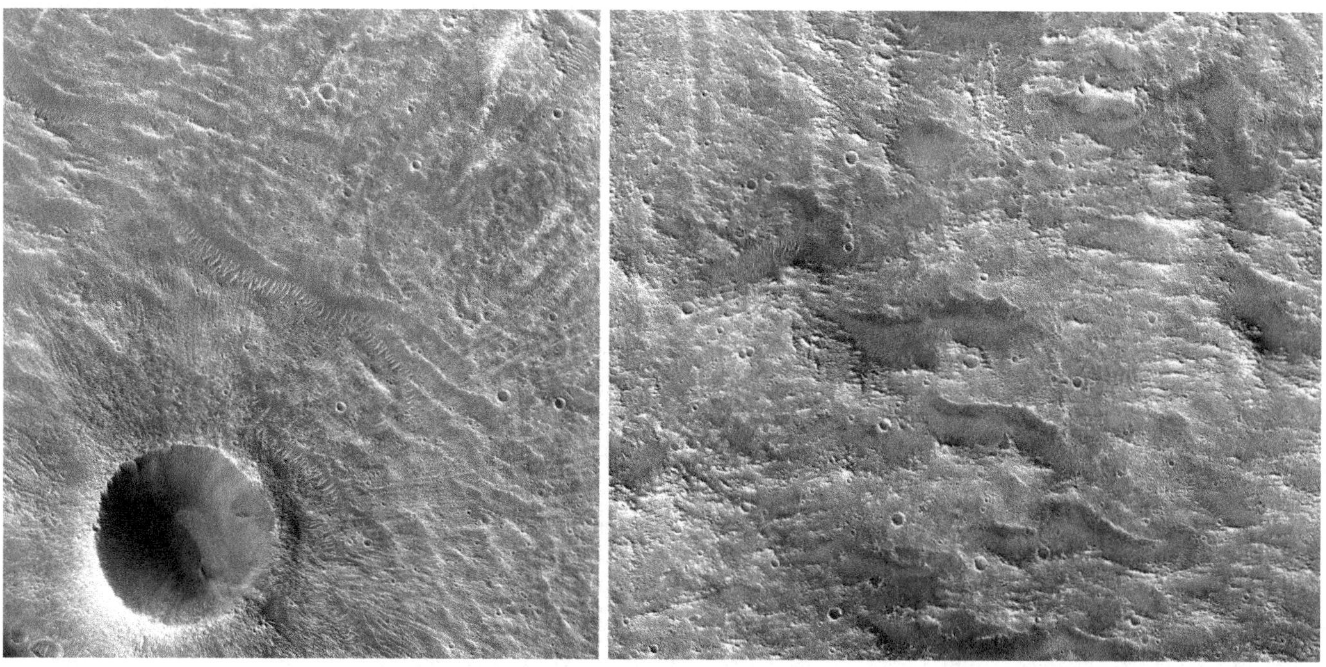

Figure 9.7 The northeast portions of the ejecta blankets of Naar crater (left) and Belz crater (right) highlight the challenges of placing a few, relatively young impact craters into chronological order through crater counts. There are several small (<100 m diameter) craters here. Which of Naar and Belz craters do you think is younger, and by how much? Images reproduced at the same scale. The largest crater in the image at left is ~750 m in diameter. HiRISE images (left) ESP_065343_2030 and (right) PSP_003776_2020.

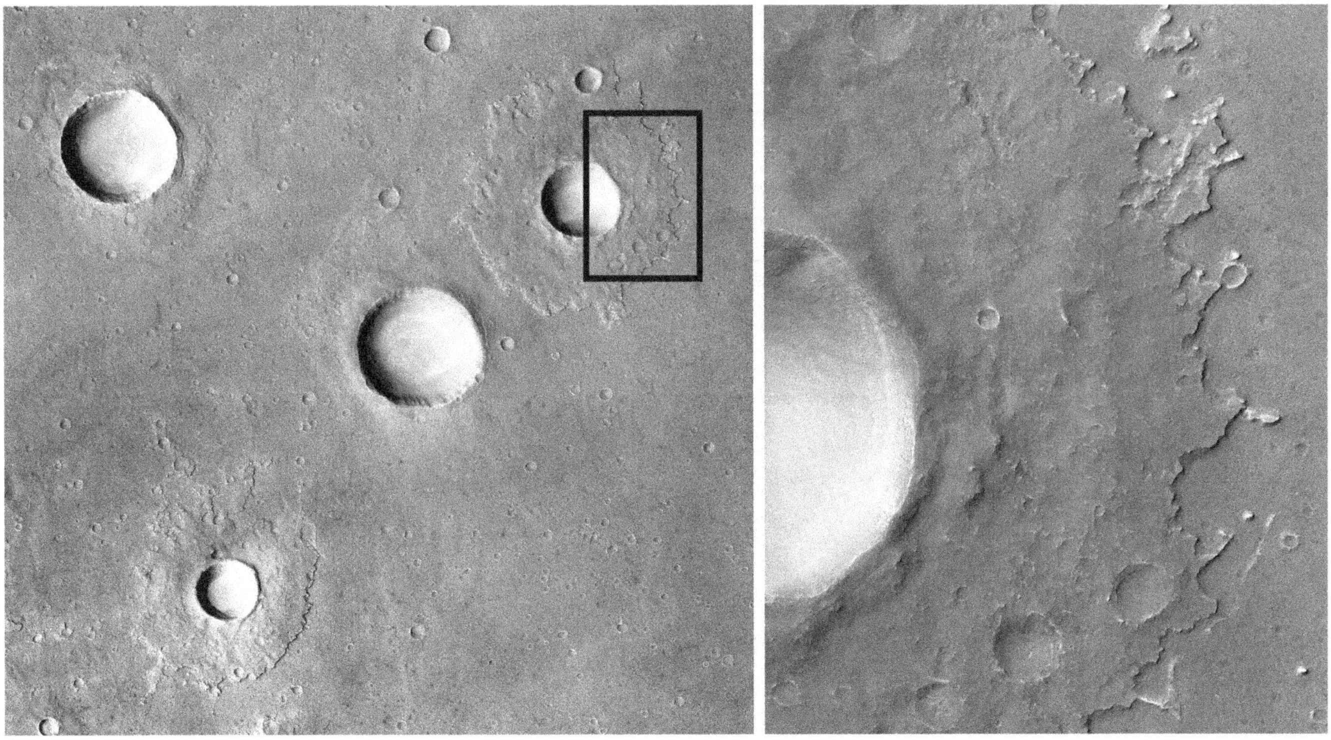

Figure 9.8 Something happens to the fluidized ejecta after emplacement, making the material more resistant to erosion than the target material. Here the surrounding plains have been eroded while the ejecta resists erosion. Note the lack of pitting or deflation features on the ejecta blankets, indicating the entire deposit is more resistant to erosion. The craters are located at 3.86°S, 296.61°E, on the western rim of Juventae Chasma at an elevation of ~2,300 m. This places them outside the region where significant volumes of ice would be expected in the target rocks. Left: CTX P06_003500_1762. Right: Higher-resolution view of the eroded ejecta of the 1.9 km diameter crater gives the suggestion that the ejecta blanket is layered. HiRISE image ESP_069355_1760.

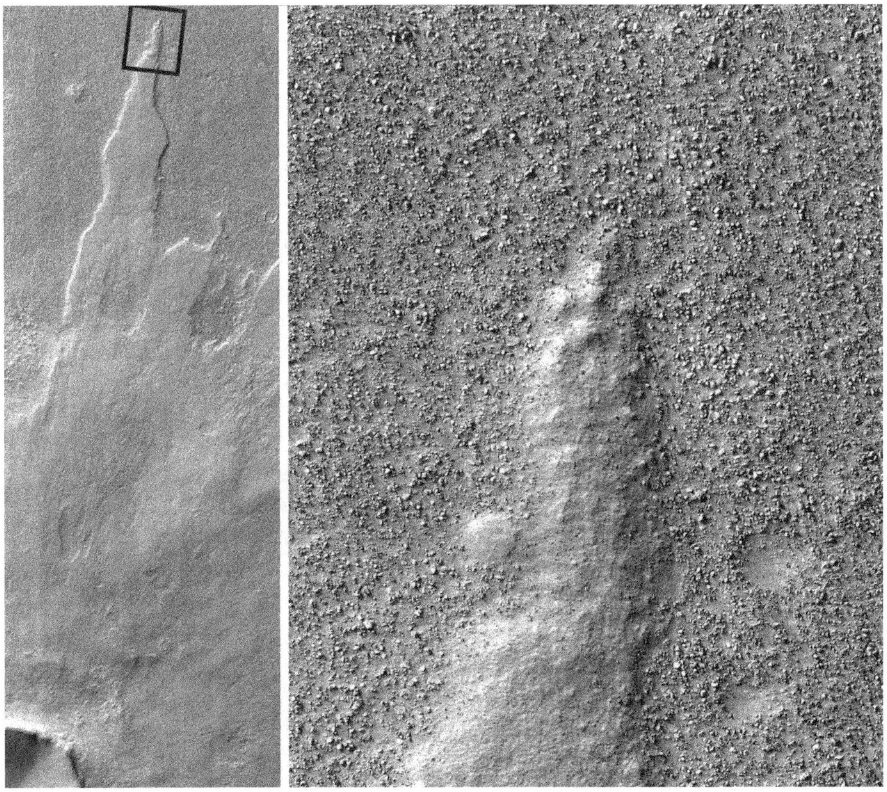

Figure 9.9 Here we see cementation of LARLE ejecta from an unnamed 2.2 km diameter crater located at 49.05°S, 135.70°E. The box in the image at left shows the location of the enlargement at the right, with the crater rim just visible at lower left. The ejecta at this location extended more than 11 km or 11 crater radii from the rim of the parent crater. Notice that the ejecta layer is almost devoid of blocks while the surrounding terrain is littered with them. Most likely the boulder field had formed prior to the emplacement of the ejecta. Both images are segments of HiRISE image ESP_016138_1305.

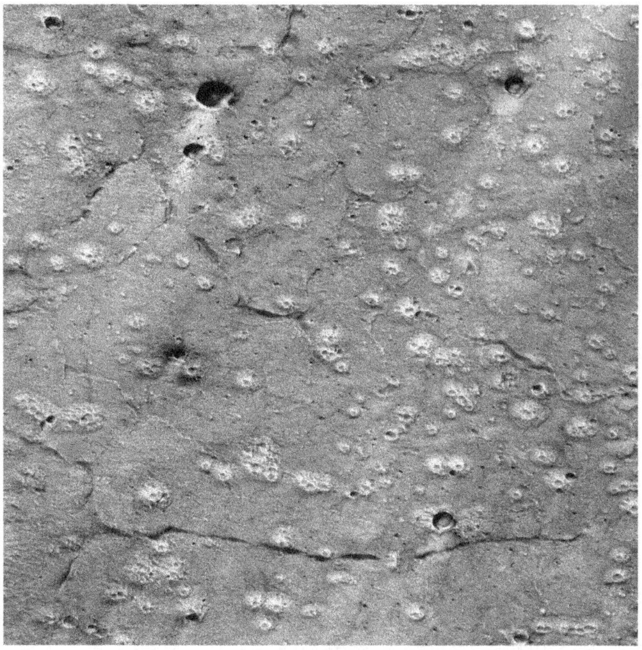

Figure 9.10 Even secondary craters disrupt the surface sufficiently to make a more resistant surface. A myriad of small secondaries that stand above the erosion surface can be seen here. These craters originate from Rauna crater at 35.27°N, 327.89°E. The largest crater in this image is ~700 m in diameter. CTX image F19_043255_2155.

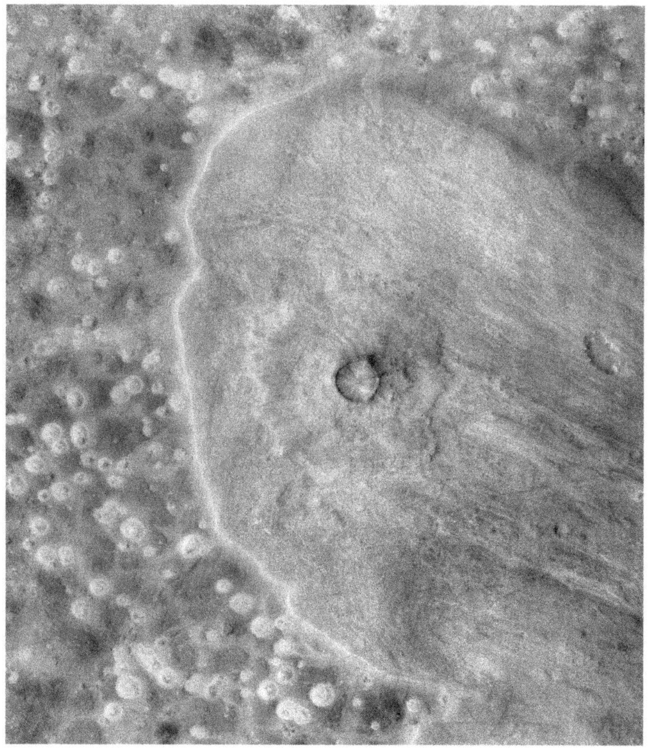

Figure 9.11 This 1.8 km diameter crater (58.92°N, 54.77°E) formed on an ejecta layer from a large (~54 km diameter) crater to the southeast of this image. This smaller crater lies ~103 km to the west of the larger crater. The persistence of fluidizing material within the target rocks over an extended period is supported by this observation. CTX frame B19_016892_2400.

ejecta blankets may in turn indicate physical attributes of the ejecta at the time of emplacement or very soon after their formation. It has been well shown that pedestal craters are characterized by a crater perched near the center of an elevated plateau surrounded by an outward-facing scarp that extends on average about three crater radii from the rim crest (Kadish et al., 2009, 2010). Such craters are typically <~2 km in diameter and are present in the middle to high latitudes. Pedestal craters are believed to have formed when an impactor was of sufficient size to excavate a substantial amount of material from beneath a surface snow an ice and that surface layer is subsequently removed (Barlow, 2006).

More challenging to explain are eroded ejecta blankets, such as the examples shown in Figure 9.8. Several craters in the diameter range of 0.7–1.9 km on the western rim of Juventae Chasma have ejecta blankets where the distal portion of the ejecta has been back-eroded, but the ejecta remains standing higher than the pre-existing terrain. This implies that the whole ejecta blanket is more resistant to erosion than the surrounding terrain. Cementation of the ejecta after emplacement might be an explanation. LARLE craters (see Section 7.4) can also illustrate the greater resistance to erosion of the ejecta layers compared to the adjacent terrain (Figure 9.9). It also appears possible for secondary craters to affect the surface materials, producing ejecta blankets that are more resistant to erosion than the pre-existing surface (Figure 9.10).

The persistence of volatiles within the region may also be addressed by searching for fluidized ejecta craters superimposed upon older ejecta layers. The example illustrated in Figure 9.11 provides an indication that the fluidizing medium is not lost from the surface after an ejecta layer is formed, but rather may persist (or be regenerated?) after the initial crater is formed. It seems possible that a "recharge of volatiles" within the top few tens of meters of the target might be possible during parts of the obliquity cycle.

9.7 ZUNIL CRATER: TWO EJECTA TYPES AT ONE CRATER

Even with the interpretations given earlier in this chapter, we do not want to give the impression that photogeology has answers to all the questions raised by the images of Martian craters. To this point, we now look at Zunil crater (10.4 km diameter, 7.71°N, 166.18°E; Figure 9.12), which was one of the first craters to be extensively investigated using images from MOC and THEMIS (McEwen et al., 2005). We focus special attention on this crater because of the very unusual bimodality of the ejecta blanket morphology; ejecta on the northeast side of the crater appear to have fluidized ejecta comparable to that seen around SLE craters of a similar size, while on the southwest side of the crater the ejecta blanket has a hummocky morphology that is more reminiscent of a fresh lunar crater or the radial craters described in Section 7.3. In addition, Zunil crater displays a diversity of rim morphology and interior structures that appear to mimic the exterior bimodality of ejecta types. Here we attempt to understand the

Figure 9.12 The crater Zunil (10.4 km diameter, 7.71°N, 166.18°E) is one of the youngest craters larger than a few kilometers in diameter, possessing many attributes associated with nearly pristine craters (see Section 3.7). In addition, it is a rare crater that displays striking azimuthal differences in its ejecta blanket. Computer-generated oblique image using a CTX-derived DEM from images G05_020211_1877 and F06_038250_1877.

inter-relationship between ejecta type, rim crest height, and crater diameter in the hope that such a comparison may reveal some underlying aspects of cratering on Mars.

The bimodality of ejecta types at Zunil crater may reveal useful information on the ejecta fluidization process and the potential correlation between such physical attributes as variations in crater rim height and rim morphology. Such an asymmetry in ejecta types is not seen at other fresh Martian craters of comparable size. Thus, the crater may provide an opportunity to investigate differences in ejecta properties and how they may relate to the role of volatiles and lithology in the impact cratering process (Kieffer and Simonds, 1980). Analysis of Zunil crater may also help to resolve the long-term issue of the potential role of the atmosphere in ejecta fluidization (Schultz and Gault, 1979; Schultz, 1992).

Zunil crater has many of the attributes of a fresh Martian crater, including pitted material on the crater floor, an extensive field of secondary craters, and thermally distinct crater rays (McEwen et al., 2005; Preblich et al., 2007). Zunil crater formed in the Cerberus Plains, which are generally interpreted to be young (<50 Ma) volcanic plains (Jaeger et al., 2010). The crater formed at an elevation of −2,750 m relative to the Mars datum, and at this elevation fluidized ejecta should dominate (Mouginis-Mark, 1979a; Barlow and Perez, 2003). The differences in the morphology of the ejecta blanket can be clearly seen in Figures 9.13 and 9.14. The southwest ejecta possess a generally hummocky morphology with less than 25 m of total relief relative to the surrounding terrain away from the outer rim crest. In contrast, the lobes seen to the northeast have ramparts that may exceed 100 m in height relative to the surrounding terrain. Striations on the surface that mark the direction of flow.

The differences in the morphology of the ejecta blanket can be clearly seen in Figures 9.13 and 9.14. The southwest ejecta have a generally hummocky morphology with less than 25 m of total relief relative to the surrounding terrain away from the outer rim crest. In contrast, the lobes seen to the northeast have ramparts that may exceed 100 m in height relative to the surrounding terrain. Striations on the surface that mark the direction of flow.

It is productive to investigate the azimuthal differences of the rim and ejecta from the Zunil crater. Such an analysis reveals that there are differences in the crater radius at a given azimuth, as well as in the height of the rim, that in some ways are correlated to the type of ejecta (fluidized or ballistic) beyond the crater rim. These differences may occur either due to differing amounts of structural uplift of the rim or to the removal of ejecta due to slumping away from the rim. Potential differences can be investigated via the use of a DEM for the crater. In Figure 9.15, we compare five transects across the southwest portion of the ejecta blanket to five transects across the NE portion of the ejecta. In Figure 9.15, “A” shows that the inner part of the southwest crater floor close to the central peak is deeper than the NE floor, but at “B” the terrace blocks make

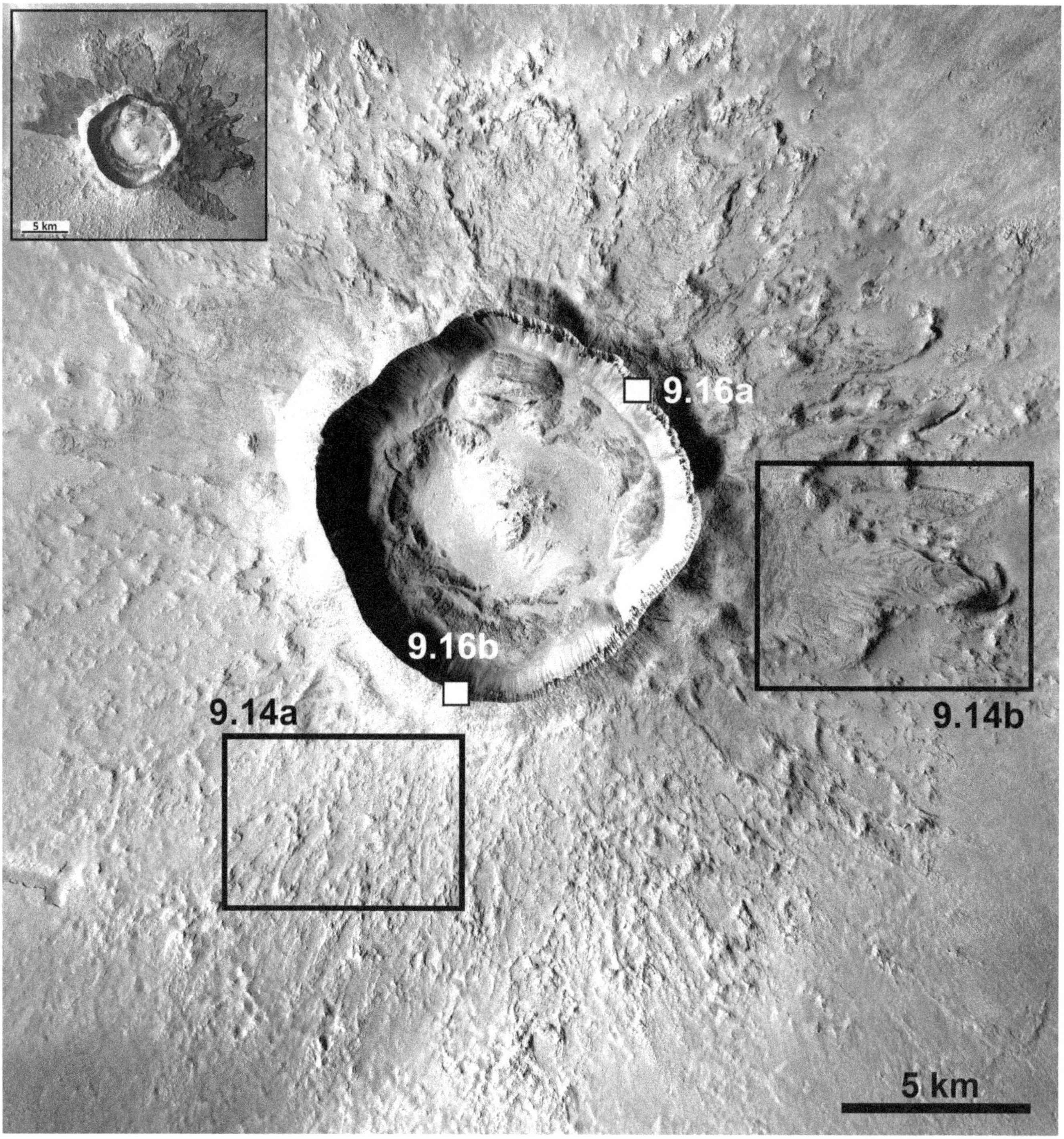

Figure 9.13 The asymmetric distribution of ejecta from Zunil crater (10.4 km diameter, 7.70°N, 166.12°E). The image at the top left shows the portions of the ejecta blanket that are fluidized (shaded area). The locations of the two detailed images of the ejecta blanket (Figure 9.14) are indicated by the boxes. The pair of white boxes on the rim crest indicate the locations of the different rim morphologies shown in Figure 9.16. CTX image G05_020211_1877.

the northeast interior higher than the southwest interior. This is also true for the proximal exterior deposits, with the fluidized ejecta higher close to the rim ("C"). At a radial distance of ~7 km does the ballistic ejecta have higher relief ("D"). Further out, at ~9 km from the rim, the relief of the fluidized ejecta is once again greater than the ballistic deposits ("E"). Only at ~14 km from the rim ("F") are the two types of ejecta of comparable thickness.

Striking differences also exist in the morphology of the rim crest of Zunil crater (Figure 9.16). The northeast portion of the rim is characterized by deep alcoves and pits. This is suggestive of extensive erosion of unconsolidated material. In contrast, the

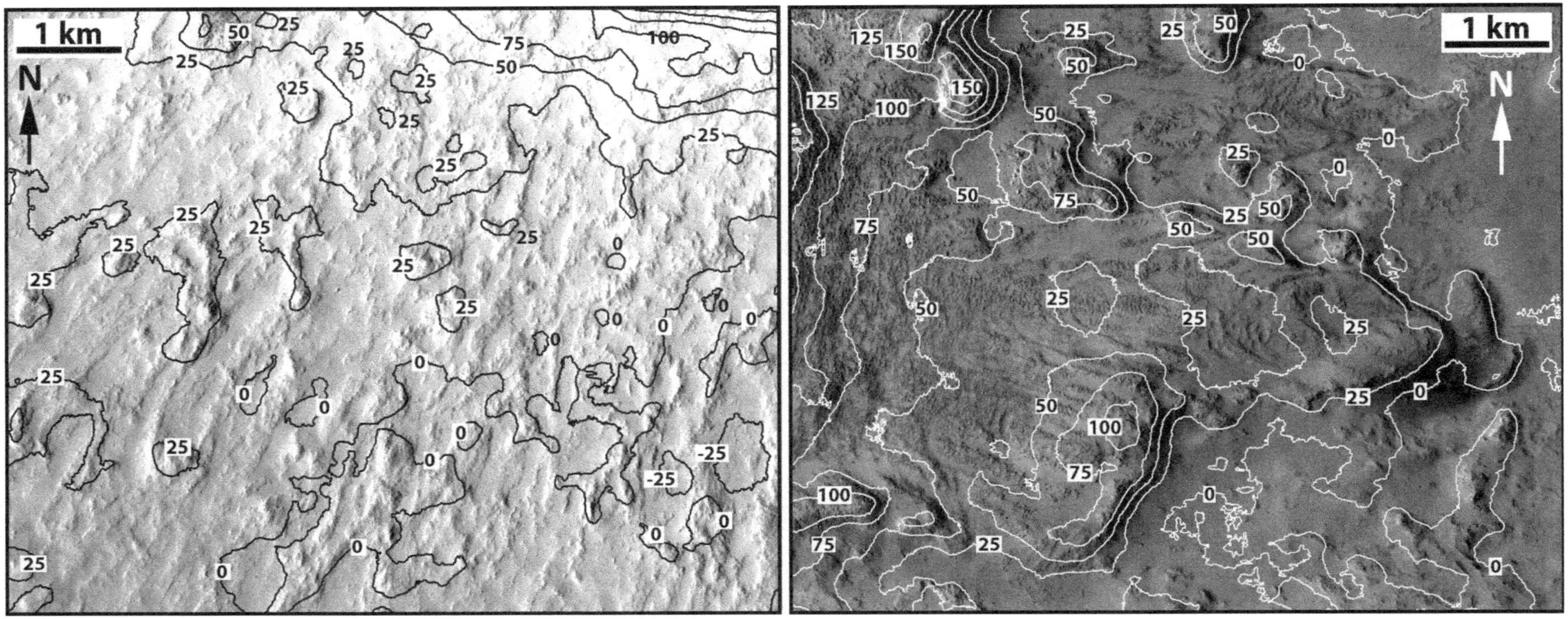

Figure 9.14 Topography relative to the local datum of the two types of ejecta present at Zunil crater. Contours in each image are in meters. See Figure 9.13 for locations. At left are ejecta to the southwest of the crater rim, which is generally hummocky and shows no sign of fluidization. A maximum of ~25 m of topographic relief occurs at this location. At right, ejecta to the east of the crater rim have the characteristic fluidized morphology associated with SLE and MLE craters, with ramparts higher than 100 m. Base images are parts of CTX image G05_020211_1877.

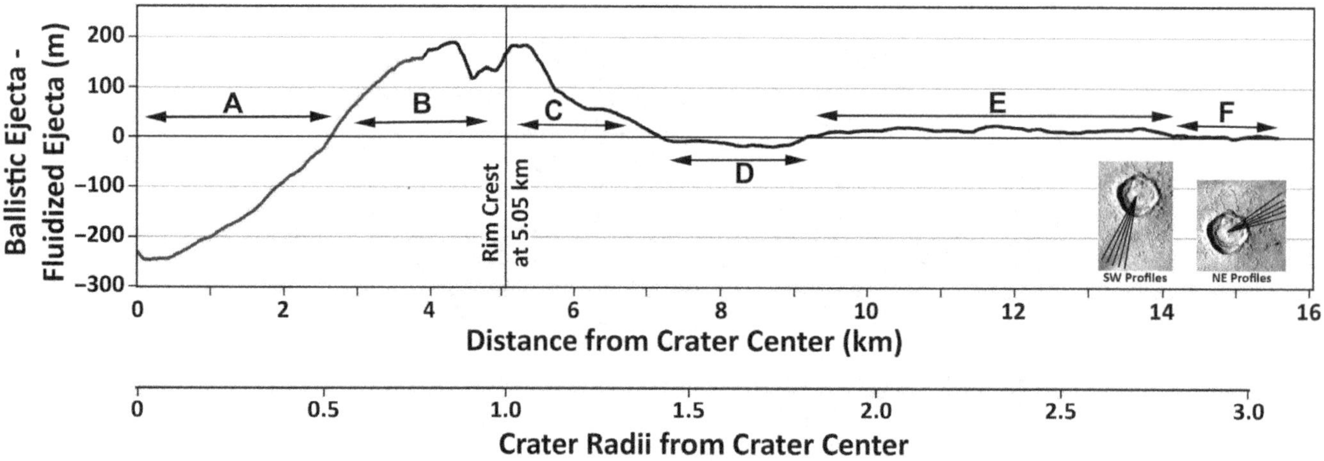

Figure 9.15 Comparison of the radial variation of topography for the SW portion (hummocky) of the ejecta blanket of Zunil crater compared to the NE (fluidized). The plot shows both the radial distance in km and crater radii. Inserts at right show where the five profiles were taken on the southwest and northwest portions of the ejecta blanket. The input data were derived from a digital terrain model generated from stereo CTX images G05_020211_1877 and F06_038250_1877.

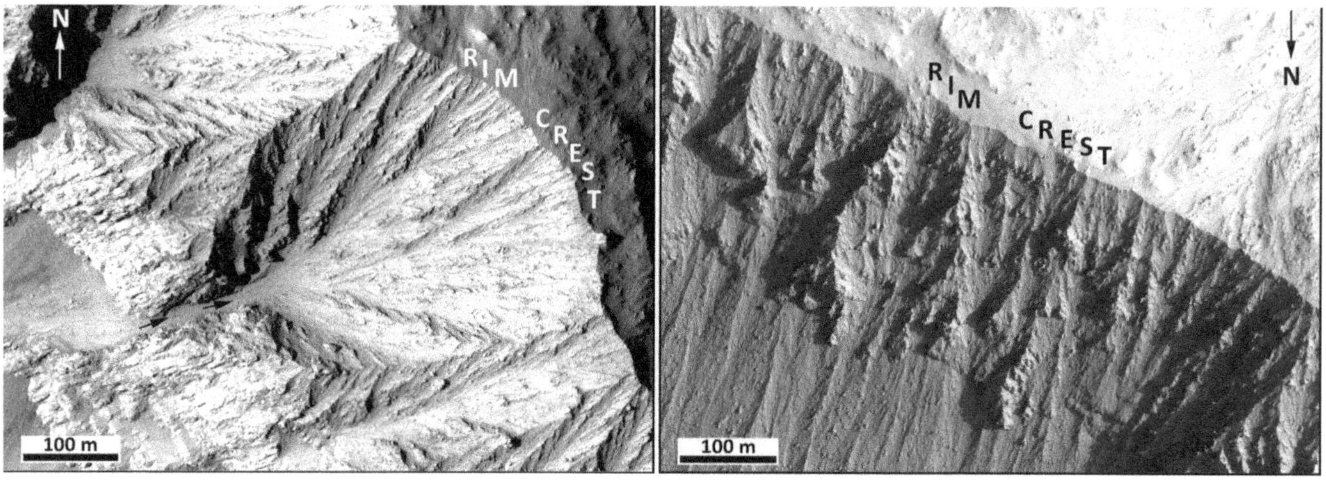

Figure 9.16 Variations in the morphology of the rim crest exist at Zunil crater. At left, a deeply incised alcove on the northeast rim. Such a morphology suggests a thick layer of impact ejecta eroded by volatiles excavated during the cratering process (Mouginis-Mark and Boyce, 2012). At right, clearly defined layers within the southwest portion of the crater wall are interpreted to be individual lava flow units uplifted during the cratering process (Sturm et al., 2016). See Figure 9.13 for locations. CTX image G05_020211_1877.

southwest rim displays layered lava flows with little sign of erosion. To further place these rim and exterior ejecta differences into a spatial context, an azimuthal analysis of the crater radius, rim height, and the location of the pitted rim and ejecta types can be utilized (Figure 9.17). The clearest correlation from such an analysis is that the pitted rim unit only occurs at azimuths where there is fluidized ejecta, and both features suggest a volatile-rich target. Even in this instance, there is no unique correlation between the two attributes.

From Figure 9.17, the radius of the crater is not correlated with the type of ejecta beyond the rim. All the pitted parts of the rim are associated with the fluidized ejecta. There is no correlation between the pitted rim and the radius of the crater. Further analysis is also possible because the DEM allows the rim height to be measured for each azimuth. Here the rim height might be expected to correlate with the type of ejecta produced, with higher rim segments associated with ballistic ejecta and slumping of the rim where the ejecta were fluidized. But once more there is no clear correlation between the parameters, raising the question as to whether the type of ejecta is a primary product of the cratering process or simply influenced by the surficial layers of the target either being volatile-rich (fluidized) or volatile-poor (ballistic).

The striking azimuthal variations in the ejecta deposits and rim morphology of Zunil crater do not correlate with the previously identified crater rays and secondary craters. This is enigmatic, because our interpretation of the eroded rim segments is that this is an indicator of unconsolidated material that formed the late-stage ejecta at these azimuths. A possible explanation could be that Zunil crater formed on top of a structural boundary between relatively competent rock (i.e., lavas) toward the southwest and less indurated material (pyroclastics,

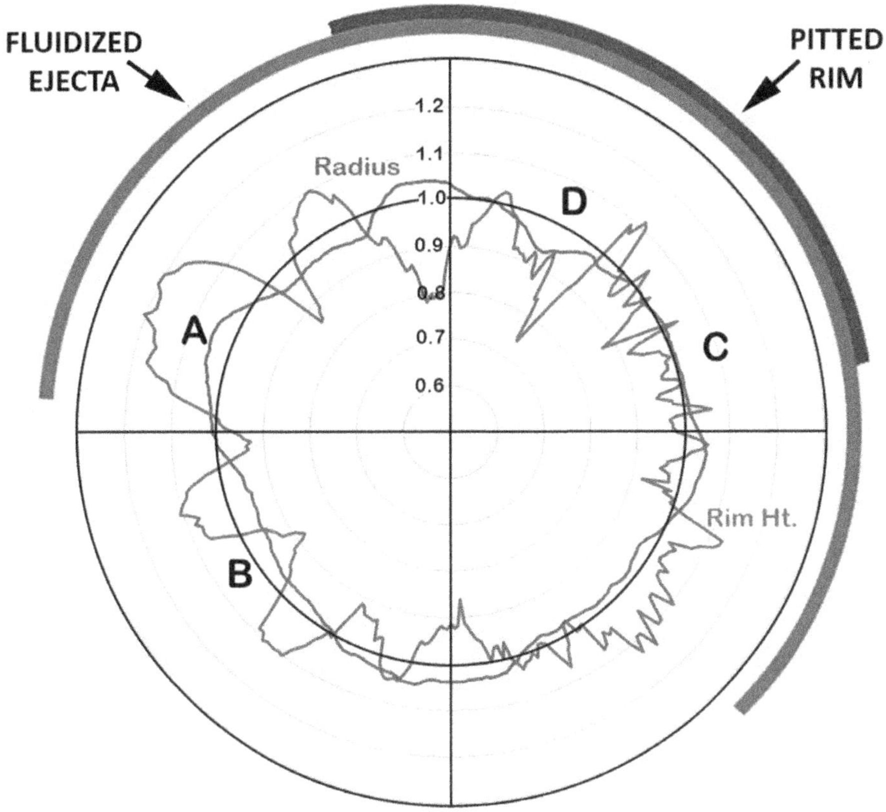

Figure 9.17 Azimuthal variations in rim radius and rim height for Zunil crater compared to the distribution of fluidized ejecta (light gray band), the pitted rim crest (dark gray), and ballistic ejecta (between 135 and 280 azimuth). Data are compared to the average radius (5,061 m) and average rim height (415 m) and indicate that the rim height varies from ~0.75 times to ~1.2 times of the average, while the crater radius varies from ~0.93× to ~1.05×. Data derived from a DEM produced from CTX images F06_038250_1877 and G05_020211_1877.

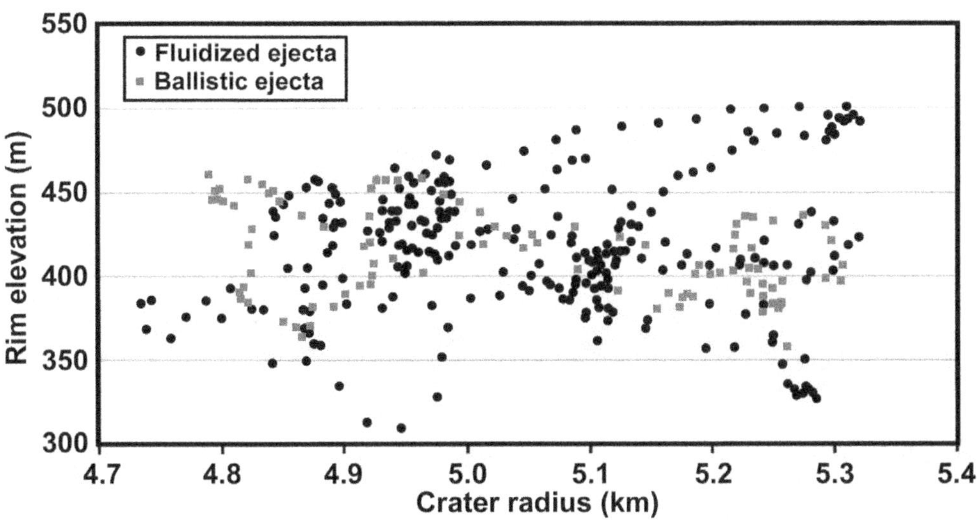

Figure 9.18 Variation of rim elevation as a function of crater radius for Zunil crater. Symbols indicate if the ejecta at each azimuth is fluidized or ballistic. Input data are the same as for Figure 9.17.

sediments, etc.) toward the north and northeast. Such a structural boundary appears to be consistent with the orientation of a fissure system that may have fed recent explosive volcanism less than 20 km northwest of Zunil crater (Horvath et al., 2021). If this structural discontinuity does exist, it seems reasonable to expect that the unconsolidated material to the northeast would be less likely to produce chains of secondary craters. The "lunar-like" hummocky ejecta to the southwest would be a sign that volatiles were not present in this area. Consequently, there may have been an asymmetric distribution of volatiles within the target materials within which Zunil crater formed. More volatiles may thus have existed within the target rocks north of the structural boundary.

One final comparison between the parts of the crater that have ballistic ejecta compared to fluidized ejecta is to compare the crater radius to the rim elevation (Figure 9.18). In this instance, it is possible that one of the types of ejecta required a lower rim elevation to exit from the growing crater cavity. However, when considering the type of ejecta, the only trend between crater radius and rim elevation appears to be the dominance of fluidized ejecta at rim elevations more than 450 m and a crater radius of more than 5.0 km.

No other large crater on Mars displays the same bimodality in its azimuthal distribution of ejecta when compared to Zunil crater. We do not have any firm answers to explain the observations, except perhaps that the fissure system that possibly created the explosive activity to the northwest (Horvath et al., 2021) included a dike that underlays the target materials where Zunil crater formed. This dike may thus have acted as an aquiclude to prevent volatiles from entering the target material on the southern side of the still-to-be-formed crater. Nevertheless, Zunil crater may have currently unrecognized properties that might allow future researchers to tackle the issue of ejecta fluidization and the correlation between ejecta types and rim topography. Once again we leave this challenge to future generations of Martian cratering researchers.

9.8 HOW MIGHT SOME OF THE QUESTIONS ABOUT FLUIDIZED EJECTA BE ANSWERED?

Here we speculate on what new data sets might be obtained for Mars that could answer some of the questions raised in this book. Short of having samples returned from individual impact craters (an unlikely event for the foreseeable future), how might we learn more about Martian craters and how they formed? Better knowledge of the topography of a crater and its ejecta blanket might provide insights that could be interpreted in terms of the rheological properties of the ejecta on emplacement. For example, the few instances where DEMs have been produced from stereo-HiRISE images, topographic profiles across the ejecta layer can reveal details of erosional processes (Figure 9.19). At lower spatial resolution, images superimposed on CTX-derived DEMs can show more clearly the stratigraphy of the ejecta layers (Figure 9.20). The greater regional coverage of these CTX-derived DEMs undoubtedly allows additional ejecta blankets to be studied in more detail.

The determination of slopes, aspects, and volumes of units might also reveal new trends that have so far been undetected. New insights might also be gained if there were more detailed information on the distribution of particle sizes and the degree of cementation within the ejecta beyond that which can be inferred from thermal inertia information. Finally, the ubiquitous dust cover undoubtedly hides important information about the interior of craters as well as the ejecta blankets.

One method to address all three of these attributes might be the flight of a synthetic aperture radar (SAR) to Mars. There have been different types of radars flow to Mars, but none that produced high-resolution images. The Mars Advanced Radar for Subsurface and Ionospheric Sounding (MARSIS) (Jordan et al., 2009) and the SHAllow RADar (SHARAD) (Seu et al., 2007) are both sounding radars that can respectively probe the subsurface of Mars to depths of hundreds of meters to a few kilometers. In contrast, imaging radars have not been flown to Mars, but have been widely used in Earth orbit, as well as the exploration of Venus (e.g., the *Magellan*, and *Venera 15* and *16* missions). An imaging radar has also flown to Saturn on the *Cassini* mission (e.g., Wood et al., 2010).

A major advantage of a SAR is the ability to operate in an interferometric mode (Massonnet and Feigl, 1998), where the spacecraft is flown in such a way that there is an almost-exact repeat of the orbit either on successive days or over longer time periods. On Earth, this method has been particularly successful at generating digital topography for almost the entire globe except the polar regions at the spatial resolution of the radar image (Farr et al., 2007). Such high-resolution (potentially, 10–30 m/pixel) digital topography for much of Mars would be a significant advance over the MOLA laser altimeter data set (Smith et al., 2001) and would undoubtedly

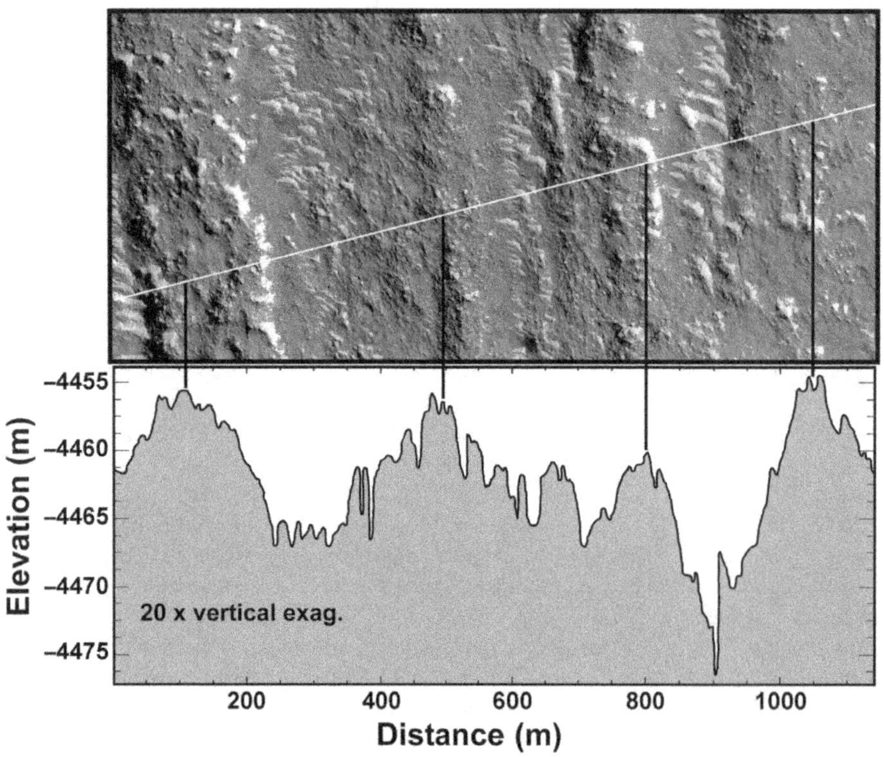

Figure 9.19 Topographic profile across a small portion of the inner ejecta layer of Bacolor crater. Vertical black lines connect similar high-relief features in both the image and the profile. The HiRISE stereo-derived topographic data show that the largest grooves are ~20 m deep and ~300–400 m wide. Such measurements hint at the erosive force required to carve such grooves, as well as the amount of material that must have been removed from the ejecta layer. Topography derived from stereo HiRISE images PSP_006750_2130 and PSP_007462_2130.

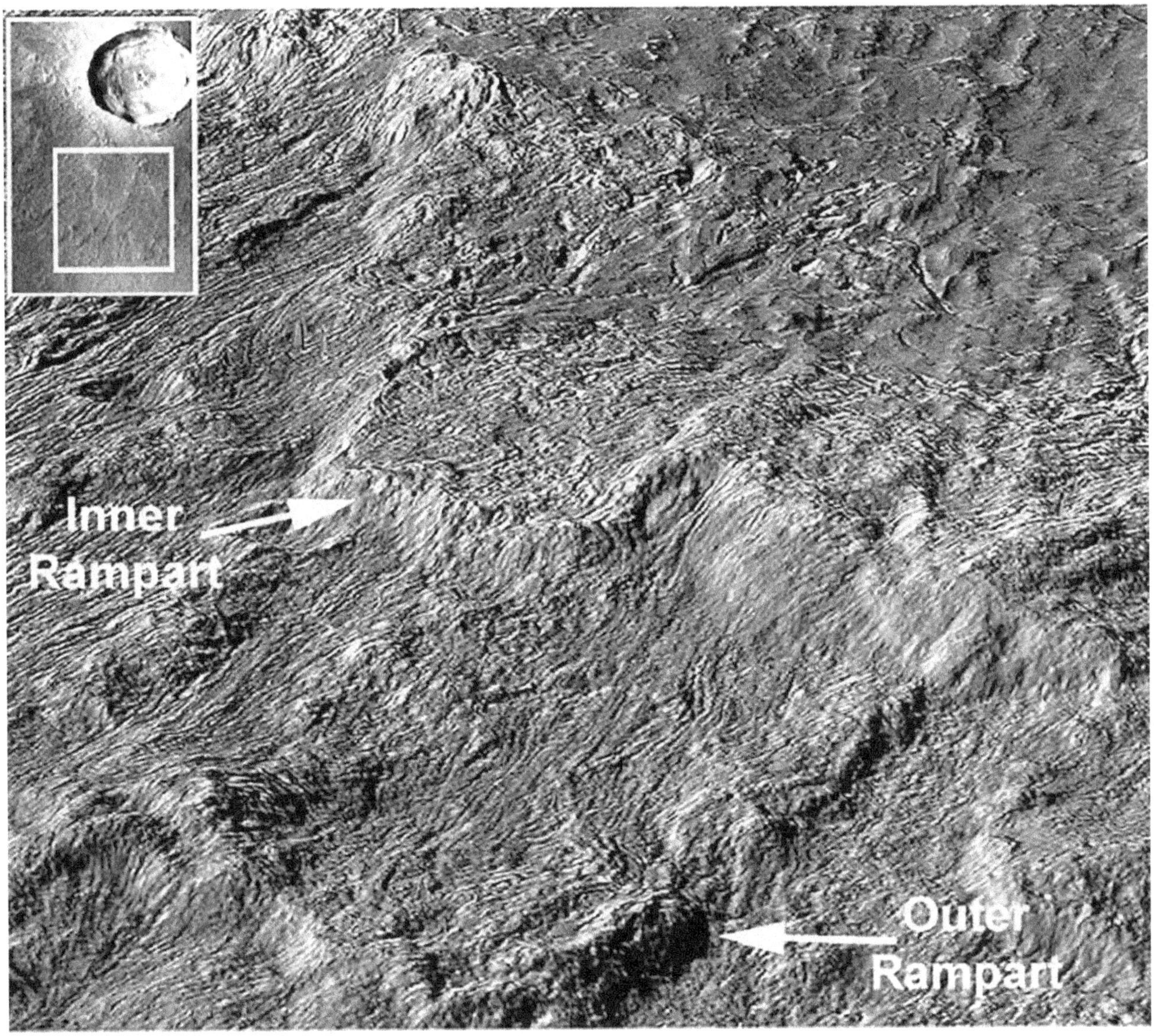

Figure 9.20 Oblique view of the ejecta layers of the Type-1 DLE crater Jaisalmer (15.0 km diameter at 33.50°N, 84.12°E), derived from stereo CTX images F23_044886_2136 and J01_045229_2136. This image shows how the grooves on the inner ejecta layer extend over the inner rampart and on to the outer ejecta layer. See also Figure 8.12 for a different view of this ejecta blanket. Insert at the top left shows the location of the ejecta layer with respect to the crater.

provide greater insight into impact crater morphology compared to DEMs derived from stereo CTX or HiRISE images (e.g., Mouginis-Mark et al., 2018) or Colour and Stereo Surface Imaging System (CaSSIS) image pairs (Tornabene et al., 2018b; Re et al., 2022; Guimpier et al., 2022) for geographically small areas. Such radar-derived high-resolution topographic data would be particularly helpful for the analysis of ejecta blankets for large (~>20 km diameter) that would extend beyond the area of coverage of individual CTX or HiRISE images (Figure 9.20).

The exciting possibility of using a SAR to "peer" beneath the layers of dust that cover much of Mars (Ruff and Christensen, 2002; Williams and Greeley, 2004) is also highly attractive. Possible differences in the near-surface structures could be detected, potentially revealing textural differences in the flow properties of the ejecta layers or the sediment fill within the crater interior. The ability of an imaging radar to penetrate a dry surface is a function of the wavelength used. The use of SAR proved to be spectacularly successful in Earth orbit during the Shuttle Imaging Radar-A (SIR-A) experiment in 1981 (McCauley et al., 1982, 1986). This experiment demonstrated that radar penetration to a few meters in the hyper-arid sands of the Sahara Desert (Figure 9.21) is possible and such penetration depths should also be possible on Mars. The SIR-A data were collected at a wavelength of 24 cm (L-Band). A Mars radar system with a depth-penetration capability of a few meters would require a longer wavelength radar than that flown to Titan (the *Cassini* SAR had a wavelength of 2.2 cm), the Moon (4.2 cm and 12.6 cm for the Mini-RF on the *Lunar Reconnaissance Orbiter*), or Venus (12.6 cm for *Magellan*), and thus a proportionally larger antenna would be needed. Operating at a longer wavelength (say, 24 cm (L-band) or 68 cm (P-band)) would make the orbital control needed for interferometric passes easier to accomplish.

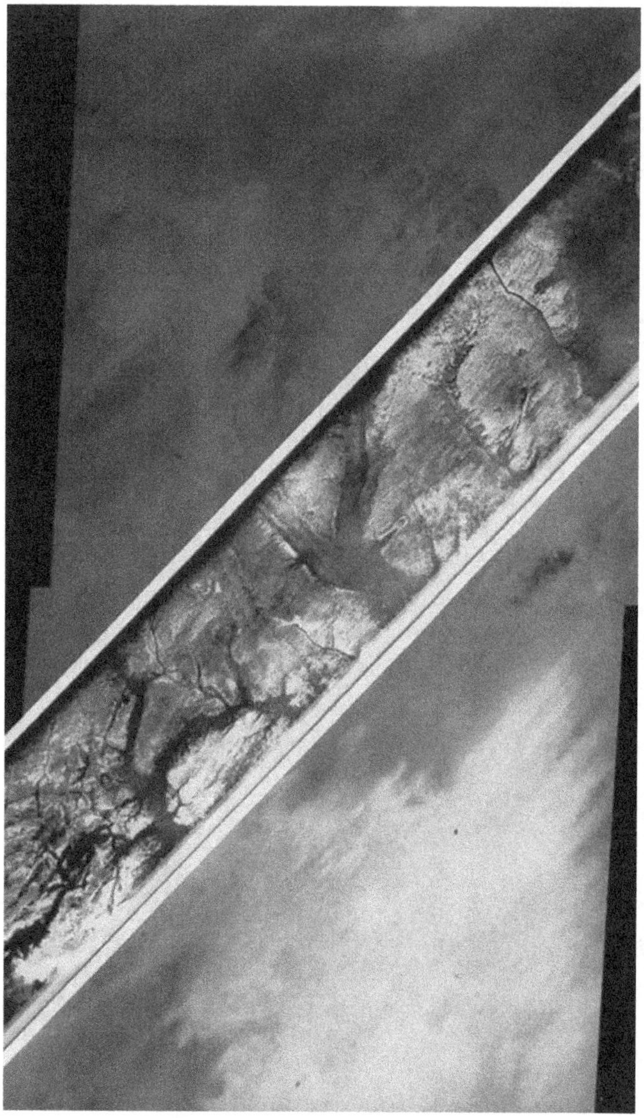

Figure 9.21 Space Shuttle Imaging Radar-A (SIR-A) radar image of the Sahara "buried rivers," superimposed upon a Landsat 4 image of the same area. The complex drainage network of these rivers (dark in this image) has been buried by about a meter of surface sand. This demonstrates the ability of L-band (23.5 cm wavelength, HH polarization with an incidence angle of ~47 degrees) radars to penetrate the surface veneer of dust and sand. USGS/NASA JPL image. SIR-A data take number 28. Width of the radar image is ~55 km. This radar experiment was flown on Space Shuttle mission STS-2 in November 1981.

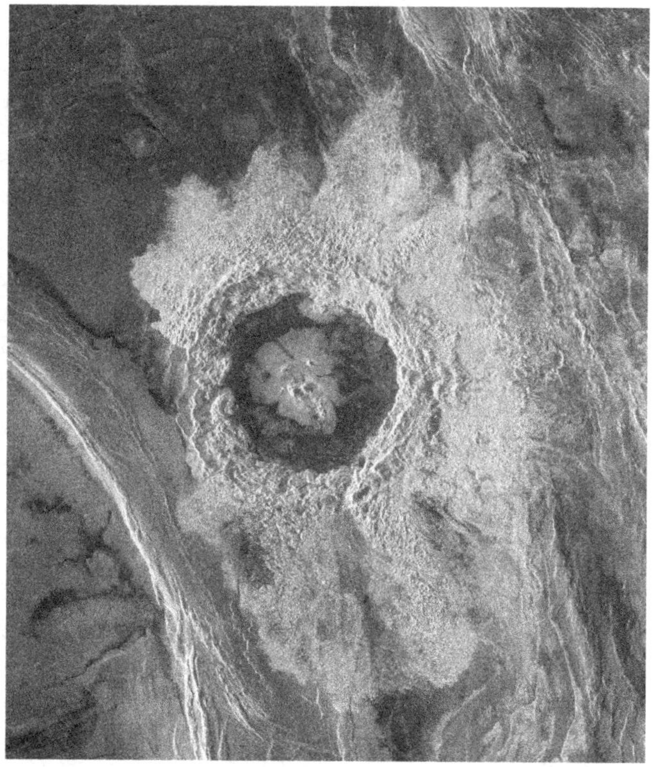

Figure 9.22 Magellan radar image of the 69 km diameter Dickinson crater on Venus (74.60°N, 177.20°E). Radar is particularly good at showing the structural and textural differences within the ejecta blanket. Different incidence angles and/or look directions would further discriminate between the roughness of different units within the ejecta blanket. The radar look direction is from the left. JPL photojournal image PIA00479.

Imaging radars are particularly sensitive to surface scatters at the scale of the radar wavelength (cf., Blom and Daily, 1984; Evans et al., 1986). Varying the angle at which the radar signal intersects the surface (the so-called incidence angle of the observation) can also be exploited to determine if the surface is rough or smooth at the radar wavelength from the "backscatter curve" of the surface. The brightness contrast between radar-dark (smooth) and radar-bright (rough) surfaces can be seen in a single *Magellan* radar image of Venus (Figure 9.22), where surface textures within the ejecta blanket are easily distinguished. The *Magellan* radar had an incidence angle that spanned from 25° to 45° and at various times of the mission, the data were obtained with the radar "looking" either east or west, so that some of the scattering properties of the surface could be determined. An important question would be whether a similar approach at Mars would give more useful textural data than the use of high-resolution optical images.

Imaging radar systems may have a single wavelength or two or three. The radar can be constructed to operate with different polarizations, wavelengths, and/or different incidence angles. Each increment in sophistication brings operational constraints as well as data rate issues. In the simplest form, a single wavelength, single polarization, fixed incidence angle would still generate unprecedented new views of Mars. Whether such a data set provides the definitive answer to the questions posed by fresh impact craters remains to be seen. Such a mission, which has not yet been selected for flight, could well be a productive step in this direction.

An alternative new mission to resolve some of the key questions about fluidized Martian craters would be a lander mission. A rover mission, employing a *Curiosity*- or *Perseverance*-like rover (see, for example, Grotzinger et al., 2015; Horgan et al., 2020), to a fresh Martian layered ejecta crater could resolve some of the most fundamental and key questions about Mars. This is because of a rover's ability to carry scientific instruments that could investigate material excavated from the subsurface. These materials could provide deep insight into the past geologic, climate, and even biological activity, as well as provide absolute age calibration that could put crater density-based chronology on a more quantitative footing.

For example, by investigating the physical properties of the ejecta, a rover could address the question of whether subsurface volatiles, like water or ice, were involved or even required for ejecta fluidization. This is important because resolving the question (even for a single crater) directly relates to the global population of layered ejecta craters of nearly all sizes and on terrains of a very wide range of ages. As argued in Chapters 6 and 7, fluidized ejecta have long been suspected to require substantial water or ice in the target materials to develop. The determination of the particle size distribution, the extent of layering observed within the deposit, and the degree of cementation of the material would all be highly informative. Considering the global distribution of these craters, conclusive proof for such volatiles would have a substantial impact on our understanding of geologic, hydrologic, and climate history of Mars; it would confirm that Mars has harbored a global aquifer for much of its history. Results from an analysis that can only be done from a rover and an *Ingenuity*-class helicopter (Balaram et al., 2021) of the ejecta flow's structures, particle size distribution, and the geochemistry of any alteration products produced by the interaction of the subsurface rock with groundwater could hold the answer to this question.

In addition, assuming the rover also carried hardware like the CODEX (Chemistry, Organics, and Dating Experiment) instrument (Anderson et al., 2020), the absolute age of the surface rock at the landing site could be obtained, as well as the age of possibly cements deposited from subsurface aqueous activity. These ages would not only provide key calibration points to make estimates of the impact flux history at Mars (e.g., Werner and Tanaka, 2011) but would also lead to a deeper understanding of the history of Mars' water. An even more exciting prospect for such a mission, depending on where the rover lands, is that the rover could investigate the material brought up from depth for signs of life. Impact-generated hydrothermal areas might well serve as incubators of early life. Because of the nature of the cratering process and ground-hugging ejecta flows, a rover mission would have ready access to samples brought up to the surface from a wide range of depths. The scientific instruments aboard the *Curiosity* and *Perseverance* rovers are well suited for this task.

Selection of the landing site would be critical to the success of the mission. Inevitably, there would be constraints on where such a mission could be conducted. First and foremost, the impact crater would have to lie at a low elevation so that a sky-crane approach to landing could be conducted (Way et al., 2007). The floor of Gale crater (where the *Curiosity* rover landed) is at an elevation of ~4,400 m below Mars datum, while Jezero crater (the *Perseverance* rover landing site) is ~2,600 m below Mars datum. It is reasonable to assume that a Mars Exploration-class rover would have to land at a crater where the ejecta blanket is at an elevation no higher than ~2.5 km below Mars datum. Many craters with the northern plains, as well as some on the highland-lowland boundary, would meet this criterion. A second constraint might be the degree of dust cover, as inferred from the Thermal Emission Spectrometer experiment (Ruff and Christensen, 2002). Dust-covered regions would be

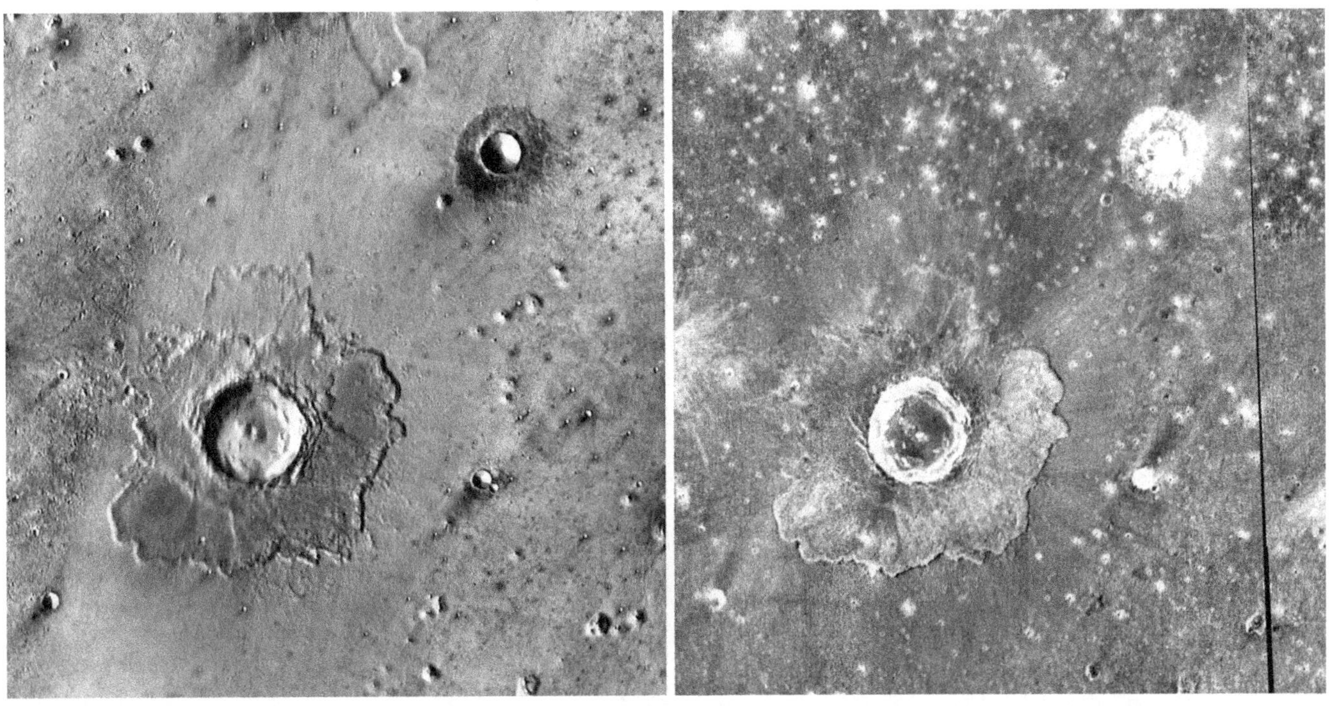

Figure 9.23 Makhambet crater (16.0 km diameter, 28.52°N, 319.44°E) would be an ideal crater to be investigated by a future Mars rover. The terrain surrounding the crater is at an elevation of ~3,850 m and is located within Chryse Planitia, which is a relatively dust-free part of the planet. THEMIS thermal data (daytime TIR at left, nighttime TIR at right) show that there are numerous rocky units within the ejecta layers. Also note the bright ejecta rays and the very bright smaller (~6 km diameter) crater at the top right.

best to avoid. Not only might rover trafficability be impaired by extensive dust deposits but also the ability to see primary morphological features would be reduced.

The rewards from a successful lander mission to a crater such as Makhambet crater (Figure 9.23) could be considerable. First and foremost, if it could be proven that ejecta layers were indeed fluidized by volatiles, then extrapolation to other craters of different ages would confirm the existence of ice near the Martian surface at numerous locations in the past. This could spur additional research into the possibility that ejecta in areas such as Chryse Planitia (see Section 9.5) could be used to investigate temporal variations in near surface volatiles. A significant enhancement to this exploration would be the determination of the age of the ejecta layer; instrumentation for such *in situ* geochronology has been advocated by Cohen et al. (2021) and would provide a firm age calibration point for the crater-counting-based chronology for the planet (Hartmann and Neukum, 2001; Werner and Tanaka, 2011).

An alternative site for a lander mission, if the elevation-based landing constraints are eased somewhat, is illustrated in Figure 9.24. Here, a very recent ~500 m diameter crater has formed on the ejecta blanket of a moderately young 10 km diameter crater. This small crater appears to be much younger than the one illustrated in Figure 9.11 because of the dark ray material that is preserved. The two craters displayed in Figure 9.24 formed at an elevation of ~1,250 m above Mars' datum, so a sky-crane landing method would have to be modified to enable this site to be reached. Nevertheless, the rewards would be great, as the smaller crater would have excavated material from the larger crater's ejecta layer. If the rover could ascend to the rim crest of the smaller crater, a vertical section through the ejecta layer should be visible, thereby enabling the thickness of this unit to be determined. Many boulders are visible on the smaller crater rim, so that the diversity of materials could be easily assessed. An added attraction to visiting this site would be the potential to obtain an age date for one of the major physiographic units on Mars, that is, the Hesperian-aged lavas.

9.9 CONCLUDING THOUGHTS

In this book we have provided the Martian geomorphologist with some of the best images of the freshest impact craters on the planet. Craters of all sizes on Mars represent probes into the subsurface that most likely contain information on the distribution of volatiles at the time of impact. While differences in interior and exterior landforms can be found, as described earlier in this chapter, the prevailing interpretation is that water or ice is responsible for most of these features. And yet the community is still challenged to quantify the amount of volatiles as well as their depth below the surface. When we look at a crater, are we thinking about 5% volatiles by volume, or 25%, or 50%? Were the volatiles concentrated within the top 10 m of the target, the top 100 m, or the top 1 km? Some progress in answering

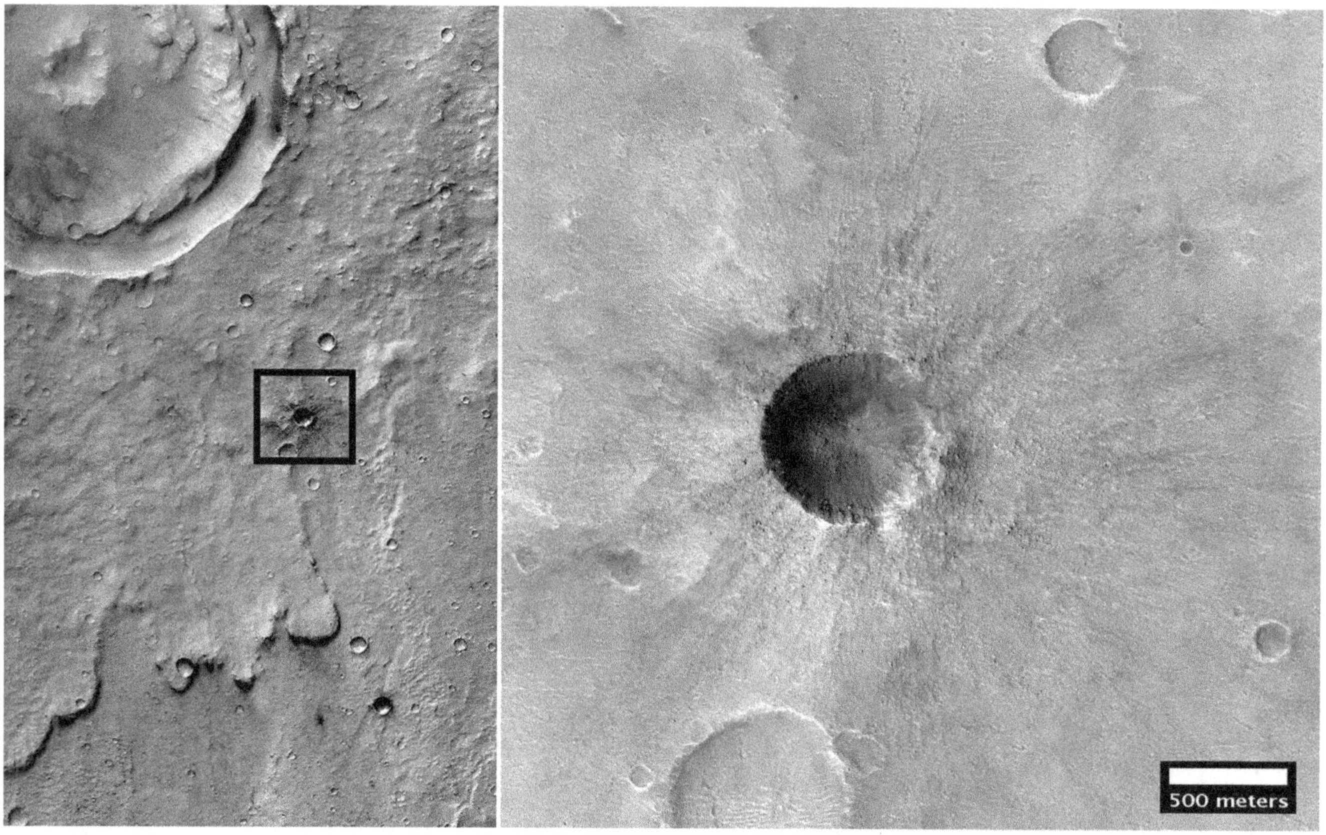

Figure 9.24 Left: Unnamed 10 km diameter crater in northern Hesperia Planum (at 11.54°S, 103.69°E) with prominent set of distal ramparts. Black box denotes the area shown at right. CTX image D16_033347_1684. Right: Very fresh crater ~500 m in diameter formed within the ejecta layer of the larger crater. Note prominent secondary craters and boulders near the rim. HiRISE image ESP_012605_1685.

these questions has recently been made using the simulations of ejecta dynamics developed by Sokolowska et al. (2024), but future numerical models need to be refined to have sufficient confidence to place numbers on the physical situation on Mars. Alas, it is not clear how image analysis alone can ever provide an answer, and numerical models can probably only go so far if virtually nothing is known about the particle size distribution and degree of cementation of the material within the flow.

In the absence of detailed field analysis of individual craters by rovers (which might even include drilling to infer emplacement processes), the most likely advances in our understanding of the details of crater formation might come from the use of as-yet-unused instrumentation such as imaging radar to "peer" beneath the dust-covered surface. We leave the task of flying a mission to resolve the questions raised in this book to the next generation of Martian crater enthusiasts. Wishing you success!

APPENDIX

The following is a listing of all the named craters included in this book. Along with the crater name, we list the diameter, the latitude and longitude of the crater. The planet where the crater is located is also included, the figure number in the book. Where more than one image of the crater appears only the "best" image is listed that shows the crater, as well as the figure number in this book. The frame number of the image is also included.

Table A.1 *Listing of named craters included in this book.*

Crater name	Diam. (km)	Planet	Latitude	Longitude	Fig. no.	Frame no.
Abedin	116.2	Mercury	*61.75°N*	*349.34°E*	2.11	JPL PIA19423
Achelous	40.0	Ganymede	*62.22°N*	*347.86°E*	2.26	JPL PIA01660
Addams	87.0	Venus	*56.09°S*	*98.81°E*	2.30	Magellan C2-MIDR 60S093
Arandas	25.1	Mars	*42.43°N*	*343.88°E*	4.3	CTX D04_028802_2227
Arima	53.6	Mars	*15.73°S*	*296.23°E*	4.6	THEMIS daytime IR mosaic
Aristarchus	40.0	Moon	*23.75°N*	*312.52°E*	2.8	LROC M1259297876LR
Bacolor	20.8	Mars	*33.01°N*	*118.58°E*	8.8	CTX P16_007462_2133
Bakhuysen	155.0	Mars	*22.88°S*	*15.71°E*	7.22	THEMIS daytime IR mosaic
Bamberg	54.0	Mars	*39.78°N*	*356.80°E*	4.8	CTX B02_010301_2200
Basho	74.6	Mercury	*32.38°S*	*189.55°E*	2.5	JPL PIA16343
Belz	10.2	Mars	*21.58°N*	*316.76°E*	9.6a	CTX B19_016909_2009
Betio	32.4	Mars	*23.07°S*	*281.30°E*	4.11	CTX B18_016805_1567
Copernicus	92.0	Moon	*9.61°N*	*339.92°E*	1.10	Apollo AS17-M-2439
Corinto	13.3	Mars	*16.95°N*	*141.70°E*	4.14	HiRISE PSP_003611_1970
Curie	114.1	Mars	*28.83°N*	*355.21°E*	4.19	THEMIS daytime IR mosaic
Danielson	64.0	Mars	*8.00°N*	*352.93°E*	5.9	CTX P18_008205_1880
Danilova	49.0	Venus	*26.30°S*	*337.15°E*	2.28	JPL PIA00086
Dickinson	69.0	Venus	*74.60°N*	*177.20°E*	9.22	JPL PIA00479
Dilly	2.1	Mars	*13.28°N*	*157.19°E*	9.4	CTX D09_030866_1933
Eberswalde	65.3	Mars	*23.83°S*	*326.38°E*	5.15	CTX B21_017911_1559
Endeavor	21.8	Mars	*2.25°S*	*354.78°E*	3.13	JPL PIA18394
Endurance	0.1	Mars	*2.01°S*	*354.50°E*	5.1	JPL PIA06866
Gaan	2.7	Mars	*38.69°N*	*356.60°E*	7.19	CTX J01_045298_2193
Gale	155.3	Mars	*5.27°S*	*137.73°E*	5.12	THEMIS daytime IR mosaic
Gasa	20.1	Mars	*35.72°S*	*129.43°E*	7.17	HiRISE PSP_009901_1440
Gratteri	7.0	Mars	*17.70°S*	*199.93°E*	8.2	CTX P07_003820_1628
Grindavik	12.0	Mars	*25.42°N*	*321.00°E*	9.6d	CTX P19_008562_2055
Gula	38.0	Ganymede	*64.39°N*	*347.34°E*	2.26	JPL PIA01660
Hale	150 × 125	Mars	*35.62°S*	*323.62°E*	5.19	THEMIS daytime IR mosaic
Hell Q	3.7	Moon	*33.00°S*	*355.55°E*	2.6	LROC M1221369684LR
Huygens basin	470.0	Mars	*13.88°S*	*55.58°E*	4.5	THEMIS daytime IR mosaic
Irharen	6.2	Mars	*34.50°N*	*140.81°E*	7.25	HiRISE ESP_023337_2150
Isabella	175.0	Venus	*29.70°S*	*204.12°E*	2.29	JPL PIA00480
Jackson	71.3	Moon	*22.05°N*	*196.68°E*	2.14	LROC M1139975629RE
Jaisalmer	15.0	Mars	*33.52°N*	*84.11°E*	9.20	CTX F23_044886_2136
Jezero	49.0	Mars	*18.49°N*	*77.37°E*	5.16	CTX P03_002387_1987
Jojutla	18.6	Mars	*81.63°N*	*189.80°E*	5.22	CTX P18_008131_2616

Table A.1 *(cont.)*

Crater name	Diam. (km)	Planet	Latitude	Longitude	Fig. no.	Frame no.
Karzok	15.6	Mars	*18.42°N*	*228.26°E*	4.16	CTX B09_013075_1985
Kokopelli	34.0	1 Ceres	*18.30°N*	*124.50°E*	2.23	JPL PIA21915
Korolev	84.2	Mars	*72.81°N*	*164.55°E*	5.23	CTX P21_009042_2528
Kotka	19.5	Mars	*19.28°N*	*169.86°E*	6.6	CTX B02_010519_2002
Lonar	11.0	Mars	*72.99°N*	*38.29°E*	8.4	CTX P22_009798_2530
Makhambet	16.0	Mars	*28.52°N*	*319.44°E*	9.23	THEMIS daytime IR mosaic
Manicouagan	100.0	Earth	*51.10°N*	*68.70°W*	2.4	Shuttle STS009-048-3139
Marcia	68.0	4 Vesta	*9.56°N*	*339.0°E*	2.25	JPL PIA16182
Mariner	170.0	Mars	*34.63°S*	*195.73°E*	3.1	Viking 562A29
Mazamba	52.3	Mars	*27.48°S*	*290.24°E*	4.7	CTX D05_029134_1523
Meteor	1.2	Earth	*35.01°N*	*111.00°W*	2.1	USGS 2007158
Michelangelo	230.0	Mercury	*44.92°S*	*250.22°E*	2.12	MESSENGER EN0241236952M
Mojave	58.5	Mars	*7.54°N*	*326.97°E*	4.21	CTX P03_002167_1877
Morteus	114.0	Moon	*70.63°S*	*354.05°E*	1.4	LROC WAC
Naar	11.4	Mars	*22.93°N*	*317.86°E*	9.6c	CTX G19_025691_2031
Nicholson	102.5	Mars	*0.28°N*	*195.47°E*	5.14	THEMIS daytime IR mosaic
Nutak	11.2	Mars	*17.43°N*	*329.73°E*	6.1d	CTX G07_020772_1967
Occator	92.0	1 Ceres	*19.83°N*	*239.34°E*	2.22	JPL PIA20350
Orientale basin	930.0	Moon	*19.56°S*	*266.25°E*	1.7	LROC WAC
Osiris	107.0	Ganymede	*38.00°S*	*193.69°E*	2.21	Voyager image
Pangboche	10.2	Mars	*17.31°N*	*226.58°E*	4.24	CTX P02_001643_1974
Poona	19.0	Mars	*23.77°N*	*307.67°E*	7.4	CTX J09_048293_2040
Puńsk	11.6	Mars	*20.62°N*	*318.87°E*	9.6b	CTX B21_017766_2007
Rustaveli basin	200.0	Mercury	*52.41°N*	*82.34°E*	2.20	JPL PIA19283
Santa Fe	20.3	Mars	*19.39°N*	*312.03°E*	4.3	CTX B02_010382_1994
Schiaparelli Basin	440 × 460	Mars	*2.50°S*	*16.62°E*	5.24	THEMIS daytime IR mosaic
Schrodinger basin	312.0	Moon	*74.73°S*	*132.92°E*	2.12	NASA Scientific Visualization Studio
Sinlap	80.0	Titan	*11.42°N*	*343.92°E*	2.31	JPL PIA16638
Slangpos	5.3	Mars	*4.73°S*	*136.81°E*	3.14	CTX D09_030735_1746
Soi	80.0	Titan	*24.37°N*	*218.99°E*	2.31	JPL PIA16638
Steinheim	11.1	Mars	*54.62°N*	*190.63°E*	6.1c	CTX P17_007736_2349
Talu	10.0	Mars	*40.34°S*	*20.08°E*	5.17	HiRISE ESP_011817_1395
Tehachapi	27.0	Mars	*36.58°N*	*329.57°E*	4.18	CTC B18_016737_2168
Telz	2.6	Mars	*21.17°N*	*111.11°E*	4.2	HiRISE ESP_018512_2015
Terby	174.0	Mars	*27.88°S*	*74.04°E*	5.10	THEMIS daytime IR mosaic
Toconao	17.7	Mars	*20.81°S*	*285.27°E*	6.13	THEMIS V05808002
Tooting	28.0	Mars	*23.21°N*	*207.76°E*	1.5	CTX P01_001538_2035
Toro	43.0	Mars	*17.04°N*	*71.81°E*	5.8	HiRISE PSP_005842_1970_MRGB
Tycho	85.0	Moon	*43.28°S*	*348.79°E*	2.9	LROC WAC
Valhalla	360.0	Callisto	*15.87°N*	*304.75°E*	2.24	JPL PIA02277
Waters	15.0	Mercury	*8.96°S*	*254.55°E*	2.17	MESSENGER 2641461
Zeami basin	129.0	Mercury	*2.96°S*	*212.59°E*	2.15	JPL PIA19267
Zumba	3.3	Mars	*28.67°S*	*226.92°E*	7.3	CTX P03_002118_1512
Zunil	10.4	Mars	*7.70°N*	*166.12°E*	9.13	CTX G05_020211_1877

REFERENCES

Abramov, O., Kring, D. A. (2005). Impact-induced hydrothermal activity on early Mars. *J. Geophys. Res.* 110(E12S09), doi:10.1029/2005JE002453, 19 pp.

Acuna, M. H. et al. (1999). Global distribution of crustal magnetization discovered by the Mars Global Surveyor MAG/ER experiment. *Science* 284, 790–793.

Allen, C. C. (1979). Large lunar secondary craters: Size-range relationships. *Geophys. Res. Lett.* 6(1), 51–54.

Alzate, N., Barlow, N. G. (2011). Central pits on Ganymede. *Icarus* 211, 1274–1283.

Anderson, F. S. et al. (2020). Pb-Pb dating of terrestrial and extraterrestrial samples using resonance ionization mass spectrometry. *Earth Space Sci.* 7, e2020EA001177, doi:10.1029/2020EA001177.

Andrews-Hanna, J. C., Lewis, K. W. (2011). Early Mars hydrology: 2. Hydrological evolution in the Noachian and Hesperian epochs. *J. Geophys. Res.* 116, E02007, doi:10.1029/2010JE003709, 20 pp.

Ansan, V. et al. (2011). Stratigraphy, mineralogy, and origin of layered deposits inside Terby crater, Mars. *Icarus* 211, 273–304.

Arvidson, R. E. (1974). Morphologic classification of Martian craters and some implications. *Icarus* 22, 264–271.

Arvidson, R. E. et al. (1976). Latitudinal variation of wind erosion of crater ejecta deposits on Mars. *Icarus* 27(4), 503–516.

Arvidson, R. E. et al. (1980). A post-Viking view of Martian geologic evolution. *Revs. Geophys. Space Phys.* 18, 565–603.

Ashley, J. W. et al. (2012). Geology of the King crater region: New insights into impact melt dynamics on the Moon. *J. Geophys. Res.* 117, 1–13, doi:10.1029/2011JE003990.

Baker, D. M. H. et al. (2011). The transition from complex crater to peak-ring basin on Mercury: New observations from MESSENGER flyby data and constraints on basin formation models. *Planet. Space Sci.* 59, 1932–1948.

Baker, V. R., Kochel, R. C. (1979). Martian channel morphology: Maja and Kasei Valles. *J. Geophys. Res.* 84(B14), 7961–7983.

Baker, V. R. et al. (1991). Ancient oceans, ice sheets and the hydrological cycle of Mars. *Nature* 352, 589–594.

Balaram, J. et al. (2021). The Ingenuity helicopter on the Perseverance rover. *Space Sci. Rev.* 217, 56, doi:10.1007/s11214-021-00815.

Baloga, S. M. et al. (2005). Emplacement of Martian rampart crater deposits. *J. Geophys. Res.* 110(E10001), doi:10.1029/2004JE002338, 12 pp.

Bamberg, M. et al. (2014). Floor-fractured craters on Mars – Observations and origin. *Planet. Space Sci.* 98, 146–162, doi:10.1016/j.pss.2013.09.017.

Bandfield, J. L. et al. (2000). A global view of Martian surface compositions from MGS-TES. *Science* 287, 1626–1630.

Baratoux, D. et al. (2002). An instability mechanism in the formation of the Martian lobate craters and implications for the rheology of ejecta. *Geophys. Res. Lett.* 29(51), doi:10.1029/2001GL013779, 4 pp.

Baratoux, D. et al. (2005). Thermal properties of lobate ejecta in Syrtis Major, Mars: Implications for the mechanism of formation. *J. Geophys. Res.* 110(E04011), doi:10.1029/2004JE002314, 14 pp.

Barlow, N. G. (1994). Sinuosity of Martian rampart ejecta deposits. *Geophys. Res. Lett.* 99, 10927–10935.

Barlow, N. G. (1995). The degradation of impact craters in Maja Valles and Arabia, Mars. *J. Geophys. Res.* 100(E11), 23307–23316.

Barlow, N. G. (2004). Martian subsurface volatile concentrations as a function of time: Clues from layered ejecta craters. *Geophys. Res. Lett.* 31(05703), doi:10.1029/2003GL019075, 4 pp.

Barlow, N. G. (2005). A review of Martian impact crater ejecta structures and their implications for target properties. In: Kenkmann, T., Hörz, F., Deutsch, A. (Eds.), *Large Meteorite Impact III, Geological Society of America Special Paper* 384, pp. 433–442, doi:10.1130/0-8137-2384-1.433.

Barlow, N. G. (2006). Impact craters in the northern hemisphere of Mars: Layered ejecta and central pit characteristics. *Meteor. Planet. Sci.* 41(10), 1425–1436.

Barlow, N. G. (2015). Characteristics of impact craters in the northern hemisphere of Mars. In: Osinski, G. R., Kring, D. A. (Eds.), *Large Meteorite Impacts and Planetary Evolution V, Special Paper* 518, *Geological Society of America, Denver, CO, USA*, pp. 31–63.

Barlow, N. G., Bradley, T. L. (1990). Martian impact craters: Correlations of ejecta and interior morphologies with diameter, latitude, and terrain. *Icarus* 87, 156–179.

Barlow, N. G., Perez, C. B. (2003). Martian impact crater ejecta morphologies as indicators of the distribution of subsurface volatiles. *J. Geophys. Res.* 108(5085), doi:10.1029/2002JE002036, 4 pp.

Barlow, N. G., Pollak, A. (2002). Comparisons of ejecta mobility ratios in the Northern and Southern Hemispheres of Mars. *Lunar Planet. Sci.* XXXIII, abs. #1322.

Barlow, N. G., et al. (2000). Standardizing the nomenclature of Martian impact crater ejecta morphologies. *J. Geophys. Res.* 105(E11), 26733–26738, doi:10.1029/2000JE001258.

Barlow, N. G. et al. (2001). Variations in the onset diameter for Martian layered ejecta morphologies and their implications for subsurface volatile reservoirs. *Geophys. Res. Lett.* 28(16), 3095–3098.

Barlow, N. G. et al. (2014). Martian low-aspect-ratio layered ejecta (LARLE) craters: Distribution, characteristics, and relationship to pedestal craters. *Icarus* 239, 186–200.

Barlow, N. G. et al. (2017). Comparison of central pits craters on Mars, Mercury, Ganymede, and the Saturnian satellites. *Meteor. Planet. Sci.* 52(7), 1371–1387.

Barnouin-Jha, O. S., Schultz, P. H. (1996). Ejecta entrainment by impact-generated ring vortices: Theory and experiments. *J. Geophys. Res.* 101(E9), 21099–21115.

Barnouin-Jha, O. S. et al. (1999a). Investigating the interactions between an atmosphere and an ejecta curtain: 1. Wind tunnel tests. *J. Geophys. Res.* 104(E11), 27105–27131.

Barnouin-Jha, O. et al. (1999b). Investigating the interactions between an atmosphere and an ejecta curtain: 2. Numerical experiments. *J. Geophys. Res.* 104(E11), 27117–27131.

Barnouin-Jha, O. S. et al. (2005). Comparing landslides to fluidized crater ejection Mars. *J. Geophys. Res.* 110(EO4010), doi:10.1029/2003JE002214, 22 pp.

Bart, G. D. et al. (2011). Global survey of lunar regolith depths from LROC images. *Icarus* 215(2), 485–490.

Beddington, C. B. et al. (2021). The relationship between thermal inertia and degradation state of craters in areas of low surface dust cover on Mars. *Icarus* 370, 114678.

Bell, J. (2008). *The Martian Surface: Composition, Mineralogy, and Physical Properties*. Cambridge Planetary Science Series, Cambridge University Press, p. 636.

Bennett, K. A., Bell, J. F. (2016). A global survey of Martian central mounds: Central mounds as remnants of previously more extensive large-scale sedimentary deposits. *Icarus* 264, 331–341.

Berman, D. C. et al. (2005). The role of arcuate ridges and gullies in the degradation of craters in the Newton Basin region of Mars. *Icarus* 178(2), 465–486.

Bernhardt, H. et al. (2016). Photogeologic mapping and the geologic history of the Hellas basin floor, Mars. *Icarus* 264, 407–442.

Black, B. A., Stewart, S. T. (2008). Excess ejecta craters record episodic ice-rich layers at middle latitudes on Mars. *J. Geophys. Res.* 113(E02015), doi:10.1029/2007JE002888, 22 pp.

Bland, M. T., Bray, V. J. (2024). The inevitability of large shallow craters on Callisto and Ganymede: Implications for crater depth-diameter trends. *Icarus* 408, 115811.

Blewett, D. T. et al. (2011). Hollows on Mercury: MESSENGER evidence for geologically recent volatile-related activity. *Science* 333(6051), 1856–1859.

Blewett, D. T. et al. (2013). Mercury's hollows: Constraints on formation and composition from analysis of geological setting and spectral reflectance. *J. Geophys. Res.* 118(5), 1013–1032.

Blom, R. G., Daily, M. (1984). Radar image processing for rock-type discrimination. *IEEE Trans Geosci. Rem. Sens.* GE-20(3), 343–351.

Boyce, J. M., Garbeil, H. (2007). Geometric relationships of pristine Martian complex impact craters, and their implications to Mars geologic history. *Geophys. Res. Lett.* 34(L16201), doi:10.1029/2007GL029731, 5 pp.

Boyce, J. M., Mouginis-Mark, P. J. (2006). Martian craters viewed by the Thermal Emission Imaging System instrument: Double-layered ejecta craters. *J. Geophys. Res.* 111(E10005), doi:10.1029/2005JE002638, 21 pp.

Boyce, J. M., Mouginis-Mark, P. J. (2020). Possible self-secondary cratering on the rim of Tooting crater, Mars. *51st Lunar Planet. Sci. Conf. abs.* No. 1106.

Boyce, J. M., Mouginis-Mark, P. J. (2021). Morphologic differences in radial grooves on Martian layered (fluidized) ejecta: Implications for emplacement processes and conditions. *Icarus* 366, 114513.

Boyce, J. M., Mouginis-Mark, P. J. (2025). A conceptual model for the formation of ramparts on Martian impact crater ejecta. *Icarus*, 425, 116336, 9 pp.

Boyce, J. M. et al. (2005). Ancient oceans in the northern lowlands of Mars: Evidence from impact crater depth/diameter relationships. *J. Geophys. Res.* 110(E03008), doi:10.1029/2004JE002328, 15 pp.

Boyce, J. M. et al. (2006). Deep impact craters in the Isidis and southwestern Utopia Planitia regions of Mars: High target material strength as a possible cause. *Geophys. Res. Lett.* 33(L06202), doi:10.1029/2005GL024462, 4 pp.

Boyce, J. M. et al. (2010). Rampart craters on Ganymede: Their implications for fluidized ejecta emplacement. *Meteor. Planet. Sci.* 45(4), 638–661.

Boyce, J. M. et al. (2012). Origin of small pits in Martian impact craters. *Icarus* 221(1), 262–275.

Boyce, J. M. et al. (2015). Origin of the outer layer of Martian low-aspect ratio layered ejecta craters. *Icarus* 245, 263–272.

Boynton, W. V. et al. (1992). Science applications of the Mars Observer gamma ray spectrometer. *J. Geophys. Res.* 97(E5), 7681–7698.

Bray, V. J. et al. (2010). New insight into lunar impact melt mobility from the LRO camera. *Geophys. Res. Lett.* 37(L21202), doi:10.1029/2010GL044666, 5 pp.

Bray, V. J. et al. (2012). Investigating the transition from central peak to peak-ring basins using central feature volume measurements from the Global Lunar DTM 100 m. *Geophys. Res. Lett.* 39(21), L21201, doi:10.1029/2012GL053693.

Breed, C. S. et al. (1979). Morphology and distribution of common "sand" dunes on Mars: Comparison with the Earth. *J. Geophys. Res.* 84(B14), 8183–8204.

Bridges, J. C., Schwenzer, S. P. (2012). The nakhlite hydrothermal brine on Mars. *Earth Planet. Sci. Lett.* 359–360, 117–123.

Bridges, N. T. et al. (2017). Martian aeolian activity at the Bagnold Dunes, Gale Crater: The view from the surface and orbit. *J. Geophys. Res.* 122(10), 2077–2110.

Brown, A. J. et al. (2008). Louth crater: Evolution of a layered water ice mound. *Icarus* 196(2), 433–445.

Buczkowski, D. L. et al. (2005). Buried impact craters: A topographic analysis of quasi-circular depressions, Utopia Basin. Mars, *J. Geophys. Res.* 110(E03007), doi:10.1029/2004JE002324, 8 pp.

Burr, D. M. et al. (2009). Floods from fossae: A review of Amazonian-aged extensional-tectonic megaflood channels on Mars. In: Burr, D. M., Carling, P. A., Baker, V. R. (Eds.), *Megaflooding on Earth and Mars*. Cambridge University Press, pp. 194–208.

Carr, M. H. (2006). *The Surface of Mars*. Cambridge University Press, p. 307.

Carr, M. H., Head, J. W. (2003). Oceans on Mars: An assessment of the observational evidence and possible fate. *J. Geophys. Res.* 108(E5, no. 5042), doi:10.1029/2002JE001963, 28 pp.

Carr, M. H., Head, J. W. (2010). Geologic history of Mars. *Earth Planet. Sci. Lett.* 294(3–4), 185–203.

Carr, M. H., Head, J. W. (2015). Martian surface/near-surface water inventory: Sources, sinks and changes with time. *Geophys. Res. Lttrs.* 42, 726–732.

Carr, M. H. et al. (1977). Martian impact craters and emplacement of ejecta by surface flow. *J. Geophys. Res.* 82(28), 4055–4065.

Caudill, C. M. et al. (2012). Layered megablocks in the central uplifts of impact craters. *Icarus* 221(2), 710–720.

Caudill, C. M. et al. (2018). Ejecta deposits of Bakhuysen crater, Mars. *Icarus* 314, 175–194.

Chadwick, D. J., Schaber, G. G. (1993). Impact crater outflows on Venus: Morphology and emplacement mechanisms. *J. Geophys. Res.* 98, 20891–20902.

Chappelow, J. E., Sharpton, V. L. (2002). An improved shadow measurement technique for constraining the morphometry of simple impact craters. *Meteor. Planet. Sci.*, 37, 479–486.

Chappelow, J. E., Herrick, R. R. (2008). On the origin of a double, oblique impact on Mars. *Icarus* 197, 452–457.

Chojnacki, M. et al. (2011). Orbital observations of contemporary dune activity in Endeavor crater, Meridiani Planum, Mars. *J. Geophys. Res.* 116(E00F19), doi:10.1029/2010JE003675, 20 pp.

Christensen, P. R. (2003). Formation of recent Martian gullies through melting of extensive water-rich snow deposits. *Nature* 422, 45–48.

Christensen, P. R. et al. (1992). Thermal emission spectrometer experiment: Mars observer mission. *J. Geophys. Res.* 97, 7719–7734.

Christensen, P. R. et al. (1998). Results from the Mars Global Surveyor Thermal Emission Spectrometer. *Science* 279, 1692–1698.

Christensen, P. R. et al. (2004). The Thermal Emission Imaging System (THEMIS) for the Mars 2001 Odyssey Mission. *Space Sci. Rev.* 110, 85–130.

Christian, J. R. et al. (2022). CRISM-based high spatial resolution thermal inertia mapping along Curiosity's traverses in Gale Crater. *J. Geophys. Res.* 127, e2021JE007076.

Chuang, F. C. et al. (2016). Zumba crater, Daedalia Planum, Mars: Geologic investigation of a young, rayed impact crater and its secondary field. *Icarus* 269, 75–90.

Cintala, M. J., Mouginis-Mark, P. J. (1980). Martian fresh crater depths: More evidence for subsurface volatiles? *Geophys. Res. Lett.* 7, 329–332.

Cintala, M. J. et al. (1976). Characteristics of fresh Martian craters as a function of diameter: Comparison with the Moon and Mercury. *Geophys. Res. Lett.* 3, 117–120.

Cintala, M. J. et al. (1977). The effects of target characteristics on fresh crater morphology: Preliminary results for the moon and Mercury. *Proc. Lunar Sci. Conf. 8th*, pp. 3409–3425.

Clifford, S. M. (1993). A model for the hydrologic and climatic behavior of water on Mars. *J. Geophys. Res.* 98, 10973–11016.

Clifford, S. M. et al. (2010). Depth of the Martian cryosphere: Revised estimates and implications for the existence and detection of subpermafrost groundwater. *J. Geophys. Res.* 115, E07001. doi:10.1029/2009JE003462.

Cohen, B. A. et al. (2021). In situ geochronology for the next decade: Mission designs for the Moon, Mars, and Vesta. *Planet. Sci. J.* 2, 145.

Conway, S. J. et al. (2012). Climate-driven deposition of water ice and the formation of mounds in craters in Mars' north pole region. *Icarus* 220, 174–193.

Cordell, B. M. et al. (1974). Martian cratering and central peak statistics: Manner 9 results. *Icarus* 21, 448–456.

Costard, F. et al. (2019). The Lomonosov crater impact event: A possible mega-tsunami source on Mars. *J. Geophys. Res.* 124, 1840–1851.

Craddock, R. A., Howard, A. D. (2002). The case for rainfall on a warm, wet early Mars. *J. Geophys. Res.* 107(5111), doi:10.1029/2001JE001505, 21 pp.

Craddock, R. A. et al. (1997). Crater morphology and modification in the Sinus Sabaeus and Margaritifer Sinus regions of Mars. *J. Geophys. Res.* 102, 13321–13340.

Craddock, R. A. et al. (2018). An assessment of regional variations in Martian modified impact crater morphology. *J. Geophys. Res.* 123, doi:10.1002/2017JE005412, 763–779.

Crumpler, L. S. et al. (2015). Context of ancient aqueous environments on Mars from in situ geologic mapping at Endeavour Crater. *J. Geophys. Res.* 120, doi:10.1012/2014JE004699, 538–569.

Daubar, I. J. et al. (2014). The morphology of small fresh craters on Mars and the Moon. *J. Geophys. Res.* 119, 2620–2639, doi:10.1002/2014JE004671.

Daubar, I. J. et al. (2016). Changes in blast zone albedo patterns around new Martian impact craters. *Icarus* 267, 86–105.

Daubar, I. J. et al. (2022). New craters on Mars: An updated catalog. *J. Geophys. Res.* 127(7), e2021JE007145.

Davis, P. A., Soderblom, L. A. (1984). Modeling crater topography and albedo from monoscopic Viking orbiter images. *J. Geophys. Res.* 89, 9449–9457.

Denevi, B. W. et al. (2012). Pitted terrain on Vesta and implications for the presence of volatiles. *Science* 338, 246–249.

De Toffoli, B. et al. (2021). Rheological and mechanical layering of the crust underneath thumbprint terrains in Arcadia Planitia, Mars. *J. Geophys. Res.* 126, e2021JE007007.

Dickson, J. L., Head, J. W. (2009). The formation and evolution of youthful gullies on Mars: Gullies as the late-stage phase of Mars' most recent ice age. *Icarus* 204, 63–86.

Ding, N. et al. (2015). The central uplift of Ritchey crater, Mars. *Icarus* 252, 255–270.

Diot, X. et al. (2014). The geomorphology and morphometry of the banded terrain in Hellas basin, Mars. *Planet. Space Sci.* 101, 118–134.

Dohm, J. M. et al. (2007). Possible ancient giant basin and related water enrichment in the Arabia Terra province, Mars. *Icarus* 190, 74–92.

Dombard, A. J., McKinnon, W. B. (2006). Elastoviscoplastic relaxation of impact crater topography with application to Ganymede and Callisto. *J. Geophys. Res.* 111, E01001, doi:10.1029/2005JE002445, 22 pp.

Dundas, C. M., McEwen, A. S. (2007). Rays and secondary craters of Tycho. *Icarus* 186(1), 31–40.

Dundas, C. M. et al. (2014). HiRISE observations of new impact craters exposing Martian ground ice. *J. Geophys. Res.* 119, 109–127.

Dundas, C. M. et al. (2019). The formation of gullies on Mars today. In: Conway, S. J., Carrivick, J. L., Carling, P. A., de Hass, T., Harrison, T. N. (Eds.), *Martian Gullies and Their Earth analogues, Geological Society American Special Publication* 467, pp. 67–94.

Dundas, C. M. et al. (2021). Widespread exposures of extensive clean shallow ice in the midlatitudes of Mars. *J. Geophys. Res.* 126, e2020JE006617.

Eberhardt, P. et al. (1973). How old is the crater Copernicus? *The Moon* 8, 104–114.

Edwards, C. S. et al. (2009). Global distribution of bedrock exposures on Mars using THEMIS high-resolution thermal inertia. *J. Geophys. Res.* 114, E11001, doi:10.1029/2009JE003363, 18 pp.

Edwards, C. S. et al. (2014). The formation of infilled craters on Mars: Evidence for widespread impact induced decompression of the early Martian mantle? *Icarus* 228, 149–166.

Elder, C. M. et al. (2012). The theoretical plausibility of central pit crater formation via melt drainage. *Icarus* 221, 831–843.

Elliott, J. R. et al. (2018). The length of lunar crater rays explained using secondary crater scaling. *Icarus* 312, 231–246.

El-Maarry, M. R. et al. (2013). Morphology and evolution of the ejecta of Hale crater in Argyre basin, Mars: Results from high resolution mapping. *Icarus* 226, 905–922.

Evans, D. L. et al. (1986). Multipolarization radar images for geologic mapping and vegetation discrimination. *IEEE Trans. Geosci. Rem. Sens.* GE-24, 246–257.

Fairen, A. G. et al. (2003). Episodic flood inundations of the northern plains of Mars. *Icarus* 165, 53–67.

Fanale, F. P. et al. (1986). Global distribution and migration of subsurface ice on Mars. *Icarus*, *67*(1), 1–18.

Farr, T. G. et al. (2007). The Shuttle Radar Topography Mission. *Revs. Geophys.* 45, 1–33.

Fassett, C. I., Head, J. W. (2007). Layered mantling deposits in northeast Arabia Terra, Mars: Noachian-Hesperian sedimentation, erosion, and terrain inversion. *J. Geophys. Res.* 112, E08002, doi:10.1029/2006JE002875, 19 pp.

Fassett, C. I. et al. (2009). Caloris impact basin: Exterior geomorphology, stratigraphy, morphometry, radial sculpture, and smooth plains. *Earth Planet. Sci. Lett.* 285, 297–308.

Feldman, W. C. et al. (2011). Mars Odyssey neutron data: 2. Search for buried excess water ice deposits at nonpolar latitudes on Mars. *J. Geophys. Res.* 116, E11009, doi:10.1029/2011JE003806.

Fishbaugh, K. E., Head, J. W. (2001). Comparison of the North and South polar caps on Mars: New observations from MOLA data and discussion of some outstanding questions. *Icarus* 154, 145–161.

Forget, F. et al. (2017). Recent climate variations. In: *The Atmosphere and Climate of Mars*. Edited by R. M. Haberle, R. T. Clancy,

F. Forget, M. D. Smith, and R. W. Zurek. Cambridge University Press, pp. 497–525.

Forsberg-Taylor, N. K. et al. (2004). Crater degradation in the Martian highlands: Morphometric analysis of the Sinus Sabaeus region and simulation modeling suggest fluvial processes. *J. Geophys. Res.* 109(E05002), doi:10.1029/2004JE002242, 12 pp.

Frey, H. (2008). Ages of very large impact basins on Mars: Implications for the late heavy bombardment in the inner solar system. *Geophys. Res. Lett.* 35, L132030, doi:10.1029/2008GL033515.

Frey, H. et al. (1999). Discovery of a 450 km diameter, multi-ring basin on Mars through analysis of MOLA topographic data. *Geophys. Res. Lett.* 26, 1657–1660.

Frey, H. V. et al. (2002). Ancient lowlands of Mars. *Geophys. Res. Lett.* 29, 1384 doi:10.1029/2001GL013832, 22 pp.

Garvin, J. B., Frawley, J. J. (1998). Geometric properties of Martian impact craters: Preliminary results from the Mars Orbiter Laser Altimeter. *Geophys. Res. Lett.* 25(24), 4405–4408.

Garvin, J. B. et al. (2000). North polar region craterforms on Mars. Geometric characteristics from the Mars Orbiter Laser Altimeter. *Icarus* 144, 329–352.

Gault, D. E., Greeley, R. (1978). Exploratory experiments on impact craters formed in viscous-liquid targets: Analogs for Martian rampart craters? *Icarus* 34, 486–495.

Gault, D. E., Wedekind, J. A. (1978). Experimental studies of oblique impacts. *Proc. Lunar Planet. Sci. Conf. 9th*, pp. 3843–3875.

Gault, D. E. et al. (1968). Impact cratering mechanics and structures. In: French B. M., Short N. M. (Eds.), *Shock Metamorphism of Natural Materials*. Mono Book Corp., pp. 87–99.

Goddard, K. et al. (2014). Mechanisms and timescales of fluvial activity at Mojave and other young Martian craters. *J. Geophys. Res.* 119, 604–634.

Goldspiel, J. M., Squyres, S. W. (2011). Groundwater discharge and gully formation on Martian slopes. *Icarus* 211, 238–258.

Gou, S. et al. (2024). Paleoenvironment implications of layered ejecta craters in the Chryse Planitia, Mars. *Icarus* 410, 115918.

Goudge, T. A. et al. (2015). Integrating CRISM and TES hyperspectral data to characterize a halloysite-bearing deposits in Kashira crater, Mars. *Icarus* 250, 165–187.

Grant, J. A., Schultz, P. H. (1993). Erosion of ejecta at Meteor Crater, Arizona. *J. Geophys. Res.* 98, 15033–15047.

Grant, J. A., Wilson, S. A. (2018). The nature and emplacement of distal aqueous-rich ejecta deposits from Hale crater, Mars. *Meteor. Planet. Sci.* 53 (4), 839–856.

Grant, J. A. et al. (2008). HiRISE imaging of impact megabreccia and sub-meter aqueous strata in Holden Crater, Mars. *Geology* 36(3), 195–198.

Grant, J. A. et al. (2016). The degradational history of Endeavor crater, Mars. *Icarus* 280, 22–36.

Greeley, R., Spudis, P. D. (1978). Volcanism in the cratered terrain hemisphere of Mars. *Geophys. Res. Lett.* 5, 453–455, doi:10.1029/GL005i006p00453.

Green, J. (1971). Copernicus as a lunar caldera. *J. Geophys. Res.* 76, 5719–5731.

Grieve, R. A. F., Cintala, M. J. (1981). Brent crater: Ontario: Observation and theory. *Lunar Planet. Sci.* XII, 362–364.

Grieve, R. A. F., Head, J. W. (1983). The Manicouagan impact structure: An analysis of its original dimensions and form. *J. Geophys. Res.* 88, A807–A818.

Grimm, R. E., Solomon, S. C. (1988). Viscous relaxation of impact crater relief on Venus: Constraints on crustal thickness and thermal gradient. *J. Geophys. Res.* 93, 11911–11929.

Grotzinger, J. P. et al. (2015). Deposition, exhumation, and paleoclimate of an ancient lake deposit, Gale Crater, Mars. *Science* 350, aac7575-1 to 12.

Guest, J. E., Murray, J. B. (1971). A large scale surface pattern associated with the ejecta blanket and rays of Copernicus. *Moon* 3, 326–336.

Guidat, T. et al. (2015). Landform assemblage in Isidis Planitia, Mars: Evidence for a 3 Ga old polythermal ice sheet. *Earth Planet. Sci Lett.* 411, 253–267.

Guimpier, A. et al., CaSSIS Team (2022). Pre-landslide topographic reconstruction in Baetis Chaos, Mars using a CaSSIS digital elevation model. *Planet. Space Sci.* 218, 105505.

Gulick, S. P. S. et al. (2013). Geophysical characterization of the Chicxulub impact crater. *Revs. Geophys.* 51, 31–52.

Hale, W. S. (1983). Central structures in Martian impact craters: Morphology, morphometry, and implications for substrate volatile distribution (abstract). *14th Lunar Planet. Sci. Conf.*, pp. 273–274.

Hamilton, C. W. et al. (2018). Episodes of aqueous flooding and effusive volcanism associated with Hrad Vallis, Mars. *J. Geophys. Res.* 123, doi:10.1029/2018JE005543, 1484–1510.

Hamilton, V. E. et al. (2003). Searching for the source regions of Martian meteorites using MGS TES: Integrating Martian meteorites into the global distribution of igneous materials on Mars. *Meteoritics Planet. Sci.* 38(6), 871–885.

Hanna, J. C., Phillips, R. J. (2006). Tectonic pressurization of aquifers in the formation of Mangala and Athabasca Valles, Mars. *J. Geophys. Res.* 111, E03003, doi:10.1029/2005JE002546.

Harris, J. K. et al. (2023). An updated catalog of rayed craters on Mars. *Earth Space Sci.* 10, e2021EA001814.

Harrison, K. P., Chapman, M. G. (2008). Evidence for ponding and catastrophic floods in central Valles Marineris, Mars. *Icarus* 198, 351–364.

Harrison, T. N. et al. (2010). Impact-induced overland fluid flow and channelized erosion at Lyot Crater, Mars. *Geophys. Res. Lett.* 37(L21201), doi:10.1029/2010GL045074, 6 pp.

Harrison, T. N. et al. (2015). Global documentation of gullies with the Mars Reconnaissance Orbiter Context Camera and implications for their formation. *Icarus* 252, 236–254.

Hartmann, W. K. (1972). Interplanet variations in scale of crater morphology – Earth, Mars, Moon. *Icarus* 17(3), 707–713.

Hartmann, W. K. (2005). Martian cratering 8: Isochron refinement and the chronology of Mars. *Icarus* 174(2), 294–320.

Hartmann, W. K., Neukum, G. (2001). Cratering chronology and the evolution of Mars. *Space Sci. Rev.* 96, 165–194.

Hartmann, W. K. et al. (2010). Do young Martian ray craters have ages consistent with the crater count system? *Icarus* 208, 621–635.

Hartmann, W. K. et al. (2014). Comprehensive analysis of glaciated Martian crater Greg. *Icarus* 228, 96–120.

Hawke, B. R., Head, J. W. (1977). Impact melt on lunar crater rims. In *Impact and Explosion Cratering: Planetary and Terrestrial Implications*, *Proc. Symp. Planetary Crat. Mech.*, Flagstaff, AZ, September 13–17, 1976, pp. 815–841, Pergamon, New York.

Hawke, B. R. et al. (2004). The origin of lunar crater rays. *Icarus* 170, 1–16.

Hayward, R. K. et al. (2007). Mars global digital dune database and initial science results. *J. Geophys. Res.* 112(E11007), doi:10.1029/2007JE002943, 17 pp.

Hayward, R. K. et al. (2014). Mars global digital dune database (MGD^3): Global dune distribution and wind pattern observations. *Icarus* 230, 38–46.

Head, J. W. (1974). Orientale multi-ringed basin interior and implications for the petrogenesis of lunar highland sample. *The Moon* 11, 327–356.

Head, J. W. (1976). Evidence for the sedimentary origin of Imbrium sculpture and lunar basin radial texture. *The Moon* 15, 445–464.

Head, J. W. (2007). The geology of Mars: New insights and outstanding questions. In: Chapman, M. (Ed.), *The Geology of Mars:*

Evidence from Earth-based Analogs. Cambridge University Press, pp. 1–46.

Head, J. W., Carr, M. H. (2003). Oceans on Mars: An assessment of the observational evidence and possible fate. *J. Geophys. Res*. 108(E5 5042), doi:10.1029/2002JE001963, 8 pp.

Head, J. W., Roth, R. (1976). Mars pedestal crater escarpments: Evidence for ejecta-related emplacement. *Symp. Planetary Crat. Mech*., pp. 50–52, The Lunar Science Institute, Houston, TX.

Head, J. W. et al. (1999). Possible ancient oceans on Mars: Evidence from Mars Orbiter Laser Altimeter data. *Science* 286(#5447), 2134–2137.

Head, J. W. et al. (2003). Recent ice ages on Mars. *Nature* 426, 797–802.

Heather, D. J., Dunkin, S. K. (2003). Geology and stratigraphy of King crater, lunar farside. *Icarus* 163, 307–329.

Henkel, H., Reimold, W. U. (1998). Integrated geophysical modelling of a giant, complex impact structure: Anatomy of the Vredefort Structure, South Africa. *Tectonophysics* 287(1–4), 1–20.

Herd, C. D. K. et al. (2024). The source craters of the Martian meteorites: Implications for the igneous evolution of Mars. *Sci. Adv*. 10, eadn2378.

Herrick, R. (2013). The shapes of simple craters in the outer solar system determined with an enhanced shadow measurement technique. 44th Lunar Planet. Sci. Conf., Abstract 7320.

Hiesinger, H., Head, J. W. (2002). Topography and morphology of the Argyre Basin, Mars: Implications for its geologic and hydrologic history. *Planet. Space Sci*. 50(10–11), 939–981.

Hiesinger, H. et al. (2016). Cratering on Ceres: Implications for its crust and evolution. *Science* 353, aaf4759, doi:10.1126/science.aaf4759.

Hill, J. R., Christensen, P. R. (2017). Well-preserved low thermal inertia ejecta deposits surrounding young secondary impact craters on Mars. *J. Geophys. Res*. 122(E6), doi:10.1029/2016JE005210, 1276–1299.

Howard, A. J. et al. (1982). Stratigraphic relationships within Martian polar cap deposits. *Icarus* 50, 161–215.

Hoover, R. H. et al. (2021). Insight into formation processes of layered ejecta craters on Mars from thermophysical observations. *J. Geophys. Res*. 126, e2020JE006801.

Horgan, B. H. N. et al. (2020). The mineral diversity of Jezero crater: Evidence for possible lacustrine carbonates on Mars. *Icarus* 339, 113526.

Horvath, D. G., Andrews-Hanna, J. C. (2024). The hydrology of the Jezero crater paleolake: Implications for the climate and limnology of the lake system from hydrological modeling. *Earth Planet. Sci. Lett*. 635, 118690.

Horvath, D. G. et al. (2021). Evidence for geologically recent explosive volcanism in Elysium Planitia, Mars. *Icarus* 365, 114499.

Hörz, F., Banholzer, G. S. (1980). Deep-seated target materials in the continuous deposits of the Ries Crater, Germany. In: Papike, J., Merrill, R. (Eds.), *Proc. Conf. Lunar Highland Crust*, Pergamon Press, pp. 211–231.

Hörz, F. et al. (1983). Bunte Breccia of the Ries: Continuous deposits of large impact craters. *Revs. Geophys. Space Phys*. 21, 1667–1725.

Howard, K. A. (1975). Geologic map of the crater Copernicus. Miscellaneous Investigations Series. United States Geological Survey.

Howard, K. A., Wilshire, H. G. (1975). Flows of impact melt at lunar craters. *J. Res. U.S. Geol. Survey* 3, 237–251.

Irwin, R. P. et al. (2013). Distribution of early, middle, and late Noachian cratered surfaces in the Martian highlands: Implications for resurfacing events and processes. *J. Geophys. Res*. 118, 278–291.

Irwin, R. P. et al. (2015). Paleohydrology of Eberswalde crater, Mars. *Geomorphology* 240, 83–101.

Jaeger, W. L. et al. (2007). Athabasca Valles, Mars: A lava-draped channel system. *Science* 317, 1709–1711.

Jaeger, W. L. et al., the HiRISE Team (2010). Emplacement of the youngest flood lava on Mars: A short, turbulent story. *Icarus* 205. 230–243.

Jakosky, B. M., Carr, M. H. (1985). Possible precipitation of ice at low latitudes of Mars during periods of high obliquity. *Nature* 315, 559–561.

Jakosky, B. M. et al. (1995). Chaotic obliquity and the nature of the Martian climate. *J. Geophys. Res*. 100(E1), 1579–1584.

Jaumann, R. et al. (2015). Quantifying geological processes on Mars – Results of the high resolution stereo camera (HRSC) on Mars Express. *Planet. Space Sci*. 112, 53–97.

Johnsson, A. et al. (2014). Evidence for very recent melt-water and debris flow activity in gullies in a young mid-latitude crater on Mars. *Icarus* 235, 37–54.

Jones, A. P. et al. (2011). A geomorphic analysis of Hale crater, Mars: The effects of impact into ice-rich crust. *Icarus* 211, 259–272.

Jones, E., Osinski, G. R. (2015). Using Martian single and double layered ejecta craters to probe subsurface stratigraphy. *Icarus* 247, 260–278.

Jones, E. et al. (2016). Insights into complex layered ejecta emplacement and subsurface stratigraphy in Chryse Planitia, Mars, through an analysis of THEMIS brightness temperature data. *J. Geophys. Res*. 121(6), 986–1015.

Jordan, R. et al. (2009). The Mars Express MARSIS sounder instrument. *Planet. Space Sci*. 57, 1975–1986.

Jozwiak, L. M. et al. (2012). Lunar floor-fractured craters: Classification, distribution, origin and implications for magmatism and shallow crustal structure. *J. Geophys. Res*. 117, E11005, doi:10.1029/2012JE004134.

Kadish, S. J., Head, J. W. (2011). Impacts into non-polar ice-rich paleodeposits on Mars: Excess ejecta craters, perched craters and pedestal craters as clues to Amazonian climate history. *Icarus* 215, 34–46.

Kadish, S. J. et al. (2008). Martian pedestal craters: Marginal sublimation pits implicate a climate-related formation mechanism. *Geophys. Res. Lett*. 35, L16104, doi:10.1029/2008GL034990.

Kadish, S. J. et al. (2009). Latitude dependence of Martian pedestal craters: Evidence for a sublimation-driven formation mechanism. *J. Geophys. Res*. 114(E10001), doi:10.1029/2008JE0023318, 25 pp.

Kadish, S. J. et al. (2010). Pedestal crater heights on Mars: A proxy for the thickness of past, ice-rich, Amazonian deposits. *Icarus* 210, 92–101.

Kegerreis, J. A. et al. (2020). Atmospheric erosion by giant impacts onto terrestrial planets. *Astrophys. J*. 897, 161 (15 pp.)

Kieffer, S. W., Simonds, C. H. (1980). The role of volatiles and lithology in the impact cratering process. *Revs. Geophys. Space Phys*. 18, 143–181.

Kirchoff, M. R., Grimm, R. E. (2017). Timing and distribution of single-layered ejecta craters imply sporadic preservation of tropical subsurface ice on Mars. *J. Geophys. Res*. 123, 131–144.

Kite, E. S. et al. (2014). Low paleopressure of the Martian atmosphere estimated from the size distribution of ancient craters. *Nature Geosci*. 7, 335–339.

Krohn, K. et al. (2016). Cryogenic flow features on Ceres: Implications for crater-related cryovolcanism. *Geophys. Res. Lttrs*. 43, 11994–12003.

Kumar, P. S. (2005). Structural effects of meteorite impact on basalt: Evidence from Lonar crater, India. *J. Geophys. Res*. 110, B12402 doi:10.1029/2005JB003662, 10 pp.

Kumar, P. S. et al. (2010). Erosional modification and gully formation at Meteor Crater, Arizona: Insights into crater degradation processes on Mars. *Icarus* 208, 608–620.

Lagain, A. et al. (2021). Model age derivation of large Martian impact craters, using automatic crater counting methods. *Earth Space Sci.* 8(2), e2020EA001598.

Lapen, T. J. et al. (2010). A younger age for AL84001 and its geochemical link to Shergottite sources in Mars. *Science* 328(5976), 347–351.

Laskar, J. et al. (2004). Long term evolution and chaotic diffusion of the insolation quantities of Mars. *Icarus* 170, 343–364.

Leith, A. C., McKinnon, W. B. (1991). Terrace width variations in complex Mercurian raters and the transient strength of cratered Mercurian and lunar crust. *J. Geophys. Res.* 96(E4), 20923–20931.

Le Mouelic, S. et al. (2008). Mapping and interpretation of Sinlap crater on Titan using Cassini VIMS and RADAR data. *J. Geophys. Res.* 113(E04003), doi:10.1029/2007JE002965, 15 pp.

Levrard, B. et al. (2004). Recent ice-rich deposits formed at high latitudes on Mars by sublimation of unstable equatorial ice during low obliquity. *Nature* 431, 1072–1075.

Levy, J. et al. (2010). Concentric crater fill in the northern mid-latitudes of Mars: Formation processes and relationships to similar landforms of glacial origin. *Icarus* 209(2), 390–404.

Malin, M. C., Dzurisin, D. (1977). Landform degradation on Mercury, the Moon, and Mars: Evidence from crater depth/diameter relationships. *J. Geophys. Res.* 82, 376–388.

Malin, M. C., Edgett, K. S. (2000). Sedimentary rocks of early Mars. *Science* 290, 1927–1937.

Malin, M. C., Edgett, K. S. (2003). Evidence for persistent flow and aqueous sedimentation on Early Mars. *Science* 302, 1931–1934.

Malin, M. C. et al. (1998). Early views of the Martian surface from the Mars Orbiter Camera on Mars Global Surveyor. *Science* 279(5357), 1681–1685.

Malin, M. C. et al. (2006). Present-day impact cratering rate and contemporary gully activity on Mars. *Science* 314, 1573–1577.

Malin, M. C. et al. (2007). Context Camera Investigation on board the Mars Reconnaissance Orbiter. *J. Geophys. Res.* 112(E05S04), doi:10.1029/2006JE002808, 25 pp.

Maloof, A. C. et al. (2010). Geology of Lonar Crater, India. *Geol. Soc. Am. Bull.* 122, 109–126.

Mandt, K. E. et al. (2008). Origin of the Medusae Fossae Formation, Mars: Insights from a synoptic approach. *J. Geophys. Res.* 113, E12011, doi:10/1029/2008JE003076, 15 pp.

Mangold, N. et al. (2012). The origin and timing of fluvial activity at Eberswalde crater, Mars. *Icarus* 220 (2), 530–551.

Marzo, G. A. et al. (2010). Evidence for Hesperian impact-induced hydrothermalism on Mars. *Icarus* 208(2), 667–683.

Massonnet, D., Feigl, K. (1998). Radar interferometry and its application to changes in the Earth's surface. *Rev. Geophys.* 36, 441–500.

McCauley, J. F. et al. (1982). Subsurface valleys and geoarcheology of the Eastern Sahara revealed by Shuttle Radar. *Science* 218, 1004–1020.

McCauley, J. F. et al. (1986). Paleodrainages of the Eastern Sahara – The radar rivers revisited (SIR-A/B implications for a Mid-Tertiary Trans-African drainage system). *IEEE Trans. Geosci. Remote Sens.* GE-24, 624–648.

McEwen, A. S. et al. (2005). The rayed crater Zunil and interpretations of small impact craters on Mars. *Icarus* 176, 351–381.

McEwen, A. S. et al. (2007). Mars Reconnaissance Orbiter's High Resolution Imaging Science Experiment (HiRISE): *J. Geophys. Res.* 112(E05S02), doi:10.1029/2005JE002605, 40 pp.

McGetchin, T. R. et al. (1973). Radial thickness variation in impact crater ejecta: Implications for lunar basin deposits. *Earth Planet. Sci. Lett.* 20, 226–236.

McGill, G. E. (1989). Buried topography of Utopia, Mars: Persistence of a giant impact depression. *J. Geophys. Res.*, 94, 2753–2759.

McSween, H. Y. (2002). The rocks of Mars, from far and near. *Meteor. Planet. Sci.* 37, 7–25.

Mellon, M. T. et al. (2000). High-resolution thermal inertia mapping from the Mars Global Surveyor Thermal Emission Spectrometer. *Icarus* 48(2), 437–455.

Melosh, H. J. (1982). A schematic model of crater modification by gravity. *J. Geophys. Res.* 87, 371–380, doi:10.1029/JB087iB01p00371.

Melosh, H. J. (1989). *Impact Cratering: A Geologic Process*. Oxford University Press, 245 pp.

Melosh, H. J. (2011). *Planetary Surface Processes*. Cambridge University Press.

Melosh, H. J., Ivanov, B. A. (1999). Impact crater collapse. *Ann. Rev. Earth Planet. Sci.* 27, 385–415, doi:10.1146/annurev.earth.27.1.385

Meresse, S. et al. (2006). Martian perched craters and large ejecta volume: Evidence for episodes of deflation in the northern lowlands. *Meteor. Planet. Sci.* 41, 1647–1658.

Michalski, J. R., Niles, P. B. (2010). Deep crustal carbonate rocks exposed by meteor impacts on Mars. *Nature Geosci.* 3, 751–755.

Mittlefehldt, D. W. (1994). ALH84001, a cumulate orthopyroxenite member of the martian meteorite clan. *Meteoritics Planet. Sci.* 29(2), 214–221.

Miyamoto, H., Sasaki, S. (2010). Two different supply styles of crater outflow materials on Venus inferred from numerical simulations over DEMs. *Icarus* 145, 533–545.

Moore, J. M. (1990). Nature of the mantling deposit in the heavily cratered terrain of northeastern Arabia, Mars. *J. Geophys. Res.* 95(B9), 14279–14289.

Moore, J. M. et al. (1998). Large impact features on Europa: Results of the Galileo Nominal Mission. *Icarus* 135, 127–145.

Moore, J. M. et al. (2001). Impact features on Europa: Results of the Galileo Europa Mission (GEM). *Icarus* 151, 93–111.

Morgan, J. V. et al. (2000). Peak-ring formation in large impact craters: Geophysical constraints from Chicxulub. *Earth Planet. Sci. Lett.* 183, 347–354.

Morris, A. R., Mouginis-Mark, P. J. (2006). Thermally distinct craters near Hrad Vallis, Elysium Planitia, Mars. *Icarus* 180(2), 335–347.

Morris, A. R. et al. (2010). Possible impact melt and debris flows at Tooting Crater, Mars. *Icarus* 209, 369–389.

Mouginis-Mark, P. J. (1978). Morphology of Martian rampart craters. *Nature* 272, 691–694.

Mouginis-Mark, P. J. (1979a). Martian fluidized crater morphology: Variations with crater size, latitude, altitude and target material. *J. Geophys. Res.* 84, 8011–8022.

Mouginis-Mark, P. J. (1979b). Ejecta emplacement of the Martian impact crater Bamburg. *Proc. Lunar and Planetary Sci. Conf. 10th*, 2651–2668.

Mouginis-Mark, P. J. (1981). Ejecta emplacement and modes of formation of Martian fluidized ejecta craters. *Icarus* 45, 60–76.

Mouginis-Mark, P. J. (1987). Water or ice in the Martian regolith?: Clues from rampart craters seen at very high resolution. *Icarus* 71, 268–286.

Mouginis-Mark, P. J. (2015). Cratering on Mars with almost no atmosphere or volatiles: Pangboche crater. *Meteor. Planet. Sci.* 50(1), 51–62.

Mouginis-Mark, P. J., Baloga, S. M. (2006). Morphology and geometry of the distal ramparts of Martian impact craters. *Meteor. Planet. Sci.* 41, 1469–1482.

Mouginis-Mark, P. J., Boyce, J. M. (2012). Tooting crater: Geology and geomorphology of the archetype large, fresh, impact crater on Mars. *Chemie der Erde* 72, 1–23.

Mouginis-Mark, P. J., Carey, D. L. (1980). Volume estimates of fluidized ejecta deposits for craters in the Northern Plains of Mars. Lunar Planet. Sci. Conf. XI (abstract), pp. 759–761.

Mouginis-Mark, P. J., Garbeil, H. (2007). Crater geometry and ejecta thickness of the Martian impact crater Tooting. *Meteor. Planet. Sci.* 42, 1615–1625.

Mouginis-Mark, P. J., Hayashi, J. N. (1993). Shallow and deep fresh impact craters in Hesperia Planum, Mars. *Earth, Moon, Planets* 61, 1–20.

Mouginis-Mark, P. J., Zimbelman, J. R. (2020). Rafted pumice: A new model for the formation of the Medusae Fossae Formation, Mars. *Icarus* 343, 113684.

Mouginis-Mark, P. J. et al. (1981). Schiaparelli Basin, Mars: Morphology, tectonics and infilling history. In: P. H. Schultz and R. B. Merrill (Eds.), *Proc. Conf. Multi-Ring Basins*, Pergamon Press, pp. 155–172.

Mouginis-Mark, P. J. et al. (2018). Determination of Mars crater geometric data: Insights from high-resolution digital elevation models. *Meteor. Planet. Sci.* 53, 726–740.

Moyano-Cambero, C. E. et al. (2013). SNC meteorites: Atmosphere implantation ages and the climatic evolution of Mars. In: Trigo-Rodriguez, J., Raulin, F., Muller, C., Nixon, C. (Eds.), *The Early Evolution of the Atmospheres of Terrestrial Planets. Astrophys. Space Sci. Proc.* Springer, p. 35. doi:10.1007/978-1-4614-5191-4_13.

Mukherjee, S. et al. (2020). Morphological and morphometric analysis of a topographic depression near Huygens basin, Mars: Identification of a putative endorheic playa. *Geomorphology* 351, 106912.

Murchie, S. et al. (2007). Compact Reconnaissance Imaging Spectrometer for Mars (CRISM) on Mars Reconnaissance Orbiter (MRO). *J. Geophys. Res.* 112(E05S03), doi:10.1029/2006JE002682, 57 pp.

Mustard, J. F. et al. (2001). Evidence for recent climate change on Mars from the identification of youthful near-surface ground ice. *Nature* 412, 411–414.

Mustard, J. F. et al. (2007). Mineralogy of the Nili Fossae region with OMEGA/Mars express data: 1. Ancient impact melt in the Isidis Basin and implications for the transition from the Noachian to Hesperian. *J. Geophys. Res.* 112(E08S03), doi:10/10.19/2006JE002834, 14 pp.

Mutch, P., Woronow, A. (1980). Martian rampart and pedestal craters' ejecta-emplacement: Coprates quadrangle. *Icarus* 41, 259–268.

Mutch, T. A. et al. (1977). The geology of the Viking Lander 2 site. *J. Geophys. Res.* 82, 4452–4467.

Nathues, A. et al. (2020). Recent cryrovolcanic activity at Occator crater on Ceres. *Nature Astron.* 4, 794–801.

Neish, C. D. et al. Cassini RADAR Team (2013). Crater topography on Titan: Implications for landscape evolution. *Icarus* 223, 82–90.

Neish, C. D. et al. (2014). Global distribution of lunar impact melt flows. *Icarus* 239(1), 105–117.

Neish, C. D. et al. (2016). Fluvial erosion as a mechanism for crater modification on Titan. *Icarus* 270, 114–129.

Neukum, G., Hiller, K. (1981). Martian ages. *J. Geophys. Res.* 86(B4), 3097–3121.

Neukum, G. et al. (2010). The geologic evolution of Mars – Episodicity of resurfacing events and ages from cratering analysis of image data and correlation with radiometric ages of Martian meteorites. *Earth Planet. Sci. Lett.* 294, 204–222.

Newsom, H. E. et al. (1986). Fluidization and hydrothermal alteration of the Suevite deposits at the Res Crater West Germany and implications for Mars. *J. Geophys. Res.* 91(B13), E239–E251.

Newsom, H. E. et al. (1996). Impact crater lakes on Mars. *J. Geophys. Res.* 101(E6), 14951–14955.

Nishiizumi, K. et al. (1991). *In situ* ^{10}Be-^{26}Al exposure ages at Meteor Crater, Arizona. *Geochim. Cosmochim Acta* 55, 2699–2703.

Niu, S. et al. (2022). Layered ejecta craters in the candidate landing areas of China's first Mars mission (Tianwen-1): Implications for subsurface volatile concentrations. *J. Geophys. Res.* 127, e2021JE007089.

Nunes, D. C. et al. (2011). Shallow radar (SHARAD), pedestal craters, and the lost Martian layers: Initial assessments. *J. Geophys. Res.* 116(E04066), doi:10.1029/2010JE003690, 20 pp.

Nyquist, L. E. (1983). Do oblique impacts produce Martian meteorites? *J. Geophys. Res.* 88, A785–A798.

Nyquist, L. E. et al. (2001). Ages and geologic histories of Martian meteorites. *Chronology Evol. Mars*, 96, 105–164.

O'Keefe, J. D., Ahrens, T. J. (1986). Oblique impact: A process for obtaining meteorite samples from other planets. *Science* 234, 346–349.

Oberbeck, V. R. (1975). The role of ballistic erosion and sedimentation in lunar stratigraphy. *Revs. Geophys.* 13, 337–362.

Oberbeck, V. R., Morrison, R. H. (1973). On the formation of lunar herringbone pattern. *Proc. 4th Lunar Sci. Conf.*, v. 1, 107–123.

Oberbeck, V. R. et al. (1975). Transport and emplacement of crater and basin deposits. *The Moon* 13, 9–26.

Ohman, T., Kring, D. A. (2012). Photogeologic analysis of impact melt-rich lithologies in Kepler crater that could be sampled by future missions. *J. Geophys. Res.* 117, 1–21, doi:10.1029/2011JE003918.

Ojha, L., Mittelholz, A. (2023). Insight into the formation mechanism of the Medusae Fossae Formation on Mars from magnetic field data. *Icarus* 395, 115471.

Ormo, J. et al. (2004). Marine-target craters on Mars? An assessment study. *Meteor. Planet. Sci.* 39(2), 333–346.

Osinski, G. R., Pierazzo, E. (2013). *Impact Cratering Processes and Products*. Wiley-Blackwell.

Osinski, G. R. et al. (2011). Impact ejecta emplacement on terrestrial planets. *Earth Planet. Sci. Lett.* 310, 167–181.

Osinski, G. R. et al. (2013). Impact-generated hydrothermal systems on Earth and Mars. *Icarus* 224(2), 347–363.

Palumbo, A. M. et al. (2020). Rainfall on Noachian Mars: Nature, timing, and influence on geologic processes and climate history. *Icarus* 347, 113782.

Palluconi, F. D., Kieffer, H. H. (1981). Thermal inertia mapping of Mars from 60°S to 60°N. *Icarus* 45, 415–426.

Pan, C. et al. (2015). Thermal and near-infrared analyses of central peaks of Martian impact craters: Evidence for a heterogeneous Martian crust. *J. Geophys. Res.* 120, 662–688.

Pan, C. et al. (2021). Evaluating flat-crater floor fill compositions and morphologies: Insights into formation process. *J. Geophys Res.* 126, c2021JE006919.

Pathare, A. V. et al. (2005). Viscous relaxation of craters within the Martian south polar layered deposits. *Icarus* 174, 396–418.

Pathare, A. V. et al. (2018). Driven by excess? Climate implications of new global mapping of near-surface water-equivalent hydrogen on Mars. *Icarus* 301, 97–116.

Pearce, G. et al. (2011). Intra-crater glacial processes in central Utopia Planitia, Mars. *Icarus* 212(1), 86–95.

Pearce, S. J., Melosh, H. J. (1986). Terrace width variations in complex lunar craters. *Geophys. Res. Lett.* 13, 1419–1422.

Peel, S. W. et al. (2019). Formation of central pits in impact craters on Mars: A statistical investigation of proposed mechanisms. *J. Geophys. Res.* 124, 437–453.

Pettengill, G. H. et al. (1971). Martian craters and a scarp as seen by radar. *Science* 174(4016), 1321–1324.

Phillips, R. J. et al. (1991). Impact craters on Venus – Initial analysis from Magellan. *Science* 252, 288–297.

Phillips, R. J. et al. (1992). Impact craters and Venus resurfacing history. *J. Geophys. Res.* 97, 15923–15948.

Piatek, J. L. et al. (2018). Characterization of thermophysical ejecta facies in well-preserved martian craters. *49th Lunar Planet. Sci. Conf.*, abs. 2691.

Pierazzo, E., Melosh, H. J. (2000). Understanding oblique impacts from experiments, observations, and modeling. *Ann. Revs. Earth Planet. Sci.* 28, 141–167.

Pieters, C. M. (1982). Copernicus crater central peak: Lunar mountain of unique composition. *Science* 215, 59–61.

Pieters, C. M. et al. (1985). The nature of crater rays: The Copernicus example. *J. Geophys. Res.* 90, 12393–12413.

Pietrek, A. et al. (2020). Morphometric characterization of longitudinal striae on Martian landslides and impact ejecta blankets and implications for the formation mechanism. *J. Geophys. Res.* 125(2), doi:10.1029/2019JE006255, 20 pp.

Pike, R. J. (1974). Depth/diameter relations of fresh lunar craters: Revisions from spacecraft data. *Geophys. Res. Lttrs.* 1(7), 291–294.

Pike, R. J. (1977). Size-dependence in the shape of fresh impact craters on the moon. In: Roddy, D. J., Pepin, R. O., Merrill, R. (Eds.), *Impact and Explosion Cratering*, Pergamon Press, pp. 489–509.

Pike, R. J. (1980a). Control of crater morphology by gravity and target type: Mars, Earth, Moon. In: Merrill, R. B. et al., eds., *Proceedings of the 11th Lunar and Planetary Science Conference*, Lunar and Planetary Institute, pp. 2159–2189.

Pike, R. J. (1980b). Formation of complex impact craters: Evidence from Mars and other planets. *Icarus* 43(1), 1–19.

Pilon, J. A. et al. (1991). The subsurface character of Meteor Crater, Arizona, as determined by ground-probing radar. *J. Geophys. Res.* 96, 15563–15576.

Plescia, J. B. (2003). Cerberus Fossae, Elysium, Mars: A source for lava and mud. *Icarus* 164, 79–95.

Plescia, J. B., Robinson, M. S. (2019). Giordano Bruno: Small crater populations-Implications for self-secondary cratering. *Icarus* 321, 974–993.

Preblich, B. S. et al. (2007). Mapping rays and secondary craters from the Martian crater Zunil. *J. Geophys. Res.* 112(E05006), doi:10.1029 /2006JE2817, 18 pp.

Putzig, N. E. et al. (2018). Three-dimensional radar imaging of structures and craters in the Martian polar caps. *Icarus* 308, 138–147.

Quaide, W. L., Oberbeck, V. R. (1968). Thickness determinations of the lunar surface layer from lunar impact craters. *J. Geophys. Res.* 73, 5247–5270.

Quantin, C. et al. (2016). Young Martian crater Gratteri and its secondary craters. *J. Geophys. Res.* 121, 1118–1140.

Rathbun, J. A., Squyres, S. W. (2002). Hydrothermal systems associated with Martian impact craters. *Icarus* 157, 362–373.

Re, C. et al. (2022). CaSSIS-based stereo products for Mars after three years in orbit. *Planet. Space Sci.* 219, 105515.

Reiss, D., Jaumann, R. (2003). Recent debris flows on Mars: Seasonal observations of the Russell Crater dune field. *Geophys. Res. Lett.* 30(1321), doi:10.1029/2002GL016704, 4 pp.

Reiss, D. et al. (2009). Regional differences in gully occurrence on Mars: A comparison between the Hale and Bond craters. *Planet. Space Sci.* 57, 958–974.

Robbins, S. J., Hynek, B. M. (2011). Secondary crater fields from 24 large primary craters on Mars: Insights into nearby secondary crater production. *J. Geophys. Res.* 116(E10003), doi:10.1029/2011JE003820, 13 pp.

Robbins, S. J., Hynek, B. M. (2012). A new global database of Mars impact craters >1 km: 2. Global crater properties and regional variations of the simple-to-complex transition diameter. *J. Geophys. Res.* 117(E05004), doi:10.1029/2011JE003967, 18 pp.

Robbins, S. J., Hynek, B. M. (2013). Utility of laser altimeter and stereoscopic terrain models: Application to Mars craters. *Planet. Space Sci.* 86, 57–65.

Robbins, S. J. et al. (2017). Craters of the Pluto-Charon system. *Icarus* 287, 187–206.

Robbins, S. J. et al. (2018a). Measuring impact crater depth throughout the solar system. *Meteor. Planet. Sci.* 53, 583–626.

Robbins, S. J. et al. (2018b). Investigation of Charon's craters with abrupt terminus ejecta, comparisons with other icy bodies, and formation implications. *J. Geophys. Res.* 123, 20–36.

Robbins, S. J. et al. (2021). Depths of Pluto's and Charon's craters, and their simple-to-complex transition. *Icarus* 356, 113902.

Roddy, D. J. et al. (1975). Meteor Crater, Arizona, rim drilling with thickness, structural uplift, diameter, depth, volume, and mass-balance calculations. *Proc. Lunar Sci. Conf.* 6th, 2621–2644.

Roddy, D. J. (1977). Large-scale impact and explosion craters: Comparisons of morphological and structural analogs. In: Roddy, D. J., Pepin, R. O., Merrill, R. B. (Eds.), *Impact and Explosion Cratering: Planetary and Terrestrial Implications*. Pergamon Press, pp. 185–246.

Rodriguez, J. A. et al. (2016). Tsunami waves extensively resurfaced the shorelines of an early Martian ocean. *Sci. Rep.* 6, 25106, doi:10.1038/srep25106.

Ronca, L. B. (1966). Meteorite impact and volcanism. *Icarus* 5, 515–520.

Roth, L. E. et al. (1989). Radar altimetry of large Martian craters. *Icarus* 79(2), 289–310.

Ruff, S. W., Christensen, P. R. (2002). Bright and dark regions on Mars: Particle size and mineralogical characteristics based on Thermal Emission Spectrometer data. *J. Geophys. Res.* 107(E12 5127), doi:10.1029/2001JE001580, 22 pp.

Ruj, T., Kawai, K. (2021). A global investigation of wrinkle ridge formation events: Implications towards the thermal evolution of Mars. *Icarus* 369, 114625.

Salese, F. et al. (2020). Estimated minimum life span of the Jezero fluvial delta, Mars. *Astrobio* 20, doi:10.1089/ast.2020.2228, 977–993.

Salvatore, M. R. et al. (2010). Definitive evidence of Hesperian basalt in Acidalia and Chyrse Planitiae. *J. Geophys. Res.* 115(E07005), doi:10.1029/2009JE003519, 16 pp.

Sapers, H. M. et al. (2017). Evidence for a spatially extensive hydrothermal system at the Ries impact structure, Germany. *Meteoritics Planet. Sci.* 52, 351–371.

Sato, H. et al. (2010). The formation of floor-fractured craters in Xanthe Terra. *Icarus* 207, 248–264, doi:10.1016/j.icarus. 2009.10.023.

Schoenfeld, A. M. et al. (2023). Geomorphological map of the Soi crater region on Titan. *J. Geophys. Res.* 128, doi:10.1029/2022JE007499, 28 pp.

Schon, S. C., Head, J. W. (2012). Gasa impact crater, Mars: Very young gullies formed from impact into latitude-dependent mantle and debris-covered glacier deposits? *Icarus* 218, 459–477.

Schon, S. C. et al. (2012). An overfilled lacustrine system and progradational delta in Jezero crater, Mars: Implications for Noachian climate. *Planet. Space Sci.* 67, 28–45.

Schultz, P. H. (1976). *Moon Morphology: Interpretations Based on Lunar Orbiter Photography*. University of Texas Press, p. 626.

Schultz, P. H. (1978). Martian intrusions – Possible sites and implications. *Geophys. Res. Lett.* 5, 457–460, doi:10.1029/GL005i006p00457.

Schultz, P. H. (1979). Impact crater and basin control of igneous processes on Mars. *J. Geophys. Res.* 84, 8033–8047, doi:10.1029/JB084iB14p08033.

Schultz, P. H. (1992). Atmospheric effects on ejecta emplacement. *J. Geophys. Res.* 97, 11623–11662.

Schultz, P. H., Gault, D. E. (1979). Atmospheric effects on Martian ejecta emplacement. *J. Geophys. Res.* 84, 7669–7687.

Schultz, P. H., Lutz-Garihan, A. B. (1982). Grazing impacts on Mars: A record of lost satellites. *J. Geophys. Res.* 87, A84–A96.

Schultz, P. H., Lutz-Garihan, A. B. (1988). Polar wandering of Mars. *Icarus* 73, 91–141.

Schultz, P. H., Orphal, D. L. (1978). Floor-fractured craters on the moon and Mars. *Meteoritics* 13, 622–625.

Schultz, P. H., Wrobel, K. E. (2012). The oblique impact Hale and its consequences on Mars. *J. Geophys. Res.* 117(E04001), doi:10.1029/2011JE003843, 21 pp.

Schultz, R. A. (1998). Multiple-process origin of Valles Marineris basins and troughs, Mars. *Planet. Space Sci.* 46(6–7), 827–829, 831–834.

Schultz, R. A., Frey, H. V. (1990). A new survey of multiring impact basins on Mars. *J. Geophys. Res.* 95, 14175–14189.

Scott, D. H., Carr, M. H. (1978). Geologic map of Mars. U.S. Geol. Surv. Misc. Invest. Map Series, I-1083.

Scully, J. E. C. et al. (2020). The varied sources of faculae-forming brines in Cere's Occator crater emplaced via hydrothermal brine effusion. *Nature Comms.* 11, 3680, doi:10.1038/s41467-020-15973-8.

Sharpton, V. L. (2014). Outcrops on lunar crater rims: Implications for rim construction mechanisms, ejecta volumes and excavation depths. *J. Geophys. Res.* 119, 154–168.

Senft, L. E., Stewart, S. T. (2008). Impact crater formation in icy layered terrains on Mars. *Meteor. Planet. Sci.* 43, 1993–2013.

Settle, M., Head, J. W. (1977). Radial variation of lunar crater rim topography. *Icarus* 31, 123–135.

Settle, M., Head, J. W. (1979). The role of rim slumping in the modification of lunar impact craters. *J. Geophys. Res.* 84, 3081–3096.

Seu, R. et al. (2007). SHARAD sounding radar on the Mars Reconnaissance Orbiter. *J. Geophys. Res.* 112(E05S05), doi:10.1029/2006JE002745, 18 pp.

Sharpton, V. L. (2014). Outcrops on lunar crater rims: Implications for rim construction mechanisms, ejecta volumes and excavation depths. *J. Geophys. Res.* 119, 154–168.

Shoemaker, E. M. (1963). Impact mechanics at Meteor Crater, Arizona. In: Middlehurst, B. M., Kuiper, G. P. (Eds.), *The Moon, Meteorites and Comets*. University of Chicago Press, 301–336.

Shorthill, R. W. et al. (1976). The environs of Viking Lander 2. *Science* 194, 1309–1318.

Shreve, R. L. (1968). The Blackhawk Landslide. *Geol. Soc. Amer. Spec. Paper 108*, 47 pp.

Simonds, C. H. et al. (1978). Petrogenesis of melt rocks, Manicouagan impact structure, Quebec. *J. Geophys. Res.* 83, 2773–2788.

Sizemore, H. G. et al. (2017). Pitted terrains on (1) Ceres and implications for shallow subsurface volatile distribution. *Geophys. Res. Lett.* 44, 6570–6578.

Skinner, J. A., Tanaka, K. L. (2007). Evidence for and implications of sedimentary dapirism and mud volcanism in the southern Utopia highland-lowland boundary plain, Mars. *Icarus* 186, 41–59.

Smith, D. E. et al. (2001). Mars Orbiter Laser Altimeter – Experiment summary after the first year of global mapping of Mars. *J. Geophys. Res.* 106(E10), 23689–23722.

Smith, E. I. (1976). Comparison of the crater morphology-size relationship for Mars, Moon and Mercury. *Icarus* 28, 543–550.

Smith, E. I., Hartnell, J. A. (1978). Crater size-shape profiles for the moon and mercury: Terrain effects and interplanetary comparisons. *Moon Planets* 19, 479–511.

Smrekar, S. E. et al. (2004). Geologic evolution of the Martian dichotomy in the Ismenius area of Mars and implications for plains magnetization. *J. Geophys. Res.* 109, E11002, doi:10.1029/2004JE002260.

Sokolowska, A. et al. (2024). Effects of surface and subsurface water/ice on spatial distributions of impact crater ejecta on Mars. *Icarus* 420, 116150.

Song, E. et al. (2013). Bulk mineralogy of lunar crater central peaks via thermal infrared spectra from the Diviner Lunar radiometer: A study of the Moon's crustal composition at depth. *J. Geophys. Res.* 118, 689–707.

Squyres, S. W. (1989). Urey prize lecture: Water on Mars. *Icarus* 79 (2), 229–288.

Squyres, S. W. et al. (2009). Exploration of Victoria crater by the Mars Rover Opportunity. *Science* 324(5930), 1058–1061.

Squyres, S. W. et al. (2012). Ancient impact and aqueous processes at Endeavour crater, Mars. *Science* 336, 570–576.

Stewart, S. T., Valiant, G. J. (2006). Martian subsurface properties and crater formation processes inferred from fresh impact crater geometries. *Meteor. Planet. Sci.* 41(10), 1509–1537.

Stöffler, D., Ryder, G. (2001). Stratigraphic and isotopic ages of lunar geologic units: Chronological standard for the Inner Solar System. *Space Sci. Rev.* 96, 9–54.

Stopar, J. D. et al. (2014). Occurrence and mechanisms of impact melt emplacement at small lunar craters. *Icarus* 243, 337–357.

Strom, R. G. et al. (1992). The Martian impact crater record. In: Kieffer, H. H., Jakosky, B. M., Synder, C. W., Matthews, M. S. (Eds.), *Mars*, University of Arizona Press, pp. 383–423.

Sturm, S. et al. (2013). The Ries impact, a double-layer rampart crater on Earth. *Geology* 41(5), 531–534.

Sturm, S. et al. (2015). The distribution of megablocks in the Ries crater, Germany: Remote sensing, field investigation, and statistical analyses. *Meteoritics Planet. Sci.* 50, 141–171.

Sturm, S. et al. (2016). Ejecta thickness and structural rim uplift measurements of Martian impact craters: Implications for the rim formation of complex impact craters. *J. Geophys. Res.* 121, 1026–1053.

Sun, V. Z., Milliken, R. E. (2015). Ancient and recent clay formation on Mars as revealed from a global survey of hydrous minerals in crater central peaks. *J. Geophys. Res.* 120, 2293–2332. doi:10.1022/2015JE004918.

Tanaka, K. L., Leonard, G. J. (1995). Geology and landscape evolution of the Hellas region on Mars. *J. Geophys. Res.* 100, 5407–5432.

Tanaka, K. L. et al. (2014). Geologic map of Mars. U.S. Geol. Surv. Misc. Invest. Map Series, SIM-3292.

Therriault, A. M. et al. (1997). Original size of the Vredefort Structure: Implications for the geological evolution of the Witwatersrand Basin. *Meteor. Planet. Sci.* 32, 71–77.

Thomson, B. J., Head, J. W. (2001). Utopia Basin, Mars: Characterization of topography and morphology and assessment of the origin and evolution of basin internal structure. *J. Geophys. Res.* 106(E10), 23209–23230.

Thomson, B. J., Schultz, P. H. (2007). The geology of the Viking Lander 2 site revisited. *Icarus* 191, 505–523.

Thomson, B. J. et al. (2011). Constraints on the origin and evolution of the layered mound in Gale Crater, Mars using Mars Reconnaissance Orbiter data. *Icarus* 214, 413–432.

Tornabene, L. L. et al. (2006). Identification of large (2–10 km) rayed craters on Mars in THEMIS thermal infrared images: Implications for possible Martian meteorite source regions. *J. Geophys. Res.*, 111, E10006. doi:10.1029/2005JE002600.

Tornabene, L. L. et al. (2012). Widespread crater-related pitted materials on Mars: Further evidence for the role of target volatiles during the impact process. *Icarus* 220, 348–368.

Tornabene, L. L. et al. (2015). "Pristine" Martian craters: Part 1 – Criteria and characteristics. *46th Lunar Planet. Sci. Conf.* abs. no. 2531.

Tornabene, L. L. et al. (2016). Visible and thermophysical characteristics of the best-preserved martian craters, part 1: Detailed morphological mapping of Resen and Noord. *47th Lunar Planet. Sci. Conf.* abs. 2879.

Tornabene, L. L. et al. (2018a). A depth versus diameter scaling relationship for the best-preserved melt-bearing complex craters on Mars. *Icarus* 299, 68–83.

Tornabene, L. L. et al. (2018b). Image simulation and assessment of the colour and spatial capabilities of the Colour and Stereo

Surface Imaging System (CaSSIS) on the ExoMars Trace Gas Orbiter. *Space Sci. Revs.* 214(18), doi:10.1007/s11214-017-0436-7, 61 pp.

Treiman, A. H. et al. (2000). The SNC meteorites are from Mars. *Planet. Space Sci.* 48, 1213–1230.

Tsibulskaya, V. et al. (2020). Surficial geology and geomorphology of Greg crater, Promethei Terra, Mars. *J. Maps* 16(2), 524–533.

Turner, S. M. R. et al. (2016). Hydrothermal activity recorded in post Noachian-aged impact craters on Mars. *J. Geophys. Res.* 121, 608–625.

Turtle, E. P. et al. (2005). Impact structures: What does crater diameter mean? *Geol. Soc. Amer. Sp. Paper* 384, 1–24.

Udry, A. et al. (2020). What Martian meteorites reveal about the interior and surface of Mars. *J. Geophys. Res.* 125, e2020JE006523.

Vaucher, J. et al. (2009). The volcanic history of central Elysium Planitia: Implications for martian magmatism. *Icarus* 204(2), 418–442.

Viola, D. et al. (2017). Subsurface volatile content of Martian double-layered ejecta (DLE) craters. *Icarus* 284, 325–343.

Voigt, J. R. C. et al. (2023). Revealing Elysium Planitia's youngest geologic history: Constraints on lava emplacement, areas, and volumes. *J. Geophys. Res.* 128(12), e2023JE007947.

Walker, J. C. G. (1986). Impact erosion of planetary atmospheres. *Icarus* 68, 87–98.

Ward, W. R. (1973). Large-scale variations in the obliquity of Mars. *Science* 181(4096), 260–262.

Warner, N. H. et al. (2009). A refined chronology of catastrophic outflow events in Ares Vallis, Mars. *Earth Planet. Sci. Lett.* 288, 58–69.

Warner, N. H. et al. (2015). Minimum effective area for high resolution crater counting of Martian terrains. *Icarus* 245, 198–240.

Watters, T. R. et al. (2006). MARSIS radar sounder evidence of buried basins in the northern lowlands of Mars. *Nature* 444, 905–908.

Watters, T. R. et al. (2007). Radar sounding of the Medusae Fossae Formation Mars: Equatoral ice or dry, low-density deposits? *Science* 318, 1125–1128.

Watters, W. A. et al. (2015). Morphometry of small recent impact craters on Mars: Size and terrain dependence, short-term modification. *J. Geophys. Res.* 120, 226–254.

Way, D. W. et al. (2007). Mars Science Laboratory: Entry, descent, and landing system performance. IEEE Aerospace Conf., doi: 10.1109/AERO.2007.352821.

Weiss, D. K., Head, J. W. (2013). Formation of double-layered ejecta craters on Mars: A glacial substrate model. *Geophys. Res. Lett.* 40, 3819–3824, doi:10.1002/grl.50778.

Weiss, D. K., Head, J. W. (2014). Ejecta mobility of layered ejecta craters on Mars: Assessing the influence of snow and ice deposits. *Icarus* 233, 131–146.

Weiss, D. K., Head, J. W. (2018). Testing landslide and atmospheric effects models for the formation of double-layered ejecta craters on Mars. *Meteor. Planet. Sci.* 53, 741–777.

Weitz, N. et al. (2018). Modeling concentric crater fill in Utopia Planitia, Mars, with an ice flow line model. *Icarus* 308, 209–220.

Wells, R. A., Fielder, G. (1967). Martian and lunar craters. *Science* 155, 354–355.

Werner, S. C., Tanaka, K. L. (2011). Redefinition of the crater-density and absolute-age boundaries for the chronostratigraphic system of Mars. *Icarus* 215, 603–607.

Wichman, R. W., Schultz, P. H. (1995). Floor-fractured craters in Mare Smythii and west of Oceanus Procellarum: Implications of crater modification by viscous relaxation and igneous intrusion models. *J. Geophys. Res.* 100, 21201–21218, doi:10.1029/95JE02297.

Williams, J.-P. et al. (2018). Dating very young planetary surfaces from crater statistics: A review of issues and challenges. *Meteor. Planet. Sci.* 53, 554–582.

Williams, K. K., Greeley, R. (2004). Laboratory and field measurements of the modification of radar backscatter by sand. *Rem. Sens. Environ.* 89, 29–40.

Williams, N. R. et al. (2015). Evidence for an explosive origin of central pit craters on Mars. *Icarus* 252, 175–185.

Williams, R. M. E., Malin, M. C. (2008). Sub-kilometer fans in Mojave Crater, Mars. *Icarus* 198(2), 365–383.

Williams, R. M. E. et al. (2011). Evidence for episodic alluvial fan formation in far western Terra Tyrrhena, Mars. *Icarus* 211, 222–237.

Wilson, L. et al. (2009). Fissure eruptions in Tharsis, Mars: Implications for eruption conditions and magma source. *J. Volcanol. Geotherm. Res.* 185, 28–46.

Wilson, S. A. et al. (2007). Geomorphic and stratigraphic analysis of Crater Terby and layered deposits north of Hellas basin, Mars. *J. Geophys. Res.* 112(E08009), doi:10.1029/2006JE002830, 39 pp.

Wohletz, K. H., Sheridan, M. F. (1983). Martian rampart crater ejecta: Experiments and analysis of melt-water interaction. *Icarus* 56(1), 15–37.

Wood, C. A. (1973). Central peak heights and crater origins. *Icarus* 20, 503–506.

Wood, C. A. et al. (1978). Interior morphology of fresh Martian craters: The effects of target characteristics. *Proc. Lunar Sci. Conf. 9th*, 3691–3709.

Wood, C. A. et al., Cassini Radar Team (2010). Impact craters on Titan. *Icarus* 206, 334–344.

Woronow, A. (1981). Preflow stresses in Martian rampart ejecta blankets: A means of estimating the water content. *Icarus* 45, 320–330.

Wulf G., Kenkmann T. (2015). High-resolution studies of double-layered ejecta craters: Morphology, inherent structure, and a phenomenological formation model. *Meteor. Planet. Sci.* 50, 173–203.

Wulf, G. et al. (2012). Structural asymmetry in Martian impact craters as an indicator for an impact trajectory. *Icarus* 220, 194–204.

Xiao, Z. et al. (2014). Comparisons of fresh complex impact craters on Mercury and the Moon: Implications for controlling factors in impact excavation processes. *Icarus* 228, 260–275.

Yue, Z. et al. (2019). Lunar regolith thickness deduced from concentric craters in the CE-5 landing area. *Icarus* 329, 46–54.

Zanetti, M. et al. (2017). Evidence for self-secondary cratering of Copernican-age continuous ejecta deposits on the Moon. *Icarus* 298, 64–77.

Zegers, T. E. et al. (2010). Melt and collapse of buried water ice: An alternative hypothesis for the formation of chaotic terrains on Mars. *Earth Planet. Sci. Lett.* 297, 496–504, doi:10.1016/j.epsl.2010.06.049.

INDEX

Barlow, N. G., 8–9, 23, 27, 32, 35–36, 38, 46–47, 49, 51, 80–81, 85–87, 89, 93, 95, 100, 102–103, 128, 130, 133, 136–137
bowl-shaped craters. *See* simple craters
Boyce, J. M., 3, 8, 18, 25–26, 35, 39, 43–44, 53, 56, 60, 75, 80, 85, 87, 91, 93, 102, 105, 107, 115–116, 119, 122, 125–126, 128, 140, 142–143
breccia, 3, 43, 53, 65, 68, 128

candidate future missions, 144
Carr, M. H., 4, 35, 37–39, 80, 84, 93
central peak, 3–8, 12, 15, 17, 19, 27, 44, 46–49, 51, 56, 67, 74, 111, 130, 132
central pits, 3, 6, 23–24, 49–53, 62, 129
climate changes
  near-surface ice, 32, 40, 87
  obliquity cycle, 30, 35–37, 39–40, 65, 84, 103, 108, 130, 134
  paleo-ocean, 38–39
complex craters, 2–4, 6–7, 15, 17, 73
crater geometry, 6
crater rays, 16, 22, 44, 96, 109–110, 137, 145
crater types
  DLE, Type-1, 39–41, 47–48, 80, 83–88, 90–91, 96–97, 101–102, 114, 116, 119, 121, 123–127, 130, 132, 134, 143
  DLE, Type-2, 80, 84–86
  LARLE, 102–103, 119, 121, 135–136
  MLE, 47, 80–82, 85–86, 88–89, 91, 96, 102, 121, 125, 130, 139
  pedestal, 33, 35, 80, 103–104, 136
  SLE, 80–81, 85–87, 89, 91, 93–94, 96, 99–100, 102, 117–118, 125–126, 130, 133–134, 136, 139
craters, Earth
  Manicouagan, 12
  Meteor, 1, 12
  Ries, 1, 12, 53
craters, Mars
  Arandas, 48, 90
  Argyre basin, 77
  Bacolor, 114, 125, 142
  Bakhuysen, 114
  Belz, 133–134
  Betio, 52
  Curie, 56–58
  Dilly, 109, 129, 131
  Eberswalde, 69
  Endurance, 64
  Gale, 42–43, 59–60, 68
  Graindavik, 133
  Gratteri, 96, 117
  Hale, 74, 109, 113
  Hellas, basin, 77
  Huygens, 49, 59, 75
  Isidis Basin, 77
  Jaisalmer, 45, 123, 125, 143
  Jezero, 72–73
  Jojutla, 77
  Karzok, 54
  Kotka, 88–89
  Lonar, 121
  Makhambet, 145
  Naar, 133–134
  Nicholson, 71
  Nutak, 84
  Pangboche, 44–45, 60, 62
  Poona, 101
  Punsk, 133
  Schiaparelli basin, 78
  Slangpos, 41–42
  Steinheim, 85, 96, 130
  Telz, 47
  Terby, 69–70
  Toconao, 94
  Tooting, 5, 10, 18, 43–45, 48–49, 56, 60, 75, 93, 104, 107, 121
  Utopia Basin, 78
  Zumba, 101
  Zunil, 44, 95, 109, 136, 140–141
craters, Mercury
  Abedin, 19
  Basho, 16
  Michelangelo, 19
  Rustaveli, 23
  Waters, 21
craters, Moon
  Aristarchus, 17
  Copernicus, 5, 8, 22, 60
  Hell Q, 16
  Orientale, 4, 6, 49
  Schrodinger, 19
  Tycho, 17, 19
craters, Venus
  Addams, 28
  Danilova, 27
  Dickinson, 144
  Isabella, 28

data types
  CTX, 30, 34
  geological map, 36
  HiRISE, 30, 33
  MOLA, 30, 32, 35
  THEMIS, 30, 34, 95
debris flows, 74, 109, 112

depth/diameter ratio, 8–10, 30, 35, 43, 45, 64
dewatering ejecta, 111
digital elevation models, 9, 30, 85, 94, 125, 128, 137, 140, 142
distal ramparts, 37, 80, 82, 89, 93–95, 102, 119, 121, 129, 139, 146

ejecta blanket, 2, 4, 6, 12, 21–22, 33, 37, 45, 80, 91, 93–94, 101, 111, 117, 124–125, 134, 136, 144
ejecta facies, 122
ejecta thickness, 4, 85, 91–92, 94, 138
eolian dunes, 31, 63–64

Head, J. W., 4, 6, 12–13, 15, 21, 35–37, 39–40, 56, 61, 74, 77–78, 80, 85, 87, 109, 112
highly fluid ejecta, 109
hydrothermal deposits, 73, 117

impact melt, 3, 12, 16–21, 23, 27, 43–45, 60, 62, 117

Medusae Fossae Formation, 40, 68, 71, 103–104
Melosh, H. J., 1, 3–4, 17, 56, 104, 107
Mouginis-Mark, P. J., 9, 18, 35–36, 49, 60, 68, 78, 87, 89, 93, 107, 119, 125, 128, 133
Mt. Sharp, 42, 59–60, 68, 71

oblique impacts, 107, 109
Osinski, G. R., 1–3, 12, 73, 134

pitted material, 20, 43–45, 53–55, 126, 137
planetary bodies
- 1 Ceres, 23–24
- 4 Vesta, 25
- Callisto, 25
- Charon, 25–26
- Ganymede, 23–24
- Mars, 2, 6, 9, 27, 30, 34, 36, 41
- Mercury, 8, 15–21, 23, 27, 112
- Moon, 2–3, 6–8, 15–18, 44, 53–54, 95
- Titan, 27, 29
- Venus, 27

polar craters, 75

rim uplift, 3–4, 12, 60, 91, 125, 137, 140
Robbins, S. J., 3, 6, 8–9, 23, 25–26, 44, 46, 56, 80, 95, 128

Schultz, P. H., 6, 37, 40, 54, 64, 80, 85, 107–108, 112, 128, 137
secondary craters, 2, 5, 8, 22–23, 29, 40, 44, 51, 56, 80, 83, 95–97, 102, 104–105, 109, 113–115, 119, 121, 124, 130, 136–137, 141, 146
simple craters, 2–3, 6–7, 12, 15–17, 22, 46–47, 54, 60, 100
SIR-A radar, 143–144
SNC meteorites, 2, 36–37, 73, 108
spacecraft
- *Cassini*, 27, 142–143
- *Curiosity*, 41–42, 59–60, 68, 71, 144–145
- *Lunar Reconnaissance Orbiter*, 4, 6, 16–20, 22
- *Magellan*, 27, 142–143
- *Mariner IV*, 31
- *Mars Odyssey*, 30, 32, 40
- *Mars Reconnaissance Orbiter*, 30, 32
- MESSENGER, 19, 21
- Opportunity, 41–42, 64
- Perseverance, 72–73, 144–145
- *Viking Lander*, 112, 115
- *Viking Orbiter*, 31, 80, 93–94, 128

stages of crater formation
- contact and compression, 2
- ejecta emplacement, 4
- excavation, 2
- modification, 3

swirl texture, 54–55
synthetic aperture radar, 142, 144

terraces, 2–6, 15, 17, 24, 43, 46, 53–54, 56–60, 62, 75, 105, 129–130, 137
thermal inertia, 30, 44–45, 63, 87
Tornabene, L. L., 8, 43–45, 53, 55, 87, 109, 129, 143
transient cavity, 2–3, 15, 60

Vastitas Borealis Formation, 78, 86–87, 97, 112, 115–116, 130